Pathology of Aging Rats

A Morphological and Experimental Study of the Age-Associated Lesions in Aging BN/Bi, WAG/Rij and (WAG × BN) F_1 Rats

Author:

Joe D. Burek, D.V.M.

Toxicology Research Laboratory
Health and Environmental Research
Dow Chemical U.S.A.
Midland, Michigan

From the Institute for Experimental Gerontology of the Organization for Health Research TNO, Rijswijk, The Netherlands. Publication supported by National Institute on Aging, Public Health Service, National Institutes of Health, U.S. Department of Health, Education and Welfare, Washington, D.C. and Health and Environmental Research, The Dow Chemical Company, Midland, Michigan.

CRC Press
Taylor & Francis Group
Boca Raton London New York

CRC Press is an imprint of the
Taylor & Francis Group, an informa business

First published 1978 by CRC Press
Taylor & Francis Group
6000 Broken Sound Parkway NW, Suite 300
Boca Raton, FL 33487-2742

Reissued 2018 by CRC Press

Library of Congress Cataloging in Publication Data

Burek, Joe D
Pathology of aging rats.

Bibliography: p.
Includes index.
1. Age factors in disease. 2. Diseases — Animal models. 3. Rats — Diseases. 4. Rattus norvegicus.
I. Title.
RB210.B87 619'.93 78-15489
ISBN 0-8493-5649-0

A Library of Congress record exists under LC control number: 78015489

ISBN 13: 978-1-315-89633-5 (hbk)
ISBN 13: 978-1-351-07543-5 (ebk)

Visit the Taylor & Francis Web site at http://www.taylorandfrancis.com and the
CRC Press Web site at http://www.crcpress.com

THE AUTHOR

Joe D. Burek, D.V.M., is Research Specialist, Toxicology Research Laboratory, Health and Environment Research, The Dow Chemical Company, Midland, Michigan and advisor in rodent pathology and laboratory animal medicine, Institute for Experimental Gerontology TNO, Rijswijk, The Netherlands.

Dr. Burek received his B.S., D.V.M., and M.S. degrees from Michigan State University in 1968, 1969, and 1970, respectively. He subsequently received postdoctoral training in pathology and laboratory animal medicine at the U.S. Army Medical Research Institute of Infectious Disease, Frederick, Maryland, and the Johns Hopkins University School of Medicine, Baltimore. He is a Diplomate of the American College of Veterinary Pathologists and the American College of Laboratory Animal Medicine.

Dr. Burek is a consulting member of the European Late Effects Project Group; a member of the Program Committee of the American College of Laboratory Animal Medicine. His professional affiliations include the American College of Veterinary Pathologists, the American College of Laboratory Animal Medicine, the American Veterinary Medical Association, and the Midwest Association of Veterinary Pathologists.

ACKNOWLEDGMENTS

I am indebted to many people who made it possible for this work to become a reality. Therefore, I wish to express my sincere gratitude to: Professor Dr. C. F. Hollander for introducing me into the field of aging research, for providing me with the needed support to finish this study, and for his collaboration and friendship; Dr. P. Zwart for his support and comments; Dr. C. Zurcher for helping me formulate many ideas through our lengthy discussions; Dr. M. J. Van Zwieten for his encouragement and criticisms; Ms. A. Nooteboom for her technical assistance and support; Mr. A. A. Glaudemans for the preparation of the photomicrographs; Dr. S. P. Meihuizen for his assistance with the electron microscopy; Mr. W. Krause for the preparation of the charts and figures; Ms. K. Richards and Ms. D. Euser-Gaanderse for their help in the final preparation of the thesis; The technical and scientific members of the staff of the REPGO-TNO Institutes for all of their assistance; The National Institute on Aging, National Institutes of Health, U.S. Department of Health, Education and Welfare, Bethesda, Maryland and The Dow Chemical Company, Midland, Michigan for their financial support.

TABLE OF CONTENTS

Part I
Introduction

Chapter 1

INTRODUCTION

I. PURPOSE

The laboratory rat (*Rattus norvegicus*) is widely used in biomedical research. Despite this fact, few studies have been published that have correlated longevity and the age-associated patterns of spontaneous neoplastic and nonneoplastic lesions in rats. Understanding the normal pattern of diseases during aging is essential to using rats in life span studies such as the late effects of irradiation, carcinogenesis, organ transplantation, toxicology, and especially aging research. In the past, studies in these fields were relatively short term, lasting only a few months. Seldom were rats used in studies lasting 2 years or longer. Recently, there has been a trend in biomedical research to conduct more life span studies using rats, and this trend will likely continue for some time. Therefore, it is imperative that the normal range of age-associated lesions and the expected longevity of specific strains of rats be established.

This study presents age-associated pathologic findings in aging rats that had been maintained under similar and well-controlled laboratory conditions at the Institute for Experimental Gerontology TNO. The rats studied were of the BN/Bi and the WAG/Rij strains and their (WAG X BN)F_1 hybrid. The pathologic findings came from 670 untreated rats that had completed their natural life spans, were necropsied, and were examined histopathologically. Longevity data were obtained from 791 rats that were part of defined cohorts that completed their natural life spans. The results of the longevity data from the 670 necropsied rats were compared with those of the 791 rats used only for longevity studies. This study was not limited to an evaluation of data from dead or moribund rats. Experiments were performed to compare the lesions found in dead rats with those found in rats of the same age that were killed. Furthermore, electron-microscopic, histochemical, and transplantation studies were done to more clearly define some lesions. For completeness, some previously published data from this institute were also included. The various parameters studied include longevity data, age-associated neoplastic and nonneoplastic lesions, incidence of tumor metastases, and multiple lesions per rat. The background of the rats and the husbandry conditions of the rat colony are also discussed. Finally, the results from this investigation were compared with data in the literature to provide a general reference on aging rats.

II. DEFINITION OF AN "OLD RAT"

A universally acceptable definition of an old or senescent rat may not be possible because there are several conflicting definitions of aging. Some investigators consider any sexually mature rat to be old. As a result, some published studies considered rats of 14 months to be "old,"[85] while in other studies a rat of about 30 months was "old."[91]

Comfort[43] and others[29,100,182] have discussed the need for complete survival or mortality data before defining the young, adult, and aged portions of a population. Briefly, complete survival data from rats are determined from a population in which all individuals complete their normal life spans. From such a study, a survival curve can be produced which should have a shape similar to that in Figure 1.1. This curve begins after weaning and does not include neonatal or preweanling mortality. The age of young adult and mature adult rats corresponds to the postweanling plateau of the curve. The age when the population enters its phase of senescence is the point where a bend is seen, followed by a period of rapid decline in the number of surviving animals.[43]

Other investigators, working with aging rats[98] and mice,[182] have defined aged rodents as those that are older than the 50% survival age. Regardless of the definition used, complete survival data are clearly required.

The 50% survival age is useful, not only in defining an aged population, but also as a reference point to compare the longevity of different strains and stocks of rats. For example, the 50% survival age of some Sprague-Dawley,[5,144,154,184] Wistar,[17,75,102,124,152] Fischer,[42,149] and Brown Norway rats[28] is about

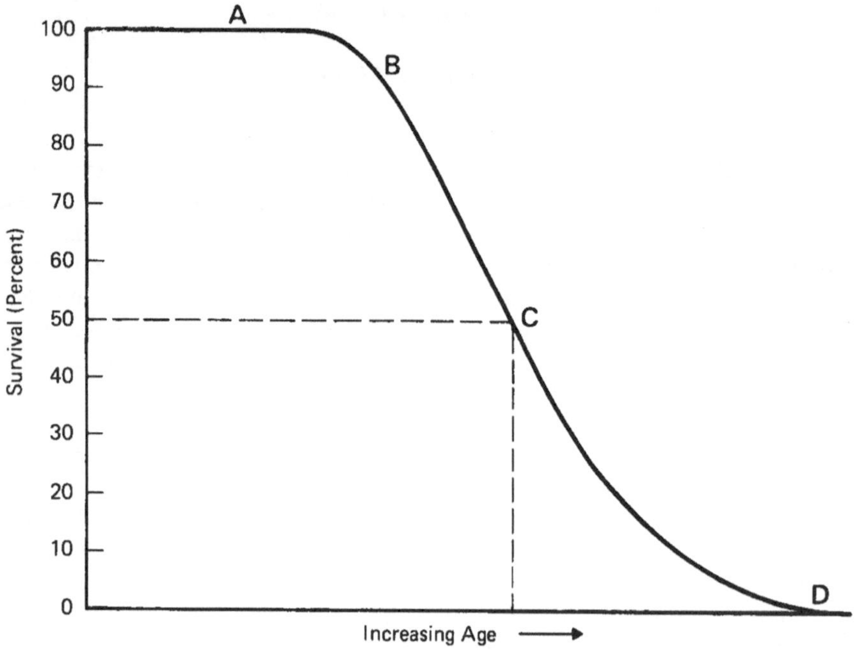

FIGURE 1.1. Percent survival with age for a population. (A) The postweanling adult population. (B) The bend in the curve followed by a rapid decline in the number of survivors. (C) The 50% survival age for the population. (D) The age when the last survivor dies.

27 to 30 months. These and other longevity studies have been recently reviewed.[29,41] If the 50% survival age is used to define an old rat, then based on the above studies, a rat of nearly 30 months is needed to study an aged rat. If the above rats are to be used in life span studies, then it can be expected that half of them will survive more than 28 months, with the oldest rats reaching ages older than 48 months.

III. LESIONS OF AGING RATS

The normal age-associated pathology of aging rats must also be defined. Several investigators have described or reviewed spontaneous nonneoplastic lesions of rats.[4,32,44,134] Tumors in rats have been discussed in detail by many investigators.[1,32,44,72,134,163,174,175] Much is known about the pathology of rats. However, few of the studies have compared both the neoplastic and nonneoplastic lesions of a strain or stock of aging rats and fewer still have studied rats that have lived out their normal life span. Much of the published data on tumor incidences for various rat strains and stocks have been derived from rats less than 2 or 2½ years old. Many of the studies included rats that were killed before the end of their natural life span, and there are few published comparisons of the lesions of rats that died spontaneously compared to those killed at the same age.

Each rat strain seems to develop its own characteristic set of lesions. Some have a high incidence of radiculoneuropathy,[12,26] while others have a very low incidence.[26] Similarly, some rat strains are plagued with a high incidence of chronic progressive glomerulonephropathy while others have a low incidence.[156] The incidence of neoplastic lesions also varies. For example, cancers of the ureter and bladder are common in BN/Bi rats[21] but are rare in other strains. Therefore, it is clearly impossible to simply list the most common lesions in aging rats. Investigators must determine the background of the neoplastic and nonneoplastic lesions in the rats they are using.

IV. FACTORS THAT CAN INFLUENCE LIFE SPAN STUDIES

Survival data alone may give misleading information because extraneous factors can alter "normal" longevity. Many factors can influence the shape of a survival or mortality curve and can change the 50%, the mean, and the maximum survival ages. Husbandry conditions need to be known before the longevity and age-associated pathology can be evaluated. Recent reviews have stressed the importance of this information.[29,39,41]

It is difficult to decide whether germ-free, specific-pathogen-free, or conventional rats are the best for life span studies; each has advantages and disadvantages. In his recent review, Weisbroth[183] reevaluated different barrier systems for laboratory rats. The major problem was the cost involved compared to the expected results. The stricter the barrier system, the greater the cost to produce and maintain rats. One wants the highest quality rats for any research, especially for life span studies, but there may be a point where the costs become too great for the potential results. Because of the need for recommendations to resolve this problem, guidelines have been published by Jonas[97] that help to clarify this question.

Specific husbandry conditions have been shown to alter longevity and pathologic findings in rats. First, infectious diseases definitely affect longevity. Paget and Lemon[124] showed that "dirty" rats do not live as long as their specific-pathogen-free counterparts. Cohen[39] has stressed how environmental influences and infectious diseases can alter longevity and the results of experiments. Chronic respiratory disease[110] is the most important infectious disease that can affect longevity, but others may be important and they have been outlined by Squire.[161]

Second, the diet can affect longevity, as clearly illustrated in the studies by Simms,[154] Sims and Berg,[153] Berg and Simms,[10,11,13] Ross[140,143] and Ross and Bras[141] and in the recent review by Cohen and Anver.[41] Dietary changes result in altered survival data and may also influence biochemical values,[57] may result in different patterns of both neoplastic[142,144] and nonneoplastic disease,[123,154] and can alter organ weights.[151]

Finally, breeding may influence the longevity of rats, as suggested by the studies of Wexler and Greenberg.[184] They showed that retired breeders do not live as long as virgins. Paget and Lemon[124] reported similar findings with earlier mortality occurring in SPF breeder rats than their SPF virgin counterparts. It is not known, however, if such observations are true for all strains of rats.

V. INTRODUCTION TO THE CHAPTERS

This study was done to evaluate two rat strains and their F_1 hybrid. All were maintained at the Institute for Experimental Gerontology TNO, The Netherlands. Evaluation of the longevity data and pathologic findings requires a knowledge of the background of the rats, the husbandry of the colony, and the literature. The following chapters provide answers to some of the questions raised in this Introduction (Chapter 1).

Chapter 2 — A general description of the BN/Bi, WAG/Rij, and (WAG × BN)F_1 rats is presented. This chapter describes their husbandry conditions, breeding history, and the methods of selecting and necropsying rats.

Chapter 3 — The longevity of each rat strain is presented. The survival of virgin female rats is compared to that of retired breeders, and the survival data from males to that of females of the same strain. These findings are compared with the survival data from the rats that are necropsied and presented in Chapters 4 and 5.

Chapter 4 — This chapter is organized into organs and organ systems. The major neoplastic and nonneoplastic lesions are described, tabulated, and often documented with photographs. The findings in each organ system are briefly discussed and are compared to previously published findings.

Chapter 5 — The rats described in Chapter 4 often had multiple neoplastic and nonneoplastic lesions. An evaluation of the number of lesions in these rats at different ages, especially the neoplastic lesions, is presented in this chapter and the age-associated risk of these rats dying with metastic neoplasms is also examined.

Chapter 6 — In this chapter, a comparison of three different figures is made in order to estimate the incidence of a lesion in the living population. These figures are (1) the percent of rats that died with a given lesion, (2) the percent of the population at risk that had the lesion, and (3) the actual percent in rats killed at specific ages. The data for Items 1 and 2 were obtained from the necropsied rats reported in

Chapter 4. The percentage in killed rats was obtained by selectively killing groups of 14- and 21-month-old rats.

Chapter 7 — Further studies on the syndrome of posterior paralysis that was observed in aging (WAG X BN)F_1 rats are reported. These studies included serial killings to determine the age of onset of the lesions, evaluation of the lesions in males and females, evaluation of the peripheral nerves, and skeletal muscle and semiserial sectioning of the entire spinal cord in order to localize lesions within the spinal cord.

Chapter 8 — Selected parameters in the F_1 hybrid rat are compared with the parent strains, and a discussion of the similarities and differences is presented in this chapter.

Chapter 9 — Potential models of certain diseases have been or are still being developed using the BN/Bi, WAG/Rij, and (WAG X BN)F_1 rats. This chapter summarizes those potential models that have been recognized and may be useful for aging or cancer research.

Chapter 10 — The importance of life span longevity and pathology data from rats are stressed in this chapter.

Chapter 11 — This chapter contains a general discussion of the age-associated lesions and their effect on aging in rats.

Part II
Background Information

Chapter 2

BACKGROUND OF THE ANIMALS AND THE RAT COLONY

I. BACKGROUND OF THE RATS

Festings and Staats[55] summarized the history and salient characteristics of several rat strains. Among those listed were the Brown Norway (BN) and the Wistar-derived (WAG) strains. These two strains are particularly emphasized here because they are the sources for the BN/Bi and WAG/Rij rats of this study.

The BN strain, BN/Bi, was originally started by Silvers and Billingham in 1958 from a brown mutation obtained from King and Aptekman. They were obtained as conventional animals from Microbiological Associates Inc. (Bethesda, Md.) in 1963. After arrival, rats were derived by hysterectomy, nursed on germ-free mothers, and raised as specific pathogen-free (SPF) rats. They have been inbred in Rijswijk since 1963 by brother × sister mating and are now in their 22nd generation. The degree of inbreeding has been frequently confirmed by reciprocal skin grafting showing permanent takes in 100% of the grafts.

The WAG/Rij is a small white WAG rat. It was obtained from Glaxo Laboratories (Greenford, Middlesex, England) in 1953. The original stock was derived from conventional animals. Later, germ-free rats were derived by hysterectomy. The germ-free rats then were used to establish an SPF breeding colony. They have been inbred since 1953 by brother × sister mating and are now in their 39th generation. As with the BN/Bi rats, the degree of inbreeding has been frequently confirmed by reciprocal skin grafting showing permanent takes in 100% of the grafts. Skin grafting between BN/Bi and WAG/Rij consistently is rejected within 10 to 12 days. They apparently differ at the major H-1 histocompatibility locus.[164,165]

The (WAG X BN)F_1 rat described in this study is the F_1 hybrid of a cross between female WAG/Rij and male BN/Bi rats. As would be expected with an F_1 hybrid, skin grafting from BN/Bi or WAG/Rij to the F_1 is successful with permanent takes in 100% of the grafts, while grafts from the F_1 to the BN/Bi or WAG/Rij parent are rejected.

II. COLONY HUSBANDRY

All rats were from the aging colony maintained at the Institute for Experimental Gerontology TNO. The source of those rats, however, was a larger production colony that produced rats for the institutes of the REP* Institutes in Rijswijk. There were three colonies in the REP Institutes: (1) a germ-free colony, (2) a SPF breeding and stock colony, and (3) the aging colony. The husbandry protocols for the various colonies are summarized in Table 2.1. A flow diagram of the rats within the REP Institutes is shown in Figure 2.1.

Bacteriologic monitoring was done on a routine basis in the SPF breeding colony. Five rats (one rat from each of five different cages) were randomly selected every month. Each rat was then killed and examined by gross necropsy. Lung tissue and feces were taken for routine bacterial cultures. In addition, culturing for mycoplasma was also attempted from the lung tissue. Individual rats from the aging colony were cultured from time to time, but routine bacteriological monitoring was not performed. If an organism was isolated from any rat, the entire colony was considered positive. Table 2.2 lists the bacteria found in 1974 and 1975. Although attempts were made to isolate other organisms, including mycoplasmas, they were not detected in the colonies.

Serologic tests for viruses latent in the SPF and aging colonies were done once a year. The tests were performed on each individual rat serum by M. C. J. van Nunen, Catholic University, Nijmegen, The Netherlands. Complement fixation tests were used to detect antibodies to murine adenovirus, lymphocytic choriomeningitis (LCM) virus, mouse hepatitis virus, reovirus-3, vaccinia virus, and Sendai virus. Hemagglutination inhibition tests were used for K-virus, polyoma virus, minute virus, Theiler gd

* REP stands for the Radiobiological Institute, Institute for Experimental Gerontology and Primate Center.

TABLE 2.1

Conditions and Standard Protocol for the Different Rat Rooms in the REP Institutes TNO

	GF	SPF breeding	SPF stock	Aging colony
Air changes per hour	Unknown	15	15	15
Pressure	+	+	+	+
Temperature (°C)	20—22	20—22	20—22	20—22
Humidity (%)	Unknown	65—70	60—65	60—65
Filters	Absolute bacterial/viral filters	Insect and dust; electrostatic filters	Insect and dust filters	Insect and dust filters
Food	Autoclaved AM II plus supplement	Autoclaved AM II plus supplement	AM II only not autoclaved	AM II only not autoclaved
Water	Autoclaved	Acidified to pH 3.0	Tap water	Tap water
Bottles	Autoclaved	Autoclaved and changed two times per week	Automatic system	Autoclaved and changed three times per week
Polycarbonate cages	Autoclaved	Cleaned and sterilized two times per week	—	Washed in 80—90°C water, changed two times per week
Wire cages	—	—	Washed one time per week with water and steam	—
Bedding	Sterilized sawdust	Sterilized sawdust	Wire cages	Presterilized sawdust
Floors	—	Cleaned daily, disinfected with Halamide®[a]	Cleaned two times per week, disinfected with Halamide®	Cleaned one time per week, disinfected with Halamide®
Light schedule	12 hr light 12 hr dark	12 hr light 12 hr dark	12 hr light 12 hr dark	12 hr light 12 hr dark
Personnel	—	Entry via four lockers, sterilized clothes, boots and gloves, caps and face mask	Entry via one locker, quarantine clothes, boots disinfected, face mask	No locker, laboratory clothes
Type of control of who enters the room	—	Restricted—only specific personnel	Limited access	Limited access

[a] Halamide = chloramine (N-chloro-p-toluensulfonamide) sodium.

VII virus, encephalomyocarditis virus, pneumonia virus of mice, and Sendai virus. Serum samples from the SPF and aging colonies tested prior to November 1974 were negative for all viruses except pneumonia virus of mice (PVM). In November and December of 1974, the entire rodent colony at the REP Institutes experienced an epizootic Sendai virus infection.[29,190] As a result, additional serology testing was done to monitor the titers to Sendai virus in the rodents. Table 2.3 summarizes the titers for all viruses tested and the number of rats examined during 1974 and 1975.

A. Germ-free Colony

Germ-free (GF) BN/Bi and WAG/Rij rats were maintained in case it might become necessary to start a new SPF breeding colony. The number of GF rats produced provided a few animals for experiments; however, they were not maintained for life span or pathology studies.

B. Specific Pathogen-free Breeding and Stock Colony

The SPF breeding rats were first given "colony-resistant flora" (CRF) and then transferred to the breeding colony. The organisms in

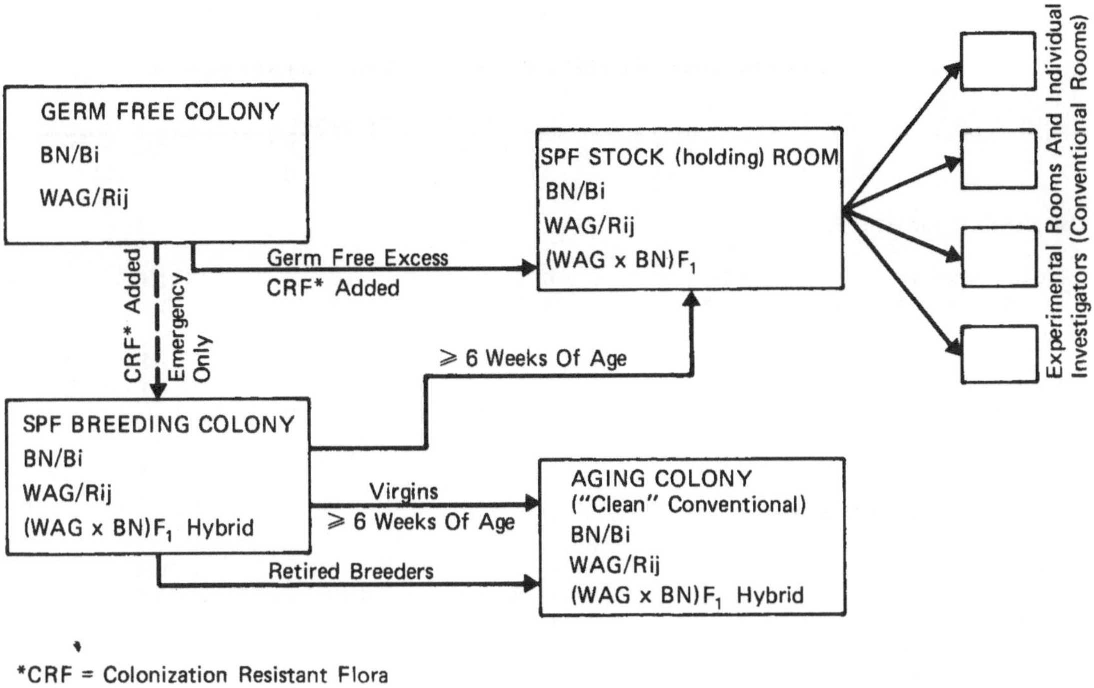

FIGURE 2.1. Flow diagram of BN/Bi, WAG/Rij, and (WAG × BN)F_1 rats within the REP Institutes TNO.

TABLE 2.2

Bacteria That Have Been Isolated from the SPF and Aging Rat Colonies

	1974		1975	
Organism	SPF	Aging	SPF	Aging
Escherichia coli	+	+	+	+
Staphylococcus aureus	−	+	−	+
S. albus	+	+	+	+
Streptococcus faecalis	+	+	+	+
Enterobacter cloacea	+	+	+	+
Hafnia sp.	+	+	+	+
Bacillus sp.	+	+	+	+
Bordetella bronchiseptica	−	−	+	+
Klebsiella pneumoniae	−	−	+	+

the CRF and the techniques used have been described[178] and included nonpathogenic strains of *Escherichia coli*, *Staphylococcus* sp., and *Streptococcus*.

In the breeding colony, rats were housed in polycarbonate cages (Makrolon®) 35 cm long × 23 cm wide × 16 cm high. One male was placed with three females. When a female became pregnant, it was removed from the breeding cage and was individually caged for parturition, nursing, and weaning. After weaning, the female was returned to the breeding cage of origin with the male and other females. The weaned rats were kept together until about 6 weeks old, at which time they were transferred to the SPF stock rooms or to the aging colony.

The young stock rats from the breeding colony were separated by sex, strain, and age. They were housed in large 63 cm long × 42 cm wide × 20 cm high wire cages. Approximately

TABLE 2.3

Summary of Serology Findings from the SPF and Aging Rat Colonies of the REP Institutes

	Number of rats examined			
	1974— 14	January 1975— 15	May 1975— 35	August 1975— 25
Complement fixation test[a]				
Murine adenovirus	<1:5	<1:5	<1:5	<1:5
Lymphocytic choriomeningitis (LCM) virus	<1:5	<1:5	<1:5	<1:5
Mouse hepatitis virus	<1:5	<1:5	<1:5	<1:5
Sendai virus	<1:5	<1:10—1280 (402)[b]	1:20—1280 (292)	1:5—160 (50)
Vaccinia virus	<1:5	<1:5	<1:5	<1:5
Reovirus-3	<1:5	<1:5	<1:5	<1:5
Hemagglutination inhibition test[c]				
K-virus	<1:10	<1:10	<1:10	<1:10
Polyma virus	<1:20	<1:20	<1:20	<1:20
Minute virus	<1:20	<1:20	<1:20	<1:20
Theiler gd VII virus	<1:20	<1:20	<1:20	<1:20
Encephalomyocarditis virus	<1:10	<1:10	<1:10	<1:10
Pneumonia virus of mice (PVM)	1:10—320 (60)	1:10—160 (49)	1:20—640 (174)	1:10—2560 (262)
Sendai virus	<1:10	Not determined	1:20—5120 (370)	1:10—320 (48)

[a] Complement fixation titers of <1:5 were considered negative.
[b] The titers for Sendai virus and PVM virus represent the range (mean) of all individual sera tested.
[c] Hemagglutination inhibition titer of <1:10 or <1:20 were considered negative.

20 rats were kept per cage until transferred to experiments.

The breeding colony was small enough so breeders could be culled from the colony on an individual basis. Female rats were culled when the number of youngborn or weaned per litter decreased and males when they failed to make females pregnant. As a result, some rats were removed from the breeding colony at about 6 months old and others remained as active breeders even when older than 15 months.

C. Aging Colony

The aging colony was maintained at the Institute for Experimental Gerontology TNO. The rats entering this colony came from the SPF stock and breeding colony of the REP Institutes (Figure 2.1) and consisted of retired breeders and virgins. The retired breeders from the breeding colony varied in age from a few months to about 15 months old when they entered the aging colony. Virgin rats were about 6 weeks old when they entered the aging colony.

The aging colony was maintained under "clean" conventional conditions.[81] It was not maintained under strict barrier isolation, but the microbiological flora was known and there was limited access into the rat rooms by certain personnel and animal caretakers (Table 2.1). *E. coli, Klebsiella pneumoniae, Bordetella bronchiseptica, Staphylococcus* sp., *Streptococcus* sp., *Enterobacter* sp., *Hafnia* sp. and *Bacillus* sp. were infrequently isolated from these rats (Table 2.2). There was also a low-level but constant infection by pinworms.

Titers to PVM and Sendai virus were detected in serum from rats as shown in Table 2.3. The longevity and pathology data presented in subsequent chapters were derived from rats that died or were killed from early 1971 until June 1976. PVM was present in the colony during most or all of this period. Sendai virus entered the colony in November or December 1974. Therefore, some rats died in the pre-Sendai period and others died or were killed after Sendai virus entered the colony.

TABLE 2.4

Composition of the AM II Diet

Ingredients	Percentage by weight
Whole wheat meal (North American, not denaturated)	24
Whole wheat meal (Western European, not denaturated)	24
Oat meal (ground)	8
Whole linseed (ground, not extracted, not pressed out)	6
Alfalfa meal (minimum crude protein 17% artificial dried)	3
Grassmeal (minimum crude protein 17% artificial dried)	2
Cornmeal (±9% crude protein)	5
Meatmeal (crude protein 70—80%)	15
Lactalbumin	4
Yeast (dried)	4
Cornoil (minimum 50% unsaturated fat)	2
Minerals	1
Vitamin A-D premix	0.1
Vitamin B premix	2.0

Analysis	Theoretical (%)	Practical (%)
Crude protein	25.2	24.8 — 25.5
Digestible crude protein	21.3	—
Real protein	21.6	—
Digestible real protein	17.7	—
Crude fat	8.0	7.8 — 8.2
Crude fiber	3.6	± 3.5
Other carbohydrates	47.3	—
Moisture	10.4	10.0 — 10.5
Ash	4.1	—

Aging rats were housed in polycarbonate cages 35 cm long × 23 cm wide × 16 cm high. The number of rats per cage varied from one to three. Sexes and strains were maintained separately.

III. DIET

The basic diet for all rats was the AM II diet (Hope Farms, Woerden, The Netherlands). The composition and the theoretical and practical analysis are summarized in Table 2.4. The GF and SPF breeding rats, however, all received autoclaved AM II diets mixed with a supplement diet (Table 2.5) The AM II pellets were mixed with the supplement food pellets at a ratio of 9 AM II to 1 supplement before autoclaving. The precise composition of this combined diet was not available. Water was provided to all animals ad libitum (Table 2.1).

IV. BREEDING PERFORMANCE AND PREWEANLING MORTALITY

The records used to calculate the survival and mortality data, presented in the subsequent chapter, began after weaning. Ideally, the data should be collected from birth, but this was not possible. To partially overcome this deficiency, some brief comments on the breeding colony are discussed.

The maximum number of litters in one generation of BN/Bi rats was about eight and was about six for WAG/Rij rats. The average litter size born per female was five for BN/Bi rats and eight for WAG/Rij rats. The average number weaned per litter for BN/Bi rats was four and for WAG/Rij rats was six. The preweanling mortality for BN/Bi rats varied from 9 to 24% depending on the number of litters per female, time of the year, and other factors. For

TABLE 2.5

Composition of Supplement Diet[a]

Analysis	Theoretical (%)
Crude protein	29.9
Digestible crude protein	25.8
Real protein	26.4
Digestible real protein	22.5
Crude fat	8.2
Crude fiber	3.1
Other carbohydrates	42.5
Moisture	9.9
Ash	4.1

[a] The precise composition of this diet is not available for publication.

TABLE 2.6

General Data from the Breeding Colony for One Average Generation of BN/Bi Rats

Litter number in one generation of breeding	0	1	2	3	4	5	6	7	8
Percent of breeding females remaining for the next litter	100	100	99	95	84	60	34	10	4
Average litter size at birth	—	4.4	5.6	5.5	5.4	5.2	4.8	3.3	6.5
Average number weaned per litter	—	3.9	5.2	5.1	4.5	4.3	3.9	0.8	3.5
Average preweanling mortality per litter (%)	—	13	9	9	16	15	16	24	46

WAG/Rij rats, the preweanling mortality ranged from 15 to 31% per litter. The average preweanling mortality per generation for BN/Bi rats was 25% and for WAG/Rij rats was 24%. These data are listed in Table 2.6 for BN/Bi rats and Table 2.7 for WAG/Rij rats.

V. WEIGHTS OF AGING RATS

Weights were determined by weighing all rats of a given age that were present in the aging colony on one day. The average weight (±1 standard deviation) in grams for virgin, male and female BN/Bi, WAG/Rij and F_1 rats are summarized in Figure 2.2. Each point plotted represents 7 to 44 rats, with the lower number of rats per point occurring in the groups older than 30 months.

Female BN/Bi, WAG/Rij, and (WAG × BN)F_1 rats of similar ages had nearly identical weights. Females older than 6 months weighed between 180 and 220 g, with the older rats weighing only slightly more than the younger. There was greater variability in the weights of male BN/Bi, WAG/Rij, and (WAG × BN)F_1 rats older than 6 months. The males as a group were significantly heavier than females of the same age.

VI. PROTOCOL FOR SICK OR DEAD RATS

A certain number of losses will occur from cannibalism or autolysis during any life span study. To reduce such losses, inspections for sick and dead rats were made twice daily during

TABLE 2.7

General Data from Breeding Colony for One Average Generation of WAG/Rij Rats

Litter number in one generation of breeding	0	1	2	3	4	5	6
Percentage of breeding females remaining for the next litter	100	99	95	90	75	46	5
Average litter size at birth	—	7.4	8.4	8.8	7.9	8.3	7.4
Average number weaned per litter	—	5.5	6.4	6.8	6.7	6.4	5.1
Average preweanling mortality per litter (%)	—	26	24	23	15	23	31

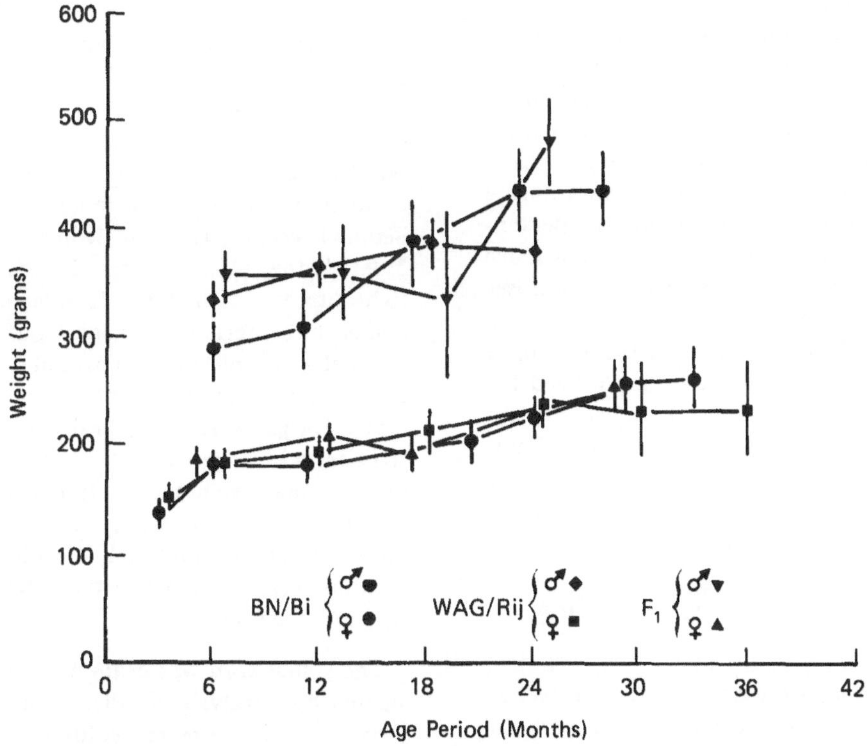

FIGURE 2.2. Mean weight (±1 standard deviation) in grams of aging, male and female, BN/Bi, WAG/Rij, and (WAG × BN)F_1 rats.

the week and once daily on weekends and holidays. When a rat was observed to be sick, less active, or to have a large lesion such as a tumor, it was removed from the aging colony, transferred to an individual cage, and moved into a "sick ward." The "sick ward" was in a location such that sick rats could be observed more frequently. Since it is desirable that aging rats live as long as possible, the sick were not killed until they became moribund. When a rat was killed, moribund, or found dead, it was submitted for necropsy. It is desirable to have complete necropsy material and well-preserved tissue for histopathology, therefore, rats that were autolyzed or partially eaten were discarded.

TABLE 2.8

Summary of the Strain, Sex, Breeding History, and Number Studied for Each Group of Necropsied Rats

Strain	Sex	History	Number examined
BN/Bi	Female	Virgins	236
BN/Bi	Male	Virgins	74
WAG/Rij	Female	Virgins	101
WAG/Rij	Male	Retired breeders and virgins	124
(WAG × BN)F_1	Female	Virgins	68
(WAG × BN)F_1	Male	Virgins	67

Strain differences were observed in the number of rats lost because of autolysis and cannibalism. Losses of WAG/Rij and F_1 rats were about 15%, while 30% of the BN/Bi rats were lost from the aging colony. Such differences are difficult to explain, but they cannot be ignored. The elimination of these rats from aging studies may have altered the age-associated patterns of lesions found in this study. Most losses from autolysis occurred in rats dying at night, weekends, or holidays, and to reduce losses during those periods would require additional personnel to examine the colony. The cost to produce rats of 2 to 3 years may, however, justify closer observations of the aging colony. In a separate publication, Burek and Hollander[29] have stressed these points and have emphasized the need for baseline pathology ata to help explain strain differences in the percentage of animals lost from autolysis and the need to consider such losses in the interpretation of aging pathology data from various rat strains.

VII. NECROPSY PROCEDURE

The necropsies were done by trained technicians under the supervision of three pathologists and followed rigid protocol. The rat was examined for external lesions and then laid on its back and pinned down. A midline incision was made from the mandibular symphysis to the anus. The skin was reflected so that mammary gland tissue, superficial lymph nodes, and other subcutaneous structures could be examined. The abdomen and thorax were opened. Each organ system and regional lymph nodes were examined for abnormalities. All gross pathologic observations were recorded and often photographed. Measurements were made and recorded for most tumors, but whole body weights and weights of specific organs or tumors were not obtained routinely.

The tissues routinely included for microscopic examination, and the minimum number of sections from each organ, are as follows: skin, 1; salivary gland, 2; lungs, 4; heart, 2; esophagus, 1; stomach, (1 section with squamous and glandular portions); duodenum, 1; ileum, 1; cecum, 1; colon, 1; liver, 3; pancreas, 2; left kidney, 1; left ureter, 1; right kidney, 1; right ureter, 1; bladder, 1; spleen, 1; lymph nodes: mesenteric — 2, superficial — 2, deep cervical — 2, anterior mediastinal — 1; thymus, 1; sternum, 1; left and right thyroid, 2; left adrenal, 2; right adrenal, 2; brain, 5; pituitary, 1; right testicle, 1; left testicle, 1; prostate, 1; accesssory male genitalia, 2; left ovary, 1; right ovary, 1; uterine horns, 2; cervix, 1; and mammary glands in females, 1. In addition, more sections were taken when indicated, as in the case of large tumors.

The spinal cord was examined only when indicated by paralysis or when lesions were observed in the vertebral column. Joints were taken for histology only when indicated by gross abnormalities. Finally, the eyes, ears, and skull (including nasal passages) were not routinely examined.

Tissues were fixed in phosphate-buffered 10% formalin or Tellyesniczky's fluid. The tissues were embedded in paraplast, sectioned with a microtome (56 μm), and were routinely stained with hematoxylin-phyloxin-saffron

(HPS). Other stains were prepared when required.

VIII. SELECTION OF NECROPSIED RATS

The groups of necropsied rats are summarized in Table 2.8. The female and male BN/Bi, female and male F_1, and female WAG/Rij rats were all virgins. The male WAG/Rij rats were a mixture of retired breeders and virgins because the breeding history was not recorded on the pathology records, but most were virgins.

The rats died during the period of 1971 to June 1976. They represent several different cohorts.* Some rats were removed from the colony for different aging studies and some were lost from studies because of autolysis or cannibalism. The remaining rats were necropsied when found dead, killed, or moribund and represented random colony deaths.

* For this study, a cohort is defined as a group of rats of the same strain, sex, and age that entered the aging colony at the same time, and all members of which were examined throughout their life spans.

Part III
Results and Discussions

Chapter 3

LONGEVITY STUDIES

I. LONGEVITY OF COHORTS* OF AGING RATS

The survival curves from cohorts of rats and from those necropsied are shown in Figures 3.1 to 3.6. A summary of the 90, 50, and 10% and maximum survival ages is presented in Table 3.1 from the cohorts of virgins, in Table 3.2 from the cohorts of retired breeders, and in Table 3.3 from the rats that were necropsied for histopathology.

A. Cohorts of Virgins

The data from complete cohorts of virgin male and female BN/Bi rats, virgin male and female F_1 rats, and virgin female WAG/Rij rats showed similar longevity which can be summarized as follows: a 90% survival age of 22 (±3) months, a 50% survival age of 31 (±2) months, and a 10% survival age of 39 (±2) months. The maximum life span observed from these cohorts ranged from 42 to 48 months.

The male WAG/Rij rats, unlike the other cohorts of virgin rats, had a shorter life span (Figure 3.4 and Table 3.1). The 90% survival age was 13 months, the 50% survival age was 24 months, the 10% survival age was 31 months, and the oldest male was 34 months. Therefore, the cohort of WAG/Rij males died 7 to 9 months earlier than rats in any of the other cohorts of virgins.

B. Cohorts of Retired Breeders

The survival curves of retired breeder female BN/Bi rats are presented in Figure 3.1 and for female WAG/Rij rats in Figure 3.3. The 90, 50, and 10% and maximum survival ages of these two cohorts are summarized in Table 3.2. Figure 3.7 shows the same data, expressed as percent mortality, for the virgin and retired breeder female BN/Bi and WAG/Rij rats.

Retired breeder BN/Bi females appeared to survive longer than virgins. The actual difference, however, was only 2 months or less over

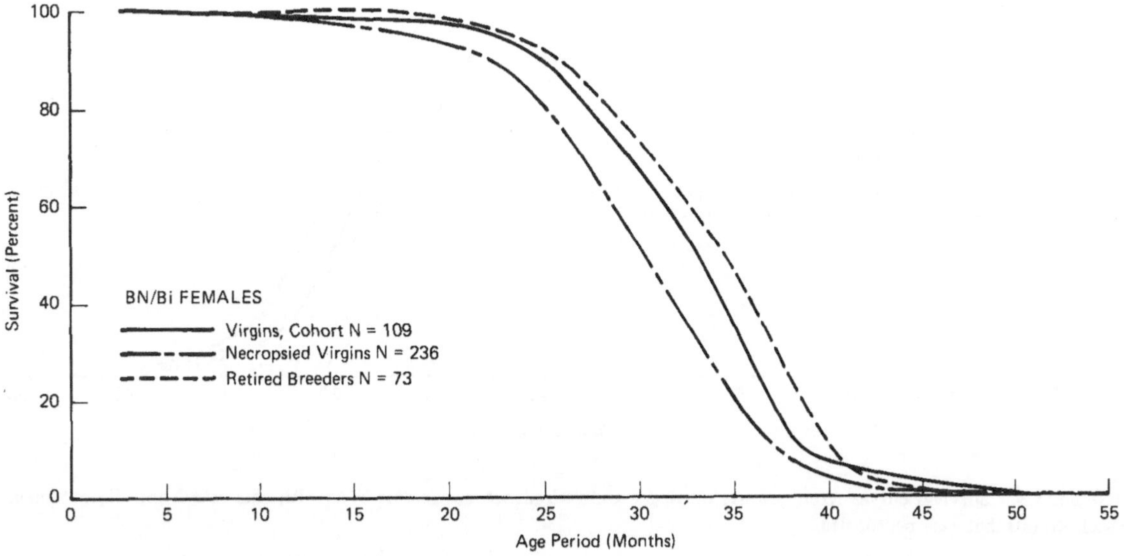

FIGURE 3.1. Survival curves from cohorts of virgin and retired breeder female BN/Bi rats and from virgin female BN/Bi rats that were necropsied.

* For this study, a cohort is defined as a group of rats of the same strain, sex, and age that entered the aging colony at the same time, and all members of which were examined throughout their life spans.

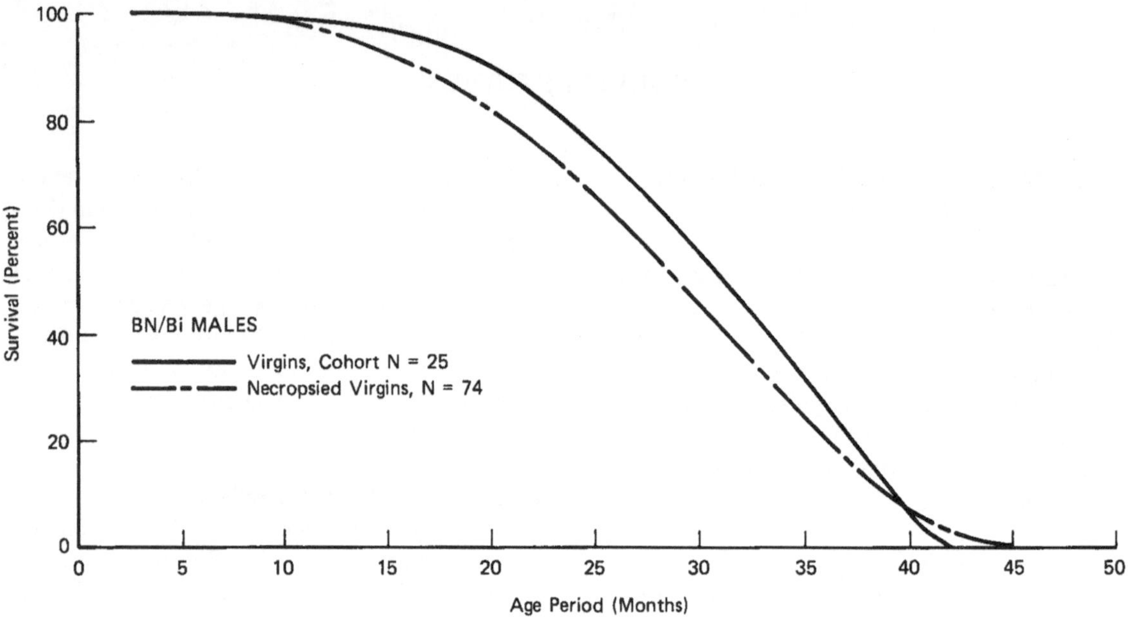

FIGURE 3.2. Survival curves from a cohort of virgin male BN/Bi rats and from virgin male BN/Bi rats that were necropsied.

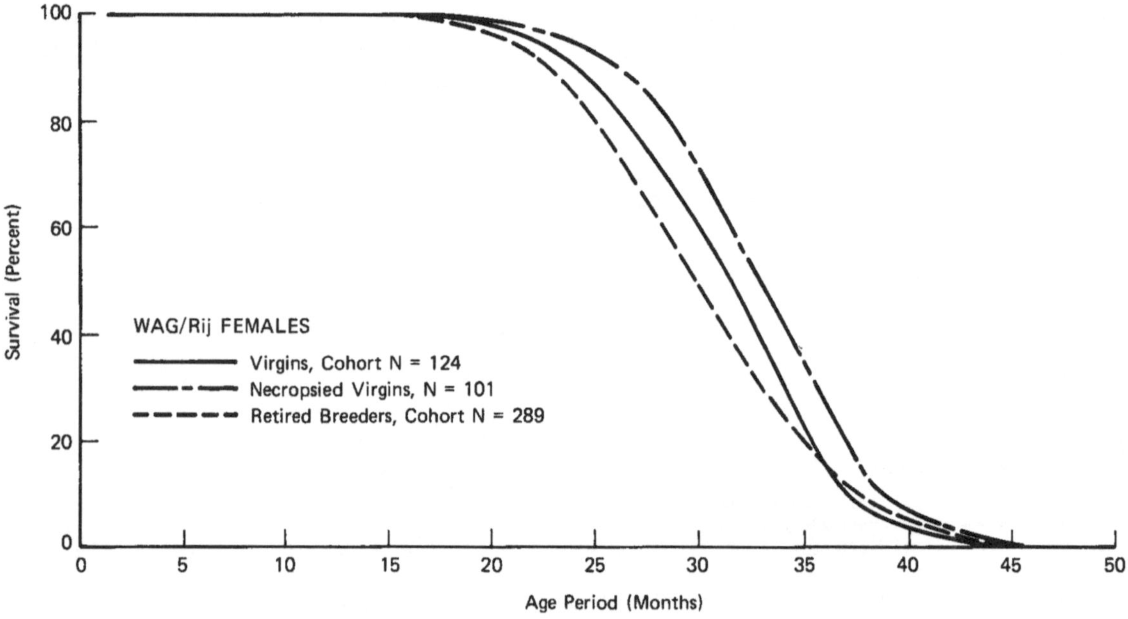

FIGURE 3.3. Survival curves from cohorts of virgin and retired breeder female WAG/Rij rats and from virgin female WAG/Rij rats that were necropsied.

the life span of both cohorts. Retired breeder WAG/Rij females, on the other hand, had a slightly shorter life span than virgin WAG/Rij females. Here again, the retired breeders and virgins differed by only 1 month at the 90 and 10% and maximum survival ages and by 3 months at the 50% survival age.

The BN/Bi female retired breeders lived

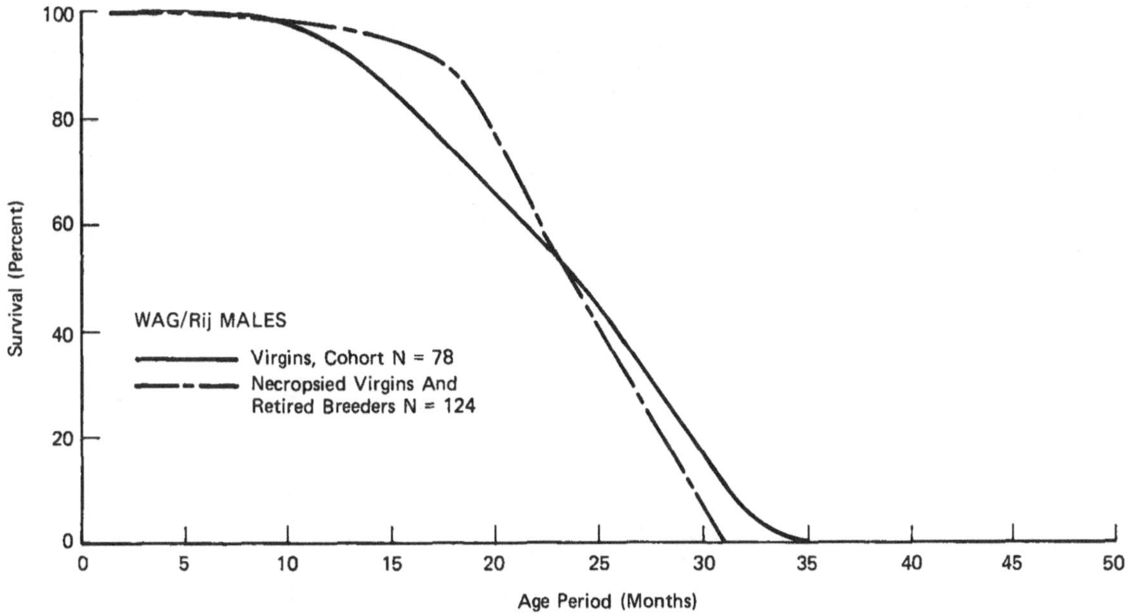

FIGURE 3.4. Survival curves from a cohort of virgin male WAG/Rij rats and from male WAG/Rij rats that were necropsied.

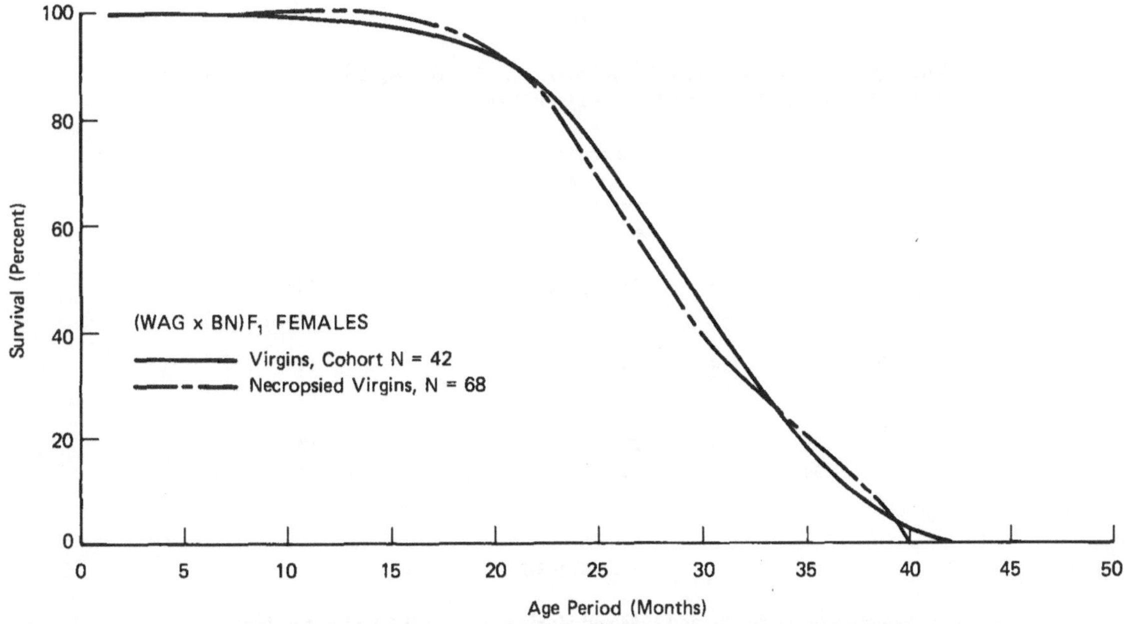

FIGURE 3.5. Survival curves from a cohort of virgin female (WAG × BN)F_1 rats and from virgin female (WAG × BN)F_1 rats that were necropsied.

longer than the WAG/Rij retired breeders as illustrated in Figure 3.7. They differed by 3 months at the 10% mortality ages, by 6 months at 50% mortality ages, and by 2 months at the 90 and 100% mortality ages.

II. LONGEVITY OF THE NECROPSIED RATS

Some differences were found between necropsied rats and cohorts of the corresponding

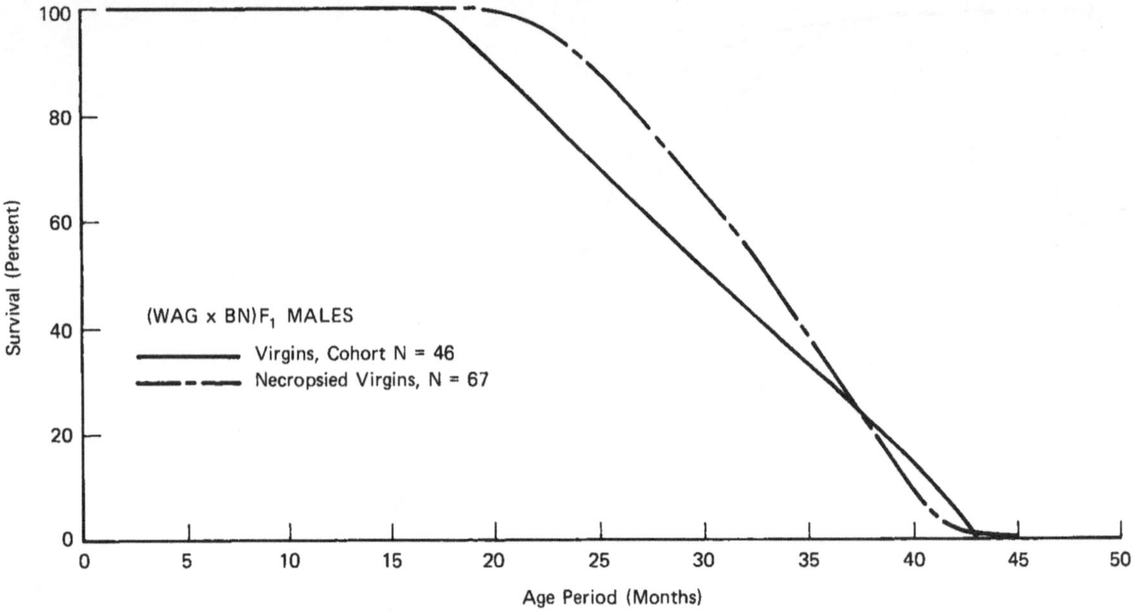

FIGURE 3.6. Survival curves from a cohort of virgin male (WAG × BN)F$_1$ rats and from virgin male (WAG × BN)F$_1$ rats that were necropsied.

TABLE 3.1

Summary of the 90, 50, and 10% and Maximum Survival Ages from Cohorts of Virgin, BN/Bi, WAG/Rij, and (WAG × BN) F$_1$ Rats

			Percent survival age (months)			
Strain	Sex	Number	90%	50%	10%	Maximum survival age (months)
BN/Bi	F	109	25	33	38	48
	M	25	21	32	39	42
WAG/Rij	F	124	24	32	37	45
	M	78	13	24	31	34
F$_1$	F	47	22	29	37	42
	M	46	19	30	41	43

TABLE 3.2

Summary of the 90, 50, and 10% and Maximum Survival Ages from Cohorts of Retired Breeder, Female BN/Bi, and WAG/Rij Rats

		Percent survival age (months)			
Strain	Number	90%	50%	10%	Maximum survival age (months)
BN/Bi	73	26	35	40	46
WAG/Rij	289	23	29	38	44

TABLE 3.3

Summary of the 90, 50, and 10% and Maximum Survival Ages from Necropsied BN/Bi, WAG/Rij, and (WAG X BN)F₁ Rats

Strain	Sex	Number	Percent survival age (months) 90%	50%	10%	Maximum survival age (months)
BN/Bi	Female	236	22	30	37	54
	Male	74	16	28	37	44
WAG/Rij	Female	101	27	34	38	46
	Male	124	18	23	27	31
F_1	Female	68	21	28	38	40
	Male	67	24	33	40	44

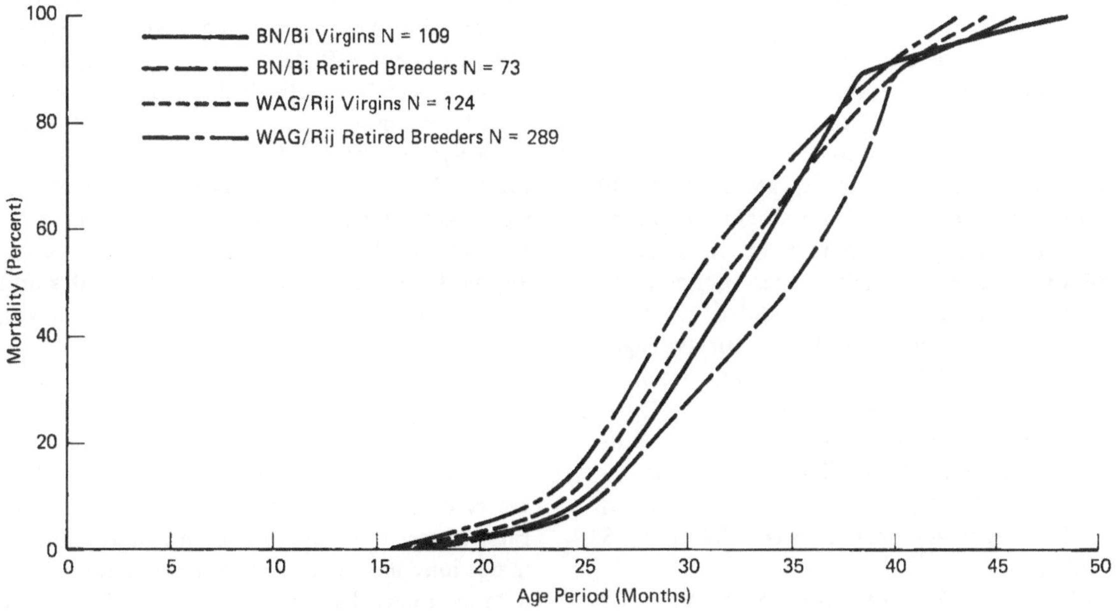

FIGURE 3.7. Mortality curves from cohorts of virgin and retired breeder female BN/Bi and WAG/Rij rats.

sex and strain. The necropsied rats showed shorter survival ages for the female (Figure 3.1) and male (Figure 3.2) BN/Bi. The necropsied groups had longer survival ages for the female WAG/Rij (Figure 3.3) and male F_1 (Figure 3.6). The necropsied male WAG/Rij (Figure 3.4) showed longer survival in the first part of the survival curve and shorter survival in the lower part. Finally, the necropsied female F_1 rats (Figure 3.5) had survival curves that were virtually identical to the cohort. The 90, 50, and 10% and maximum survival ages of all groups of necropsied rats are summarized in Table 3.3.

III. DISCUSSION

The longevity of aging rats is influenced by many variables such as the strain or stock, sex, husbandry conditions, breeding history, and diet. The importance of these factors and their influence on longevity have been stressed in recent reviews.[29,41] The need to understand these variables when evaluating conflicting experimental results from different laboratories has also been stressed.[29,56,57,71] It is difficult, therefore, to make direct comparisons of longevity data from different laboratories and among different strains or stocks of rats.

The published longevity data on male Fischer 344 rats provide an excellent example of the variable results that can be obtained from different laboratories. Early studies by Jay[96] showed that the 50% survival of Fischer 344 rats was only 12 months, suggesting that it was a short-lived strain. Chesky and Rockstein[38] found that the 50% survival age of their Fischer 344 males was about 22 to 23 months. Most recently, Coleman et al.[42] demonstrated that barrier-reared Fischer 344 male rats had a 50% survival of about 29 months. Therefore, the same rat strain, maintained at different laboratories, can have 50% survival ages ranging from 12 to 29 months.

Life span data are not available for males and females and for retired breeders and virgins of all the commonly used strains and stocks of rats. However, most published studies indicate that the 50% survival age is about 27 to 30 months for most strains and stocks of rats as previously discussed (Chapter 1). In general, SPF rats had 50% survival ages of about 28 months. Conventional rats tended to have slightly earlier mortality with 50% survival ages between 12 and 27 months. The data obtained in this study from the aging colony at the Institute for Experimental Gerontology were similar. The male and female BN/Bi, male and female (WAG X BN) F_1, and female WAG/Rij had 50% survival ages between 29 and 33 months.

Although the 50% survival age was about 28 months for most SPF strains and stocks of rats, there were a few exceptions. Simms[154] and Berg[14] showed that male Sprague-Dawley rats did not live as long as females. Kociba et al.[99] found that male Sprague-Dawley (Spartan substrain) rats had earlier mortality than females. The males of the WAG stock studied by Gsell[75] and Schlettwein-Gsell[152] had a 50% survival age of 23 months while the females under similar conditions survived 27 months. In earlier studies, Boorman and Hollander[20] showed the 50% survival for female WAG/Rij rats to be about 30 months. In contrast, the WAG/Rij males had a value of 22 months.[81] These data were derived from groups that contained both retired breeders and virgins, however. The virgin male WAG/Rij rats (Table 3.1 and Figure 3.4) of this study had a 50% survival age of 24 months, while the virgin females, in the colony at the same time, had a 50% survival age of 32 months (Table 3.1 and Figure 3.3).

The maximum life span reached in each group is another good reference point. The maximum age of male WAG/Rij reached in the virgin male cohort was 34 months. The oldest in the necropsy group was 31 months. This is in sharp contrast to all of the other rat groups (Tables 3.1 to 3.3) where the maximum ages were between 42 and 48 months in the cohort studies and between 40 and 54 months from those rats that were necropsied.

Clearly, for some rat strains or stocks, males do not live as long as females. This is not true for all rats. The female and male BN/Bi and the (WAG X BN)F_1 rats in this study had nearly identical survival data (Table 3.1). There may have been somewhat earlier mortality (90% survival age) in males of the BN/Bi and F_1 rats, but the 50 and 10% and maximum survival ages were nearly identical for both sexes. Similarly, the data of Paget and Lemon[124] did not show any major differences between SPF males and females of their Alderly Park strain (WAG) rats.

Comfort[43] alluded to the fact that males of an animal species generally have shorter life spans than females. However, this is apparently not a universal rule. The findings of similar longevity data in male and female BN/Bi and F_1 rats suggest that, in some cases, males and females may have similar life spans. Others have demonstrated that in some strains of mice the male is actually the longer lived of the two sexes. This was true for C57/B1/6J mice.[104] Storer[168] found that the males had longer life spans than females in eight mouse strains, males and females had similar life spans in four strains, and females lived longer than males in ten strains. Therefore, as suggested by Walford[182] and Kunstyr and Leuenberger,[104] survival curves are needed to evaluate the longevity of each sex and mouse strain used in aging studies. This statement is equally true for rats.

Few studies have compared longevity data from retired breeders and virgins. In the study by Wexler and Greenberg,[184] female retired breeder Sprague-Dawley rats did not live as long as virgins. Paget and Lemon[124] reported that their SPF breeding rats did not live as long as SPF virgins. The data summarized in Figure 3.7, however, do not show any major differences in the mortality of retired breeder and virgin female WAG/Rij or from retired breeder and virgin female BN/Bi rats. The retired breeder BN/Bi actually lived slightly longer than the virgins. Therefore, generalizations about retired breeders having earlier mortality than virgins are by no means universal.

Finally, the survival data from the necropsied rats were not identical, but were similar to the data from complete cohorts. It is possible that the slight differences reflect a selection process, but this cannot be demonstrated.

Chapter 4

AGE-ASSOCIATED PATHOLOGY

I. INTRODUCTION

The data presented in this chapter are summarized in several different ways. The major lesions are described morphologically and many are documented with photographs. One reason for this is to provide a general reference for investigators who only sporadically evaluate pathologic findings in aging rats. Another, perhaps more important, reason is to document the lesions clearly so that investigators who work with various aging rat strains can compare their observations with those published in this report. At times, pathologists disagree on the specific terminology for a given lesion. Therefore, the criteria used to classify lesions are given so that comparisons can more readily be made.

The incidence of the more frequently found lesions is expressed by two methods:

1. The incidence in the total population that died is summarized in tables. It was calculated by dividing the total number of dead rats into the total that died with a lesion and multiplying by 100. The incidence is therefore expressed as a percent of the total population.
2. The age-associated incidence of many lesions was calculated by life table techniques.[148] The age-associated incidences were determined on the basis of 6-month periods. The total number of rats of a given group that died with a lesion was divided by the total number of rats alive at the beginning of that time period multiplied by 100. The results of the calculations for each age period are plotted and summarized in a figure showing the age-associated percent incidence based on the population at risk.

Calculations to determine the incidence of lesions were made using the data obtained from the 670 aging rats listed in Table 2.8. Additional studies such as electron microscopy and transplantation of some tumors were not necessarily done on lesions from these specific rats. In order to perform such studies, it was necessary to have fresh material. Therefore, when needed, untreated aging rats that died or were killed from different groups were used to obtain the necessary material to conduct additional studies. When other rats were studied, care was taken to assure that the lesions were similar to those seen in the aging rats.

In a few cases, some tissues were not found at the histopathologic examination. For example, the parathyroid glands were examined only to the extent that they were found in the routine sections of the thyroid glands. At times, an organ such as the pituitary gland was lost during the processing of tissues. Attempts were made, however, to account for these tissues, and multiple sections of organs were often obtained in order to assure uniformity. Also, an earnest effort was made at the time of trimming tissues as well as on the final glass slides to account for all grossly observed lesions.

II. ORGANS AND ORGAN SYSTEMS

A. Thyroid Gland

1. Nonneoplastic Lesions

Nonneoplastic lesions in the thyroid glands were infrequent and were not tabulated for this study. These lesions were sporadic or incidental findings and included such findings as ultimobranchial duct cysts which were lined by stratified squamous epithelium, "colloid" cysts that were solitary dilated follicles filled with pink colloid material, and focal periarteritis of thyroid arteries. There also appeared to be an age-associated variability in the size of follicles; however, more detailed studies would be needed to prove this.

2. Neoplastic Lesions

a. Medullary Thyroid Carcinoma

Hyperplasia of the parafollicular or C cells of the thyroid was very common in this series, but, as is true for many proliferative lesions of the endocrine system, it is difficult to prove where hyperplasia ends and tumor begins. In this study, hyperplasias consisted of diffuse or focal collections of a few parafollicular cells or larger focal or multifocal nodules of parafollicular cells that were bounded by the follicle basement membrane (Figures 4.1 and 4.2). When the nests of C cells appeared to be extending beyond the basement membrane and infiltrating between and destroying follicles, the diagnosis of medullary thyroid carcinoma was made. This is schematically illustrated in Figure 4.3 which shows the differences between normal, hyperplastic, and carcinoma of C cells as defined in this study.

Grossly, the tumors ranged in size from small, barely visible thickenings of the thyroid gland to nodules greater than 10 mm in diameter. Most were firm, gray to white, and unilateral; however, a few were bilateral. The microscopic features were similar in all groups of rats with the cells round to polygonal in shape, their cytoplasm staining pale pink, and the cells tending to form solid nests or clusters which were often separated by a connective tissue stroma. A few tumors had cells that were more fusiform in shape and at times appeared to pallisade. Mitoses, although present in some tumors, were not common. Compressed and partially obliterated follicles were frequent within or directly adjacent to the tumors, suggesting destruction and invasion of the surrounding thyroid tissue by even the smallest neoplasms. The smallest tumors were not observed grossly, but were microscopically identical to the largest.

A hyaline stroma containing amyloid (Figure 4.4) was present in some carcinomas. The amyloid stained with congo red and crystal violet and exhibited the appropriate birefringence with polarized light. The amount of amyloid varied considerably, but when present it was

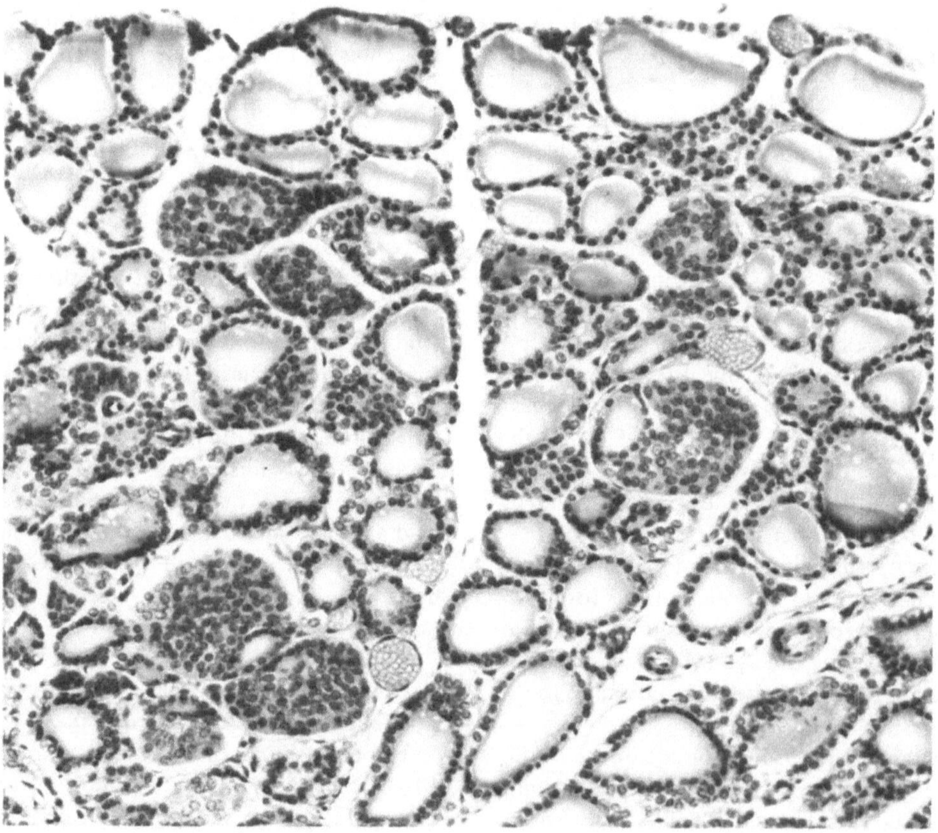

FIGURE 4.1. Thyroid gland from a female WAG/Rij rat with multifocal parafollicular cell nodular hyperplasia. (HPS; magnification × 210.)

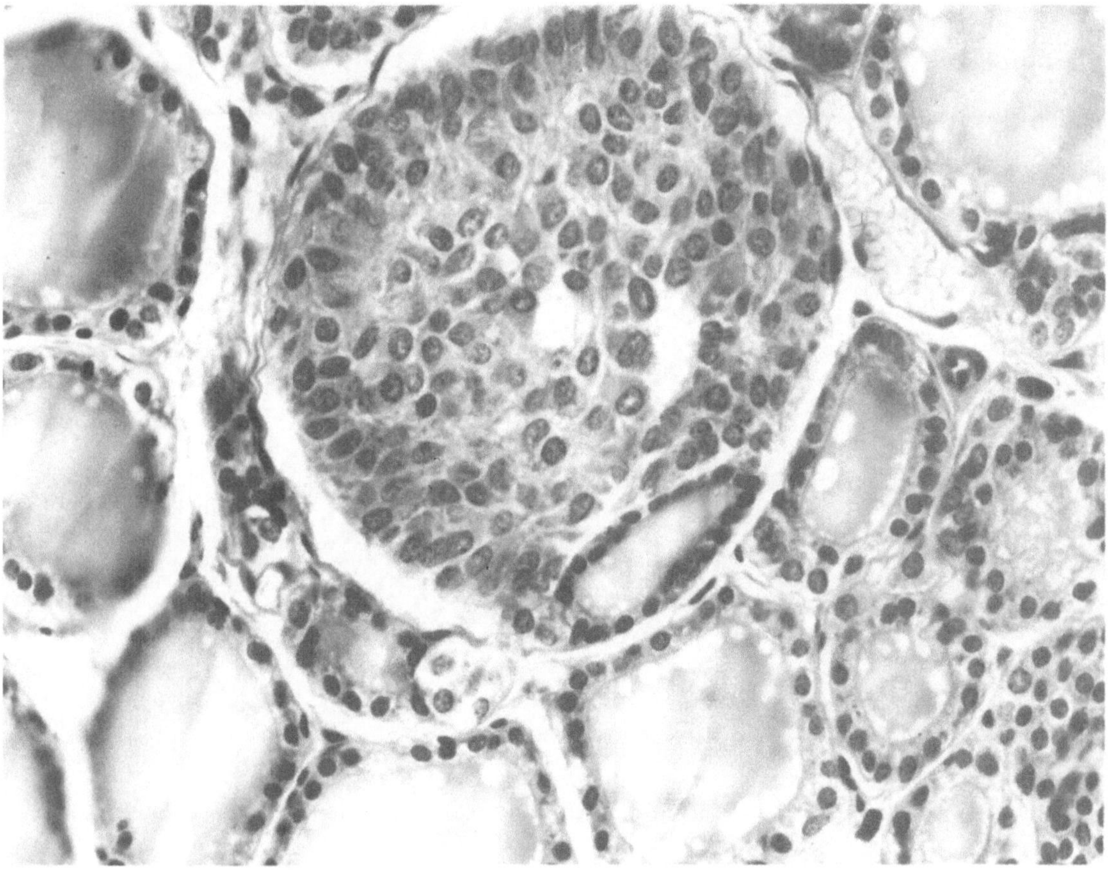

FIGURE 4.2. Nodular hyperplasia of parafollicular cells in the thyroid gland of a female WAG/Rij rat. (HPS; magnification × 500.)

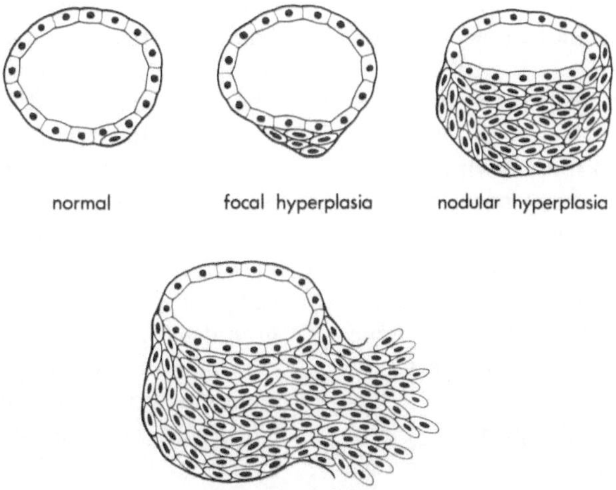

FIGURE 4.3. Schematic illustration of the differences between normal, hyperplastic, and carcinomatous medullary thyroid cells.

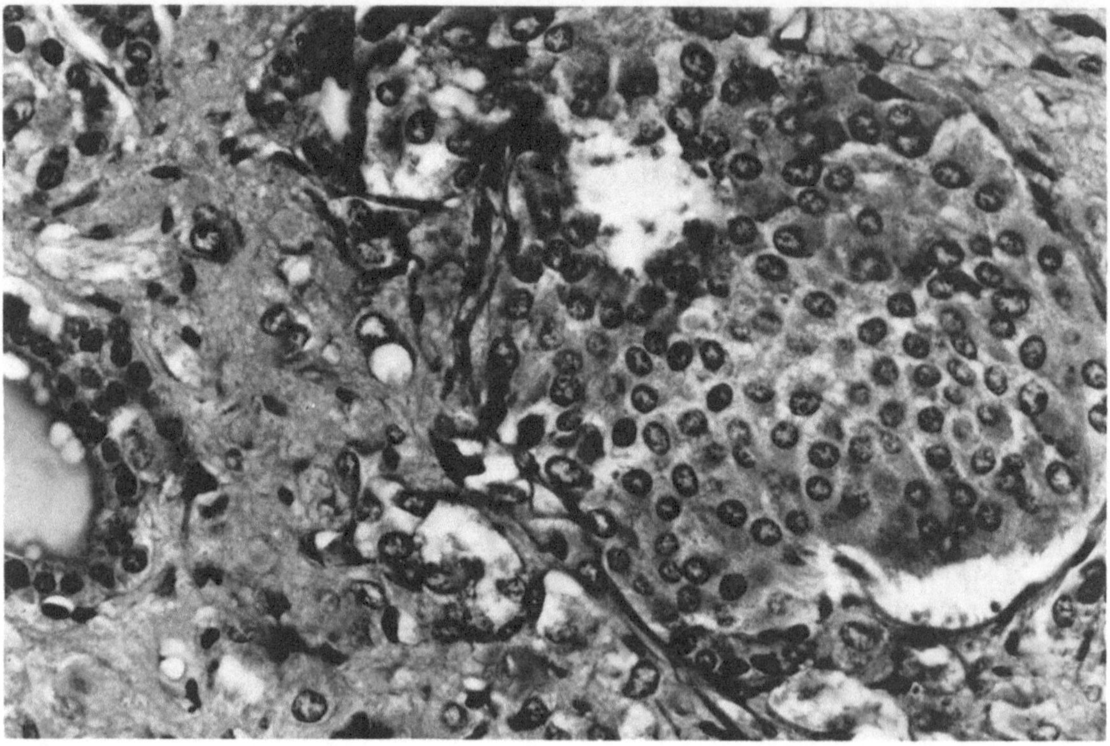

FIGURE 4.4. A medullary thyroid carcinoma from a male (WAG × BN)F_1 rat with nests of tumor cells (↑) and hyaline stroma containing amyloid (↑↑). (HPS; magnification ×465.)

usually only in slight amounts and multiple sections were often required before it could be found. Only a few tumors had extensive amyloid deposition. In the majority of tumors, amyloid was not observed, but special staining and multiple sections were not done on all tumors.

Most medullary thyroid carcinomas destroyed normal thyroid follicles, but they remained well circumscribed masses with little or no infiltration into tissues surrounding the thyroid gland. A few large tumors did have extensive local invasion beyond the thyroid gland into the adjacent cervical region. Some apparently invaded blood vessels, but blood-borne tumor metastases were not observed in rats with such lesions.

Distant metastases from the primary tumor were found in approximately 10% of the rats (Table 4.1), with the deep cervical lymph nodes the only site suggesting metastases via the lymphatics. The metastatic tumor in the deep cervical lymph nodes ranged from a few clusters of cells to masses of cancer cells that obliterated the normal lymph node architecture (Figures 4.5 and 4.6) The cellular characteristics of tumors that metastasized were similar to those that did not metastasize; that is, cellular morphology could not be used as an indication for the likelihood of metastases for these cancers. The occurrence of distant metastases was directly correlated with the size of the primary tumor and the age of the rat. Metastases were found in rats having the largest tumors (>8 mm) and in rats older than 24 months of age.

Ultrastructurally, the cytoplasm of the tumor cells contained numerous membrane-bound granules (Figure 4.7), along with variable numbers of mitochondria, free and clustered ribosomes, endoplasmic reticulum, and vacuoles. The nuclei varied in shape. Most were round to slightly elliptical while others were pleomorphic and contained vacuoles, invaginations, and irregularly clumped chromatin.

A total of 16 medullary thyroid carcinomas from WAG/Rij rats have been transplanted un-

TABLE 4.1

Summary of the Incidence of Medullary Thyroid Carcinomas and Metastases of Medullary Thyroid Carcinomas in Aging BN/Bi, WAG/Rij, and (WAG × BN)F_1 Rats

Strain	Sex	No. examined	No. with medullary thyroid carcinoma	%	Mean age (range) in months	No. medullary thyroid carcinomas with metastases	Age (in months) of rats with metastatic medullary thyroid carcinomas
BN/Bi	Female	236	15	6	33 (17—38)	2	35, 38
	Male	74	7	9	27 (15—34)	0	—
WAG/Rij	Female	101	47	47	35 (26—46)	5	35 (32—39)
	Male	124	41	33	23 (9—29)	1	29
F_1	Female	68	11	16	31 (17—38)	3	25, 27, 28
	Male	67	20	29	34 (22—42)	3	28, 30, 38

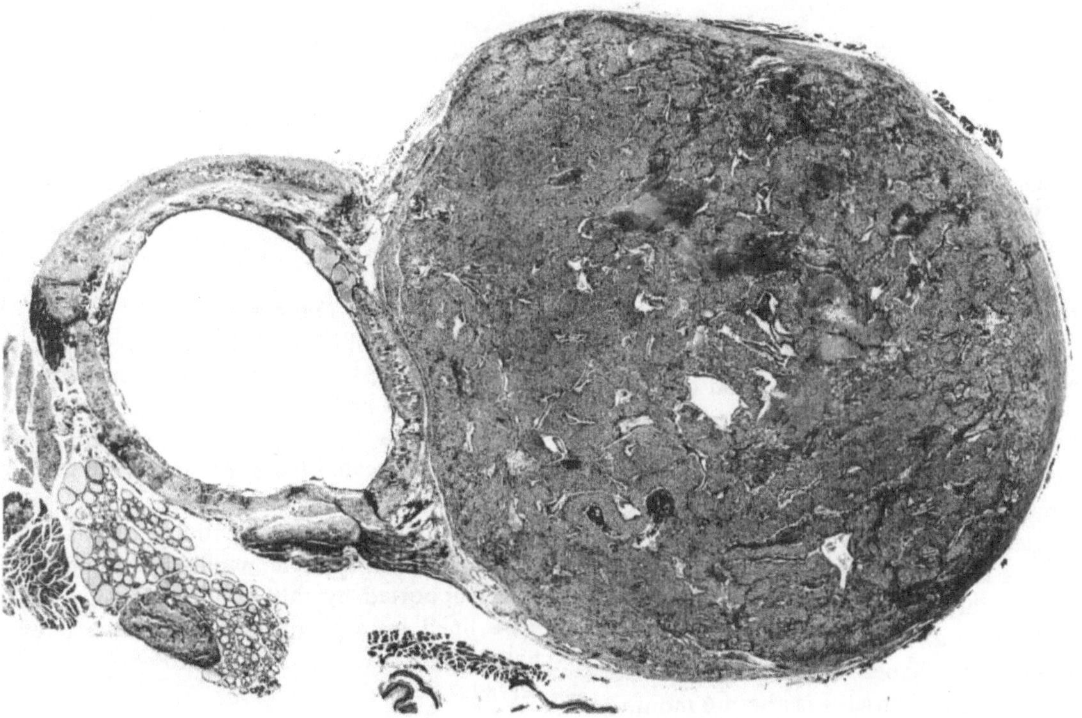

FIGURE 4.5. A large medullary thyroid carcinoma from a female (WAG × BN)F_1 rat. The opposite thyroid and parathyroid glands in the lower left of the photograph are normal. (HPS; magnification × 80).

der the capsule of the kidney or occasionally subcutaneously into approximately 90 rats with nearly 100% success. Fifty-three rats with transplanted medullary thyroid carcinomas have been examined histopathologically. The morphology of the transplanted tumors was similar to that of the primary neoplasms. Amyloid was also present in a few transplanted cancers, but, like the primary tumors, it was often present in very small amounts. As a general rule, transplanted medullary thyroid carcinomas grew slowly, requiring a minimum of 6 to 9 months before a large enough (>8 mm) lesion developed at the site of transplantation. Some

FIGURE 4.6. Deep cervical lymph node from the same rat in Figure 4.5. The lymph node is nearly replaced by metastatic medullary thyroid carcinoma. (HPS; magnification × 95.)

transplants took longer than 12 months to develop a tumor 1 cm in diameter.

The incidence of medullary thyroid carcinomas was greatest in female (47%) and male (33%) WAG/Rij rats, followed by male (29%) and female (16%) F_1 and lowest in male (9%) and female (6%) BN/Bi rats (Table 4.1). The age-associated incidence as calculated by life table methods is shown in Figure 4.8. The male WAG/Rij apparently had an earlier onset of tumors, with the peak age-associated incidence of 40% occurring in the oldest age group (25 to 30 months). The age of onset was about 6 months later in the other groups of rats, with only a few tumors present prior to 24 months. The female WAG/Rij had the highest peak age-associated incidence of 64%. The male and female F_1 had peak age-associated tumor incidences of 33 and 30%, respectively. Female BN/Bi had a peak incidence of 18%. The cancers in male BN/Bi had an apparent peak age-associated incidence of 12% during the period of 31 to 36 months. Those dying in the oldest age period (>37 months) had a 0% incidence.

b. Follicular Cell Tumors

Only two follicular cell tumors were observed. The first was in a 33-month-old female WAG/Rij rat and was a cystic papillary follicular adenoma of the thyroid. The tumor was sharply demarcated from the surrounding normal tissue. The outer edge consisted of large follicle-like structures, while the center was microcystic with multiple papillary fronds that were supported by thin connective tissue and were lined by a single layer of well-differentiated epithelial cells. No mitoses were observed.

The second was an undifferentiated giant cell (pleomorphic) carcinoma found in a 27-month-old female WAG/Rij rat. Grossly the tumor measured 2 × 1.5 × 1 cm, was soft, and appeared white with multiple small hemorrhagic foci. Microscopically it was surrounded by a dense connective tissue capsule and was highly vascular. The cells were pleomorphic (Figures 4.9 and 4.10). Some were large giant cells with a single nucleus while a few were multinucleated. The nuclei were variable in size and

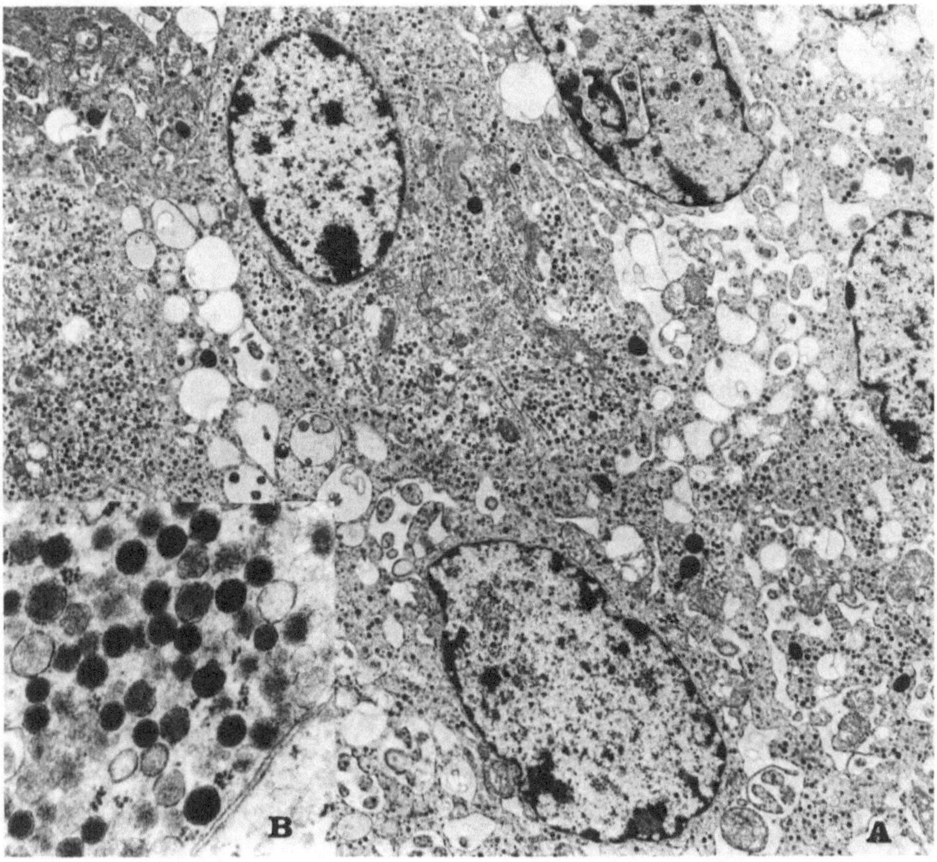

FIGURE 4.7. (A) Numerous cytoplasmic granules are present in cells of a medullary thyroid carcinoma. (Magnification × 6800.) (B) Shows the cytoplasmic granules in greater detail. (Magnification × 41,200.)

FIGURE 4.8. Percent incidence of medullary thyroid carcinomas with age in 236 female and 74 male BN/Bi, 101 female and 124 male WAG/Rij, and 68 female and 67 male (WAG × BN)F_1 rats.

shape. Mitoses were numerous and often bizarre. Neoplastic cells invaded the tumor capsule at several locations, but distant metastases were not observed.

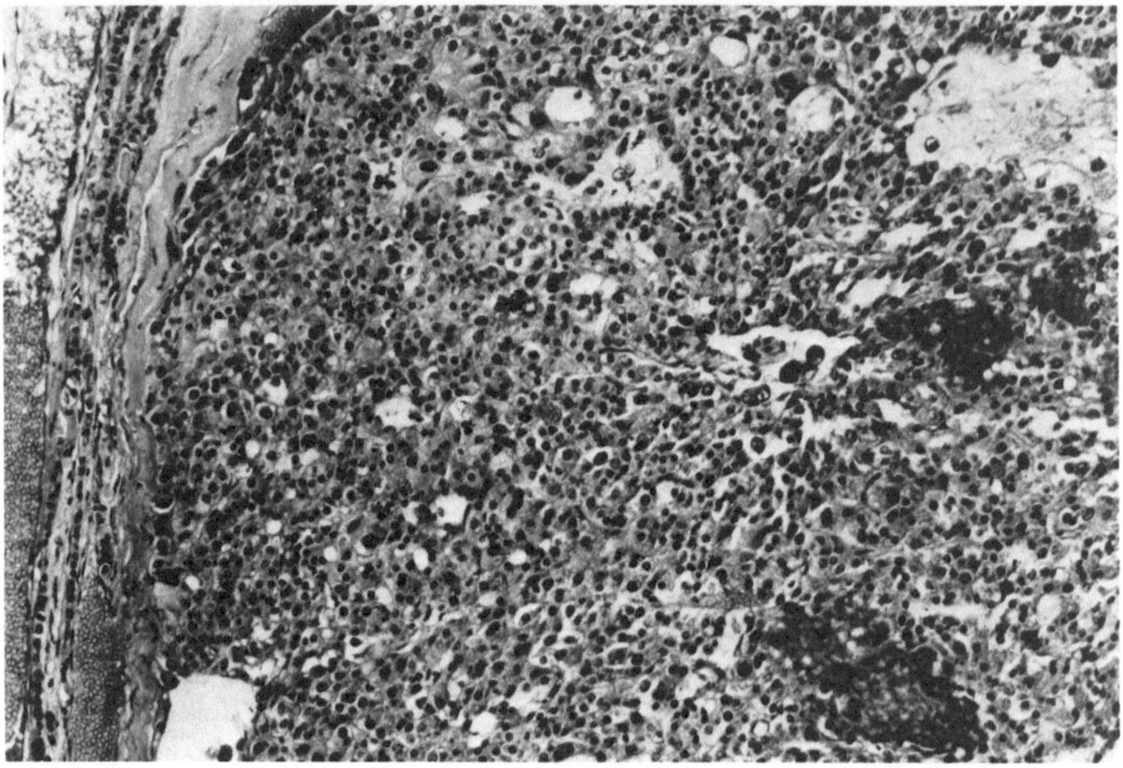

FIGURE 4.9. An undifferentiated giant cell (pleomorphic) carcinoma in the thyroid gland of a 27-month-old-female WAG/Rij rat. (HPS; magnification × 195.)

3. Discussion

Nonneoplastic age-associated lesions of the thyroid glands, with the exception of parafollicular cell hyperplasias, were not common in the aging BN/Bi, WAG/Rij, or (WAG X BN)F_1 rats. As reviewed by Russfield,[147] an increase in reticular and collagenous connective tissue follicular size and decreased height of the follicular epithelium are observed in aging rats. She also observed that ultimobranchial rests were common in the thyroid of some "old rats." Such changes appear to be present in the BN/Bi, WAG/Rij, and (WAG X BN)F_1 rats, but more detailed investigations are clearly needed before a definite age-associated pattern can be proven.

Medullary thyroid carcinomas were described in WAG/Rij rats by Boorman et al.[18,21,24] They described the light microscopic and ultrastructural features of the tumors and confirmed the parafollicular cell as the cell of origin. The presence of both intracellular and extracellular amyloid was also confirmed by electron microscopy. Calcitonin has been isolated from medullary thyroid tumors of WAG/Rij rats,[33] and the structure of the calcitonin has been studied.[34]

Medullary thyroid carcinomas occurred in the BN/Bi, WAG/Rij, and (WAG X BN)F_1 rats in this study, suggesting that it is a common tumor in rats. In support of this, others have also observed these tumors as spontaneously occurring in several different rat strains and stocks. For example, Lindsey et al.[108] and Lindsey and Nichols[109] described these tumors in Long-Evans, Sprague-Dawley, Fischer, Wistar, Buffalo, and Osborne-Mendel rats. Kroes et al.[102] have seen them in random-bred Wistar rats. Anver and Cohen[5] reported them in retired breeder female Sprague-Dawley rats. Kociba et al.[99] have seen them routinely in control and experimental Sprague-Dawley (Spartan substrain) rats. Coleman et al.[42] observed them in aging male Fischer 344 rats.

There was a direct correlation with age (Figure 4.8) and the presence of a C cell tumor in

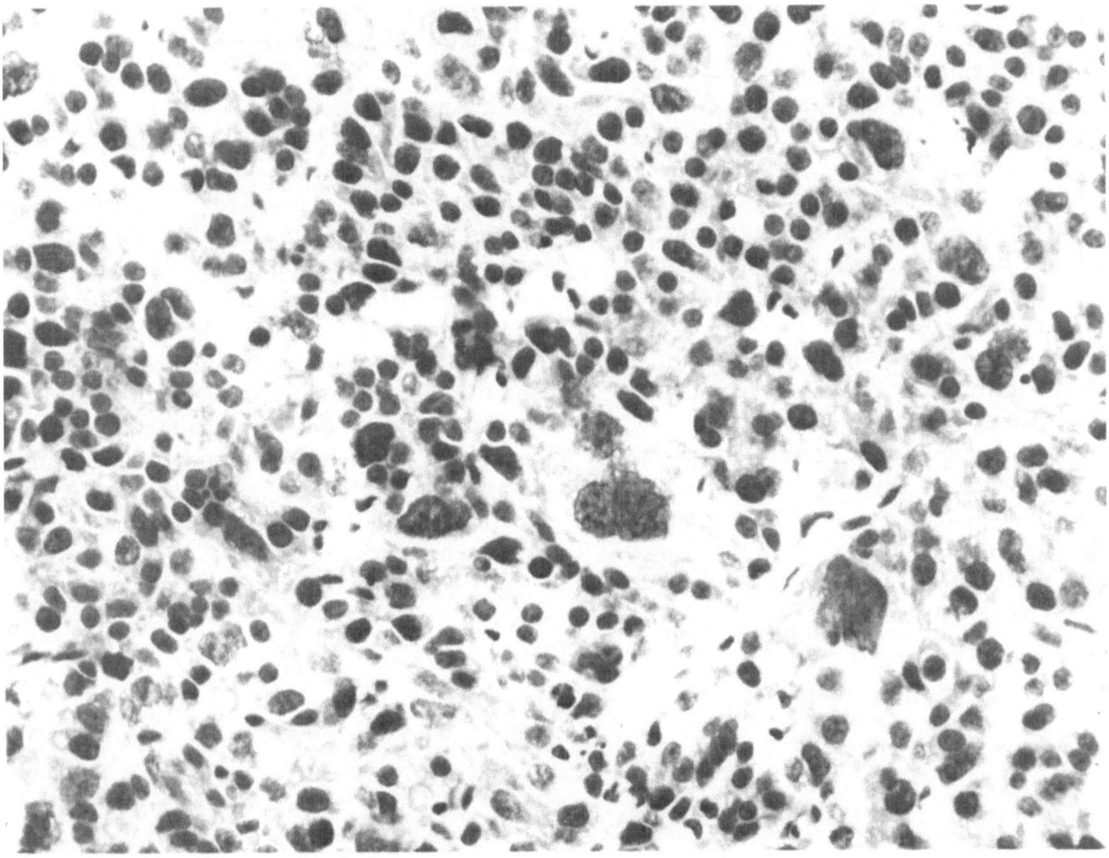

FIGURE 4.10. Higher magnification of the giant cell carcinoma shown in Figure 4.9. Note the cellular pleomorphism and the large bizarre tumor cells. (HPS; magnification × 500.)

the BN/Bi, WAG/Rij and (WAG X BN)F_1 rats of this study, with nearly half of the tumors found in rats older than 30 months. The male WAG/Rij rats had an earlier onset of medullary thyroid carcinomas, however, compared to the other groups in this study. The mean age was 23 months compared to mean ages between 27 and 35 months for the other groups. The incidence varied among the different strains, however. The WAG/Rij rats had a high incidence, the F_1 rats were intermediate, and the BN/Bi rats had a relatively low incidence.

The biological behavior of medullary thyroid carcinoma in the host and their transplantability support the view that these are malignant tumors. Metastases were found only in older rats and in those with the largest tumors, and large tumors were seldom found in the younger rats. The transplanted tumors grew in syngeneic recipients, but grew slowly, requiring 6 months to 1 year before large tumors were available for additional studies. These findings suggest that the medullary thyroid carcinomas are apparently slow-growing tumors that increase in frequency and size with age. Although metastases occurred in approximately 10% of the medullary thyroid cancers, they occurred late in the progression of this neoplasm.

In view of the high incidence of clear-cut carcinomas, it is possible that the diagnosed hyperplasias may actually represent adenomas or carcinoma *in situ*. This controversy cannot be resolved without further studies to determine the biological behavior of such lesions. Hyperplasias were common and occurred in rats with and without tumors. Some hyperplasias consisted of only a few cells, while others were nodular, and some small carcinomas appeared to have originated from areas of nodular hyperplasia. In general, the observations of this study support the view of Squire and Goodman[163] that the small localized lesions should be considered as carcinomas based on their progressive growth, invasion, and destruction of thyroid follicles and their potential for metastases. Also, this classification is in general agreement with DeLellis et al.[51] for similar lesions in man. Finally, both parafollicular and intrafollicular C cells occur in man, but most tumors appear to be of intrafollicular origin.[51] This also seems to be true for rats.

Despite the relatively high incidence of tumors of parafollicular cell origin, spontaneous tumors of the thyroid follicular cells were not common. Only two tumors were found that were not medullary thyroid carcinomas. This apparently low frequency of follicular tumors in rats was alluded to by Squire and Goodman.[163] On the other hand, others have induced follicular tumors in the thyroid glands of rats as reviewed by Napalkov.[120]

B. Parathyroid Gland

1. Nonneoplastic Lesions

a. Interstitial Fibrosis

Fibrosis was recognized in parathyroid glands from two female BN/Bi, one male F_1, and two female WAG/Rij rats. The ages of the rats ranged from 30 to 39 months. The capsule and the interstitial tissue of the involved gland were thickened by an accumulation of hyaline and fibrous connective tissue (Figure 4.11). The stroma stained positive for collagen with the routine HPS and the Massons trichrome stains. Congo red stained sections viewed with polarized light did not show amyloid in the hyaline stroma, however.

This lesion appeared to be a specific condi-

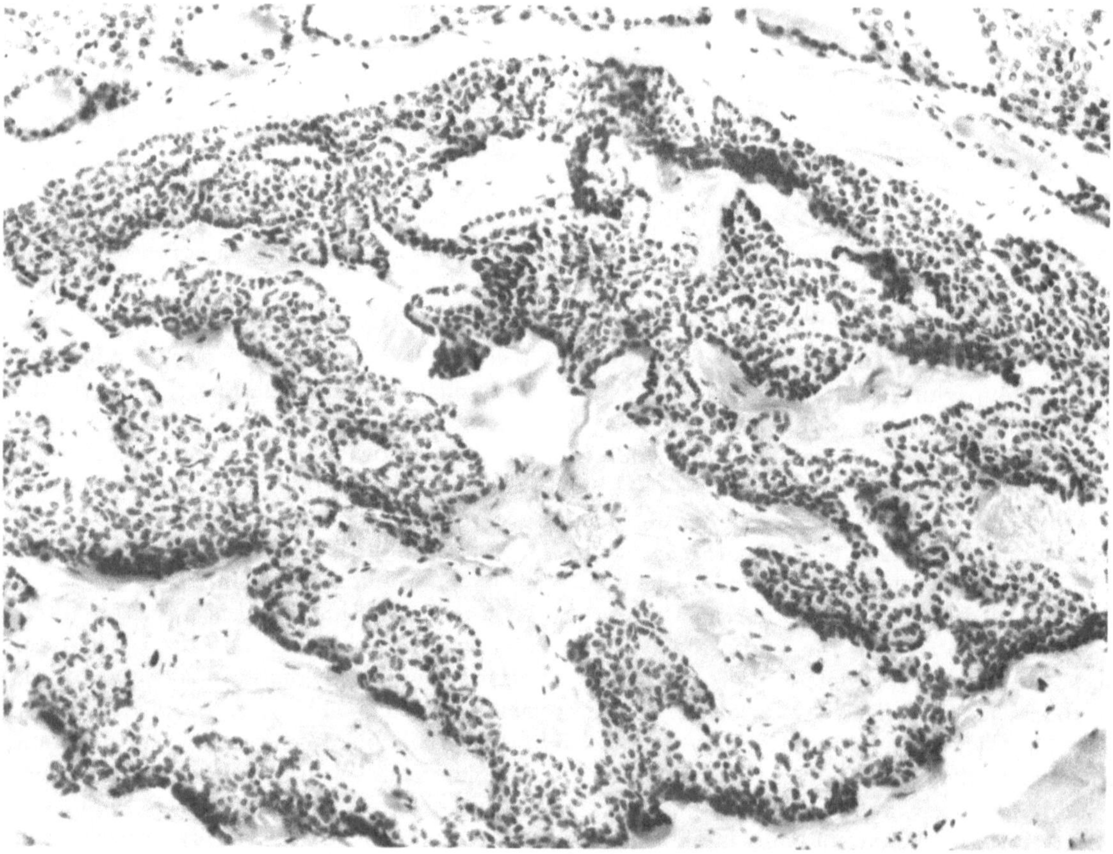

FIGURE 4.11. Interstitial and capsular fibrosis of a parathyroid gland from a male (WAG × BN)F_1 rat. (HPS; magnification ×210.)

tion that was usually confined to one parathyroid gland. It was rare in this series, with an incidence of 0 to 2% in the different rat groups. The available evidence indicates that the material is collagenous connective tissue, but its significance and pathogenesis are unknown.

b. Cystic Parathyroid Glands

Cystic change was another sporadic lesion found in parathyroid glands from rats. The cysts were lined by flattened to low cuboidal epithelial cells. The dilated and cystic acini were filled with pink proteinaceous material that often contained crystal-like concretions (Figure 4.12). Cystic parathyroid glands were observed in two female (ages 20 and 28 months) and one male (age 26 months) BN/Bi rats and one female (26 months) WAG/Rij. It was not observed in male WAG/Rij or F_1 rats.

2. Neoplastic Lesions

a. Parathyroid Adenomas

The only neoplastic lesions in parathyroid glands of this series were small adenomas. They were not common (Table 4.2) and consisted of nests of cells only slightly different from the normal tissue. The cells were usually larger than normal with pink, finely granular cytoplasm and round, lightly vesicular nuclei (Figure 4.13). Compression of the normal adjacent parathyroid tissue was usually present, but this was not prominent.

3. Discussion

Lesions of the parathyroid glands were not common in rats of this study. Part of this may be explained by the fact that the only parathyroids studied were those present in routine sections. Parathyroid glands were not observed in

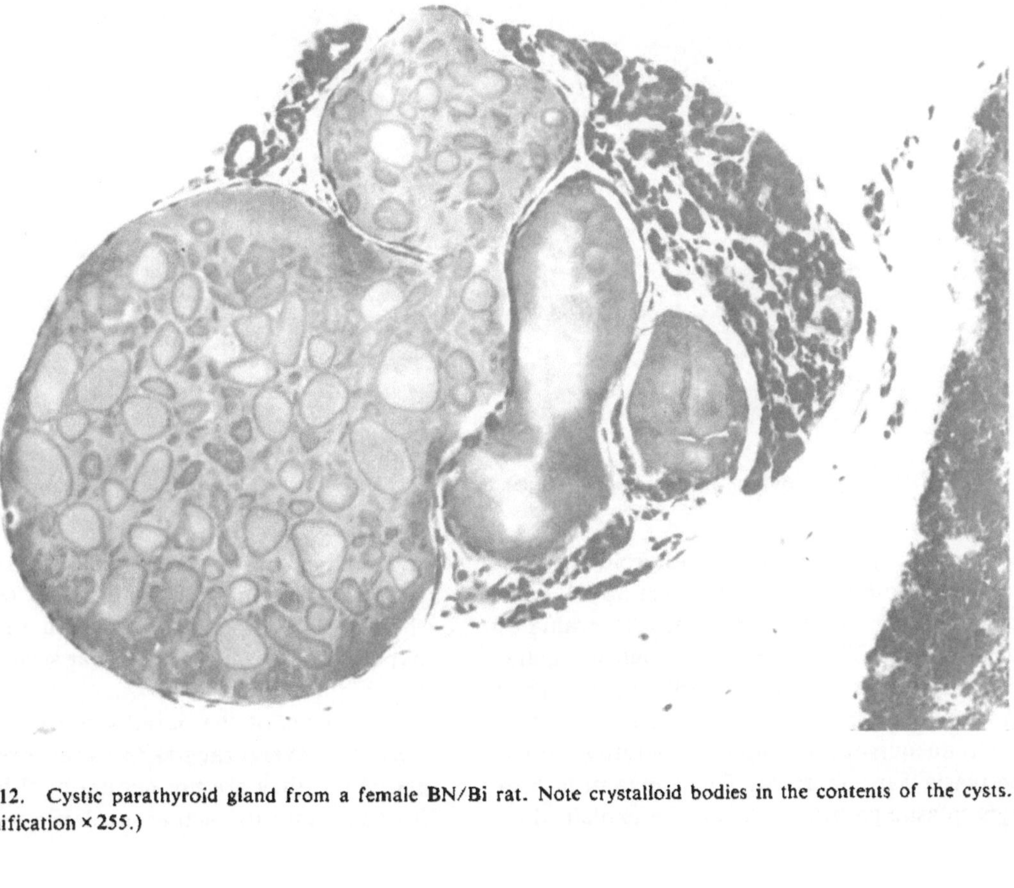

FIGURE 4.12. Cystic parathyroid gland from a female BN/Bi rat. Note crystalloid bodies in the contents of the cysts. (HPS; magnification × 255.)

TABLE 4.2

Summary of the Incidence of Parathyroid Adenomas in Aging Male and Female BN/Bi, WAG/Rij, and (WAG × BN)F_1 Rats

Strain	Sex	No. examined	No. with parathyroid adenoma	%	Mean age (range) in months
BN/Bi	Female	236	1	<1	32
	Male	74	0	—	—
WAG/Rij	Female	101	2	2	32, 37
	Male	124	3	2	25 (19—28)
F_1	Female	68	0	—	—
	Male	67	0	—	—

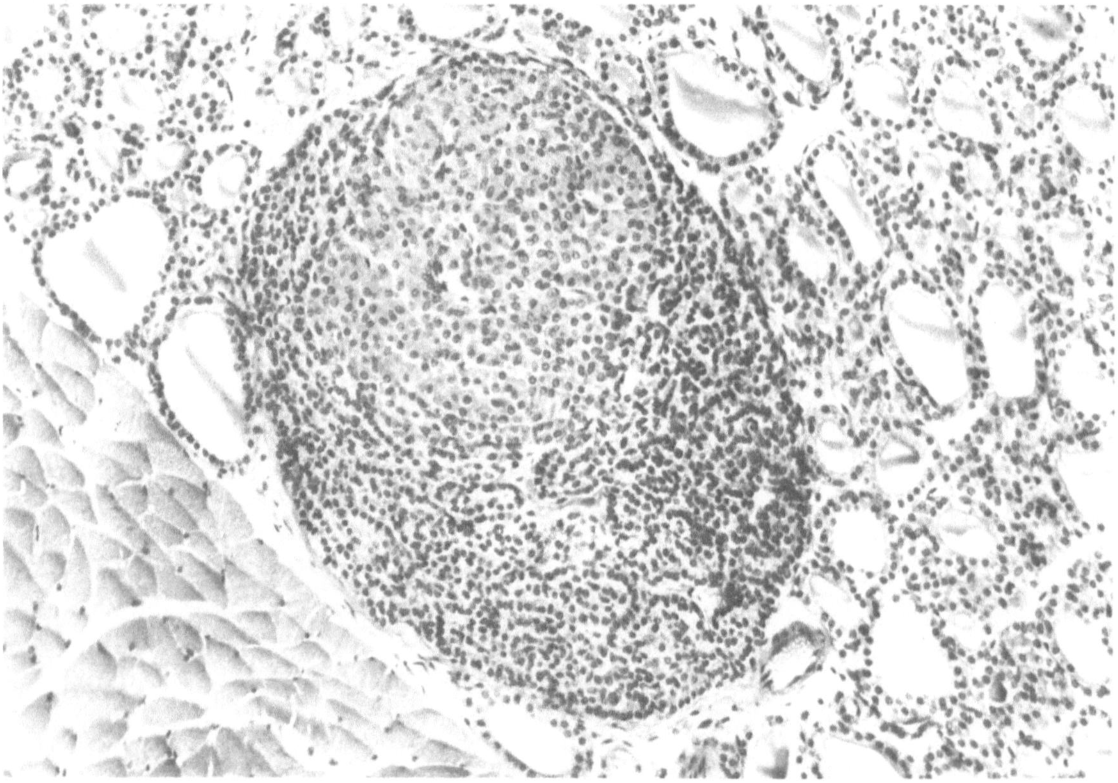

FIGURE 4.13. Adenoma in the parathyroid gland from a female WAG/Rij rat. (HPS; magnification × 195.)

some rats and only one parathyroid was observed in others. Only in exceptional cases could two or more parathyroids be evaluated from one rat. Therefore, the incidence of parathyroid lesions would probably be higher if more detailed studies were conducted.

No unequivocal examples of parathyroid hyperplasia were observed. This apparent lack of hyperplastic parathyroids may be explained by the relatively insignificant renal disease in these rats (see Section II. K). This is in contrast to the reports of others[13,109,147] who have described hyperplastic parathyroids in rat strains with severe kidney disease.

Adenomas of the parathyroid have been reported in several rat strains as reviewed by others,[156] but they are not common.[163] It is clear, however, that the actual incidence in different strains and stocks of rats will vary depending upon the degree of enthusiasm one uses to look for all of the parathyroid glands in each rat. In this series of BN/Bi, WAG/Rij, and F_1 rats, parathyroid adenomas were rare in the routine material occurring in less than 3% of rats.

C. Adrenal Gland

1. Nonneoplastic Lesions

a. Foci of Cellular Alteration in the Adrenal Cortex

Foci of cellular alteration represented a wide range of lesions found in the adrenal cortex and were the most frequent age-associated lesions observed. The smallest foci consisted of only a few cells, while the larger foci occupied most of the cortex and occasionally formed nodules. They were found in all zones of the adrenal cortex, but were most frequent in the zona fasciculata.

The foci consisted of collections or groups of cells that differed in tinctorial and textural appearance from the surrounding normal tissue (Figures 4.14 and 4.15). Some altered cells were finely vacuolated while others were coarsely vacuolated and were usually larger than the surrounding normal cells. The color of the cytoplasm varied from pale pink to strongly acidophilic. Foci of basophilic cells were infrequently observed. The foci were usually multiple, were distinct from the normal tissue, and often disturbed the normal architecture, but zones of compression were not present. They were unilateral in some rats, but were usually bilateral.

Foci of cellular change were very common in male and female BN/Bi rats, occurring in nearly 100% of those over 24 months old. They were less frequent in male and female F_1 rats and in male and female WAG/Rij rats, but were still present in the majority of rats that died. The foci were age associated, the oldest rats having the greatest risk of dying with the lesions.

b. Foci of Basophilic Change in the Adrenal Medulla

Many of the rats had what appeared to be normal adrenal medullas, but with individual clusters or foci of cells that were more basophilic and slightly larger than the remaining medullary cells. These cells were morphologically similar to cells in pheochromocytomas,

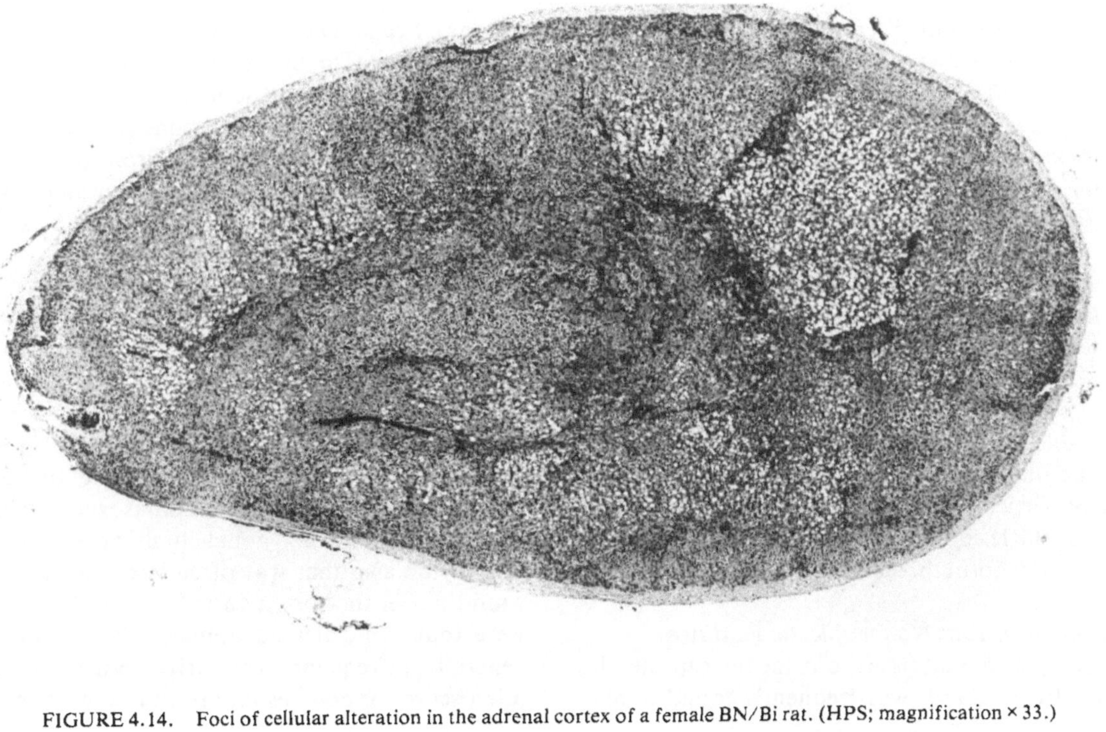

FIGURE 4.14. Foci of cellular alteration in the adrenal cortex of a female BN/Bi rat. (HPS; magnification × 33.)

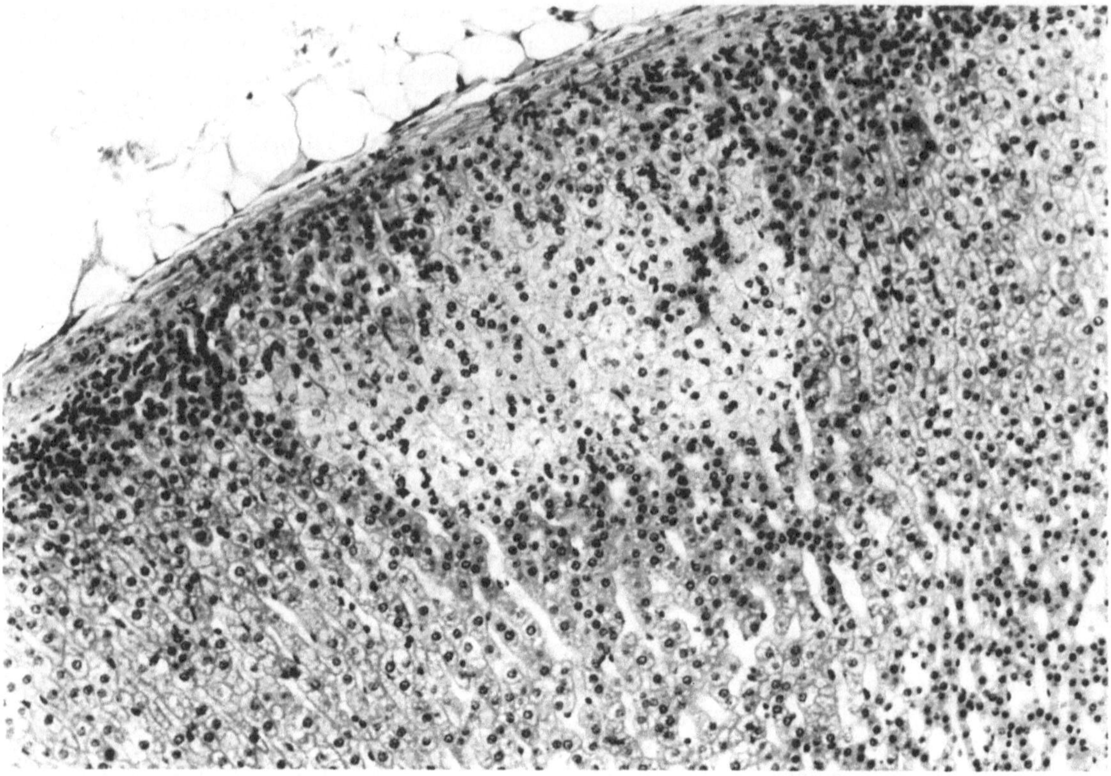

FIGURE 4.15. Solitary focus of enlarged, pale cells at the junction of the zona glomerulosa and zona fasiculata. (HPS; magnification × 195.)

but did not form discrete tumor-like nests or proliferations and mitoses were not observed. Their significance is unknown and, like the foci of cellular change of the adrenal cortex, they were more frequent in the oldest rats.

c. Blood-filled Sinuses and Cysts

The cortex of about 10% of the F_1 rats and less than 5% of the BN/Bi and WAG/Rij rats had dilated blood-filled sinusoids and blood-filled cysts (Figure 4.16). These lesions were located in the zona fasiculata and zona reticularis, consisted of endothelial-lined spaces of variable size, and contained blood. Such lesions occasionally produced compression and necrosis of cortical epithelial cells around the periphery of the lesion, and some blood-filled sinuses contained fibrin thrombi.

d. Miscellaneous Nonneoplastic Findings

Ectopic adrenal tissue, outside the capsule of the adrenal gland, was frequently found in all groups of rats. The most common site was directly adjacent to the adrenal capsule, but it was occasionally found some distance from the adrenal gland. Other observations such as focal extramedullary hematopoiesis, iron (hemosiderin) pigment, lipofuscin-like pigment, and focal necrosis were observed in some aged rats, but they were not specifically studied.

2. Neoplastic Lesions

a. Cortical Adenomas

Adenomas occurred in all zones of the adrenal cortex, but were most frequent in the zona fasiculata. The morphologic features of the cells were similar to the cells in the foci of cellular change. Cells in the adenomas were larger than normal and polygonal in shape and had pink cytoplasm that was often vacuolated and round nuclei that often varied in size. Mitoses were found in a few adenomas, but they were generally infrequent. Their size ranged from microscopic to nodules that produced gross en-

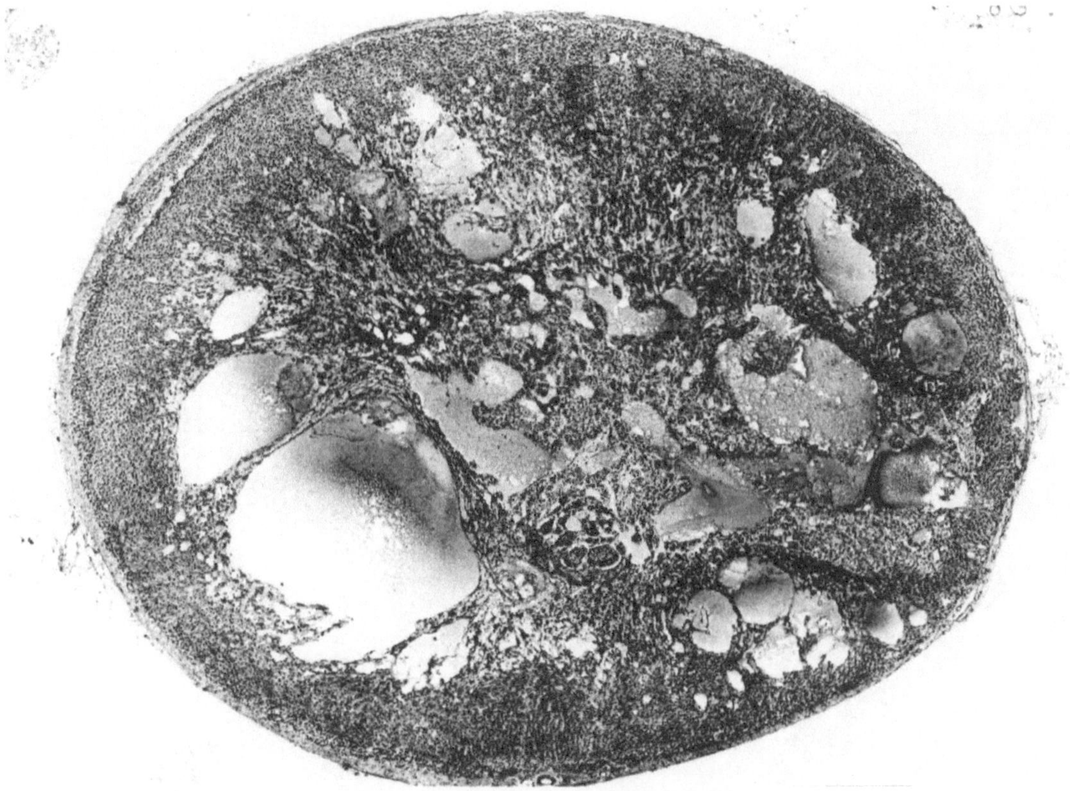

FIGURE 4.16. Multiple dilated and cystic blood-filled spaces in the adrenal cortex and adrenal medulla. (HPS; magnification × 36.)

largement of the adrenal. In all cases, compression of adjacent normal tissue was present (Figure 4.17). The highest incidence of adenomas occurred in female WAG/Rij (40%), followed by male F_1 (31%), female F_1 (22%), female BN/Bi (19%), male BN/Bi (12%), and male WAG/Rij (6%) (Table 4.3).

The age-associated incidence is shown in Figure 4.18. The peak age-associated incidence was in the oldest groups of rats for each strain, with the female WAG/Rij having the highest peak incidence at 60% and the male WAG/Rij the lowest at 15%. The peak incidence in female BN/Bi was 38%, in male BN/Bi it was 20%, and for both male and female F_1 rats the peak was 40%.

b. Cortical Carcinomas

Nearly all the adrenal cortical carcinomas were grossly visible as enlargements of the adrenal gland which ranged in size from 5 mm to over 2 cm in diameter. The histological features of the carcinomas were similar, consisting of large polyhedral cells with pink vacuolated cytoplasm and distinct cell boundaries (Figure 4.19). The sizes of the cells often varied, however, among different areas of the same tumor. The nuclei also varied in size, but were usually round. The number of mitoses in each tumor was also variable, with few in some tumors to very numerous in others. In different areas of a single tumor the pattern often differed with sheets of cells in some areas, lobular patterns in others, and cords of cells in still others. Most carcinomas were highly vascular with many blood-filled vascular spaces and often with blood-filled cystic structures. Necrosis was another common finding in the larger tumors.

The incidence of adrenal cortical carcinomas in the rats of this study and the number of carcinomas that had distant metastases are shown in Table 4.4. Female BN/Bi and female F_1 rats had the highest incidence of 9 and 7%, respectively. The other groups of rats had incidences

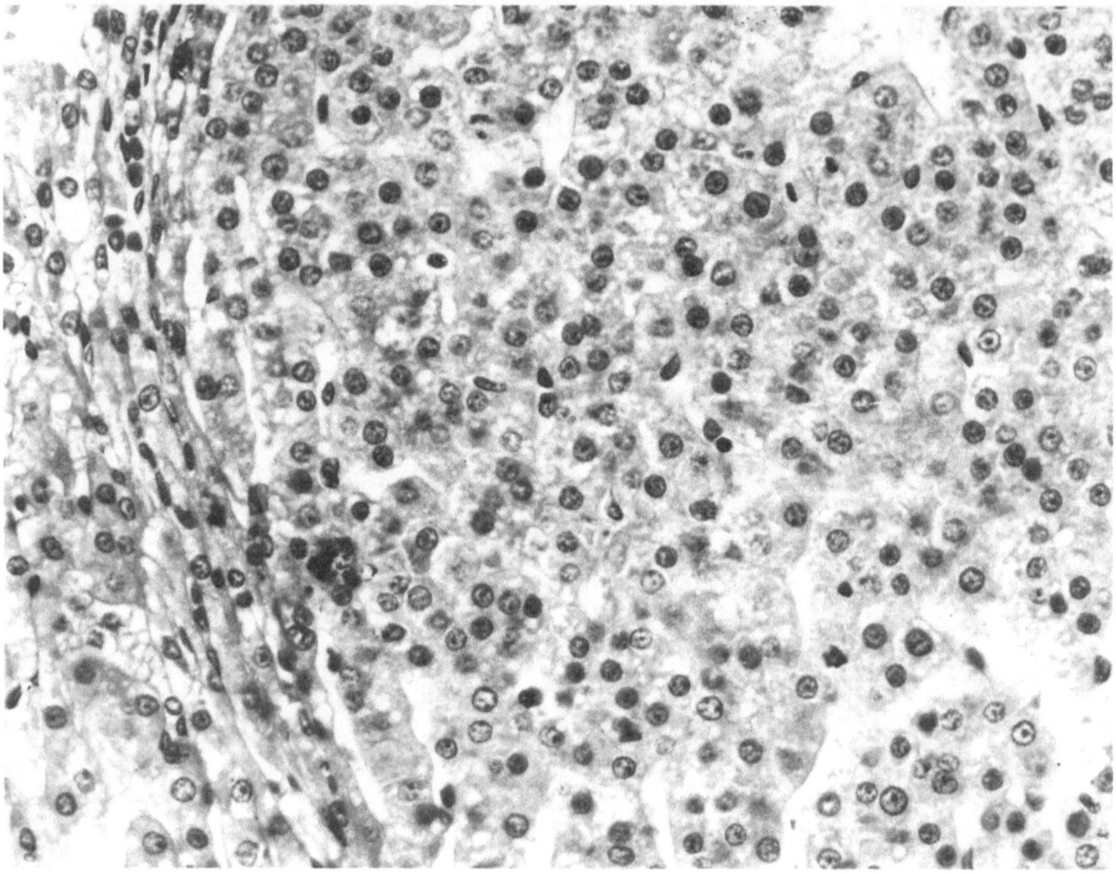

FIGURE 4.17. Junction of an adrenal cortical adenoma with the normal adrenal cortex showing zones of compressed glandular tissue at the periphery of the adenoma. (HPS; magnification × 500.)

TABLE 4.3

Incidence of Adrenal Cortical Adenomas in Aging BN/Bi, WAG/Rij, and (WAG × BN)F_1 Rats

Strain	Sex	No. examined	No. with cortical adenoma	%	Mean age (range) in months
BN/Bi	Female	236	45	19	33 (23—54)
	Male	74	9	12	33 (27—43)
WAG/Rij	Female	101	40	40	35 (29—46)
	Male	124	7	6	22 (9—29)
F_1	Female	68	15	22	30 (23—38)
	Male	67	21	31	35 (27—44)

of 1 to 3%. Also, 13 of the 22 (59%) tumors in female BN/Bi and 2 of 5 (40%) in female F_1 rats had distant metastases. Similarly, both of the carcinomas in male WAG/Rij rats had metastasized. If all of the tumors are taken as a whole, then 34 adrenal carcinomas were diagnosed and 17 had metastases. Therefore, 50% of the rats that died with an adrenal cortical carcinoma had distant metastases of that tumor. The most common sites for the metastases

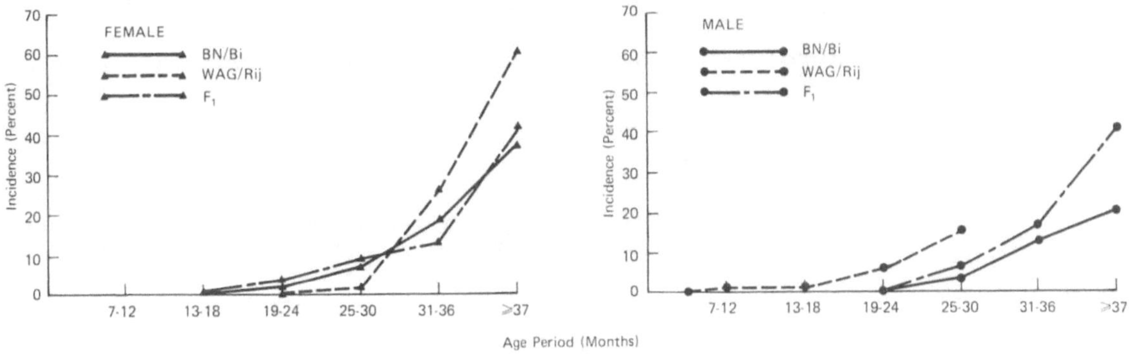

FIGURE 4.18. **Percent incidence with age of adrenal cortical adenomas in 236 female and 74 male BN/Bi, 101 female and 124 male WAG/Rij, and 68 female and 67 male (WAG × BN)F_1 rats.**

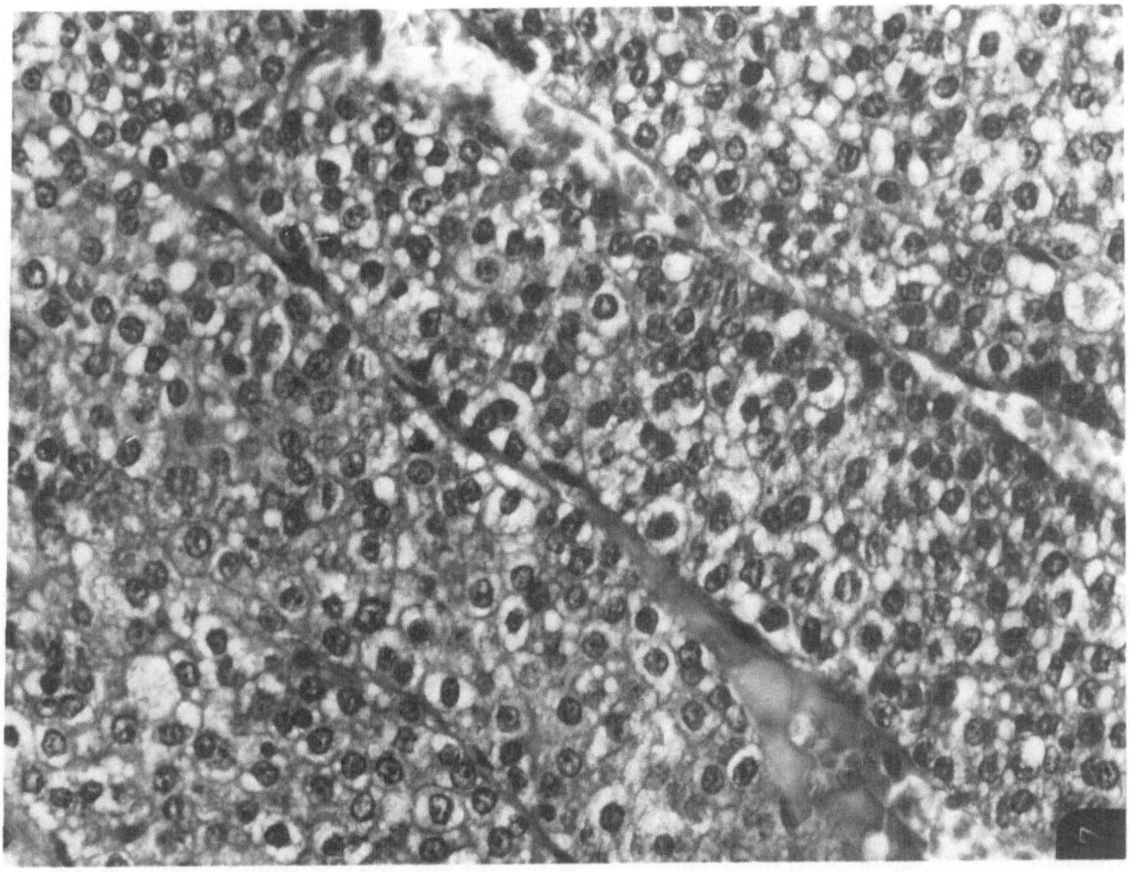

FIGURE 4.19. High magnification of an adrenal cortical carcinoma. (HPS; magnification × 500.)

were the liver only, both the liver and the lung, and, less commonly, the lung alone or other sites.

The age-associated risk of metastases for these tumors was evaluated in the female BN/Bi rats, which had the largest numbers of cortical carcinomas, and the results are shown in Table 4.5. Although the numbers in each individual group are small, there appeared to be a trend for the older rats to have a greater risk of

TABLE 4.4

Summary of the Incidence of Adrenal Cortical Carcinoma and Metastases of Adrenal Cortical Carcinomas in Aging BN/Bi, WAG/Rij, and (WAG × BN) F_1 Rats

Strain	Sex	No. examined	No. with adrenal cortical carcinoma	%	Mean age (range) in months	No. adrenal carcinomas with metastases	Age (in months) of rats with metastasis of adrenal cortical carcinoma
BN/Bi	Female	236	22	9	31 (19—40)	13	19, 25, 27, 31, 31, 32, 32, 34, 35, 35 37, 38, 40
	Male	74	1	1	32	0	—
WAG/Rij	Female	101	2	2	34, 39	0	—
	Male	124	2	2	24, 25	2	24, 25
F_1	Female	68	5	7	29 (23—35)	2	26, 34
	Male	67	2	3	40, 42	0	—

TABLE 4.5

Adrenal Cortical Carcinomas in Female BN/Bi Rats: Comparison of Number of Tumors and Number of Metastases with Age

Age in months	No. alive at beginning of period	No. dead during period	No. of dead rats with carcinoma	No. of rats with metastatic carcinoma	% of carcinomas with metastases
7—12	236	2	0	0	—
13—18	234	9	0	0	—
19—24	225	26	3	1	33
25—30	199	73	4	2	50
31—36	126	98	11	7	64
37—54	28	28	4	3	75
Total	0	236	22	13	59

dying with a metastatic adrenal carcinoma. The percent of metastatic tumors increased from 33% in the younger age period of 19 to 24 months to 75% in the oldest age period of 37 to 54 months.

An attempt was made to correlate the size of the tumor with the occurrence of metastases, but no correlation was found. Some of the smaller tumors (<1 cm in diameter) had metastasized, while some of the larger tumors (>1.5 cm in diameter) had not. The reverse situation was also true. Therefore, no clear correlation of tumor size with metastases was apparent in the rats of this study.

Adrenal carcinomas from two different female BN/Bi rats were transplanted subcutaneously into syngeneic recipients. One tumor was transplanted into five young BN/Bi rats and the other into six. In all cases, the transplanted tumors grew very rapidly and were larger than 2 cm in diameter within 2 to 4 months after transplantation. All recipients died spontaneously or were killed moribund between 4 and 6 months of age with massive tumor growth.

Finally, nearly all of the rats that died with adrenal cortical carcinomas had severe atrophy of the opposite adrenal gland. Similar severe atrophy was not recognized in rats dying without cortical carcinomas and was not seen in any of the rats with cortical adenomas. These observations would suggest that most of the carcinomas secreted steroids.

c. *Pheochromocytomas of the Adrenal Medulla*

The only tumors recognized in the adrenal medulla were pheochromocytomas. Most were small tumors that produced only slight or no obvious gross enlargement of the adrenal gland. The tumor cells generally were well differentiated, with basophilic cytoplasm and dark round nuclei. The most common pattern showed small nests or clumps of cells (Figures 4.20 and 4.21), but areas with cords and trabecular growth were seen in some. These tumors, like the cortical tumors, were usually vascular, and microinvasion or growth into vascular lumina was sometimes observed.

Only two BN/Bi rats (a 20-month-old male and a 39-month-old female) had metastases of a pheochromocytoma to the lung (Table 4.6).

The incidence of pheochromocytomas is shown in Table 4.6. The highest incidence was in male F_1 (12%), followed by male BN/Bi (8%), female WAG/Rij (8%), and female BN/Bi (7%) rats. Female F_1 had a lower incidence (3%), with male WAG/Rij rats the least (1%).

3. Discussion

The rat adrenal gland develops a number of age-associated neoplastic and nonneoplastic lesions. Many of the histologic interpretations that must be made on this organ are controversial, however, because of a lack of knowledge about the biological significance of these lesions. For example, foci of cellular alteration represented a range of lesions with individual cells generally larger than normal (hypertrophy). Because the lesions are sometimes nodular, they could be interpreted as areas of hyperplasia (increased numbers of cells). In most of the areas of altered cells, however, it was impossible to determine if there was an actual increase in the number of cells. Similarly, the adenomas of this study could be diagnosed as nodular hyperplasia. They clearly represent benign proliferative lesions, but it was impossible to determine which diagnosis was the most accurate.

The problem of differentiating hyperplasia, adenoma, and carcinoma in the adrenal cortex of rats has been stressed by Hollander and Snell.[84] They also reviewed the literature on rat adrenal gland tumors. They assumed that if a focus or nodule of cells differed morphologically from the normal cells and if the lesion compressed normal tissue, the lesion was an adenoma. Similarly, they were unable to differentiate focal hyperplasia in the adrenal medulla

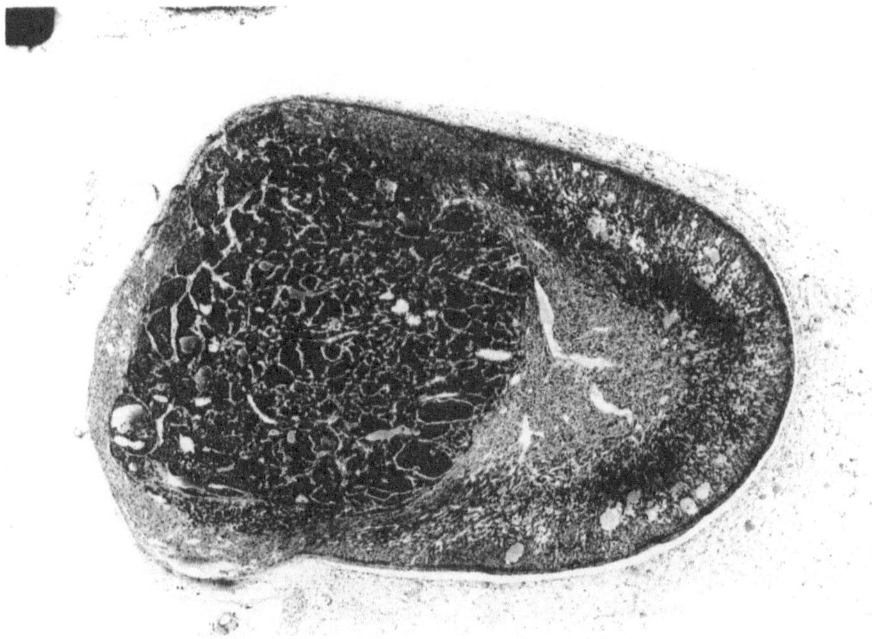

FIGURE 4.20. Pheochromocytoma arising in the adrenal medulla. (HPS; magnification × 16.)

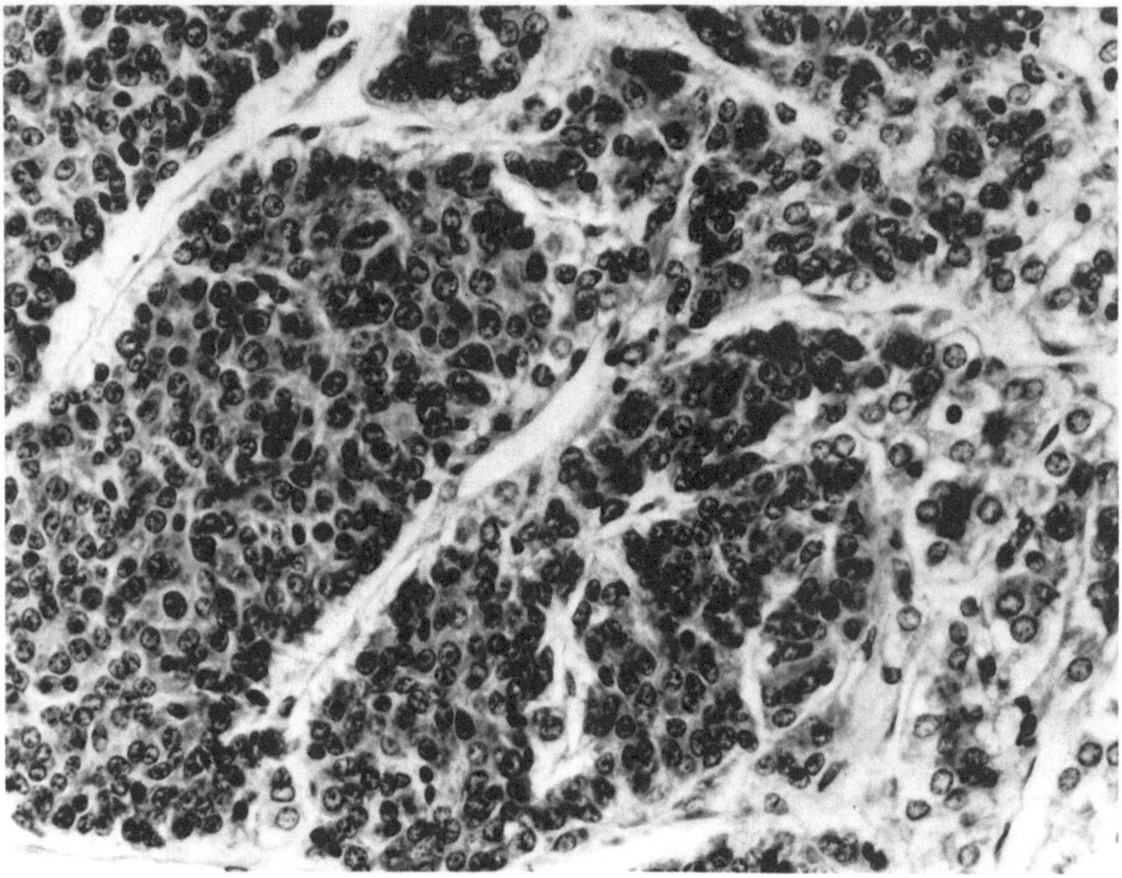

FIGURE 4.21. Higher magnification of the pheochromocytoma from Figure 4.20 showing the nests of tumor cells. (HPS; magnification × 500.)

TABLE 4.6

Incidence of Pheochromocytomas in Aging BN/Bi, WAG/Rij, and (WAG × BN)F_1 Rats

Strain	Sex	No. examined	No. with pheochromocytoma	%	Mean age (range) in months	No. with metastatic pheochromocytoma	Age (in months)
BN/Bi	Female	236	16	7	34 (28—39)	1	39
	Male	74	6	8	30 (20—42)	1	20
WAG/Rij	Female	101	8	8	36 (33—43)	0	—
	Male	124	1	1	19	0	—
F_1	Female	68	2	3	25, 26	0	—
	Male	67	8	12	33 (22—39)	0	—

from small pheochromocytomas. They therefore diagnosed focal proliferative lesions of the chromaffin cells as pheochromocytomas. The same criteria were used in this study of BN/Bi, WAG/Rij, and (WAG × BN)F_1 rats. As a result, no cases of hyperplasia were recognized in the adrenal cortex or medulla. Regardless of the criteria used, the incidence of cellular alteration, adenomas (or nodular hyperplasia), cortical cell carcinomas, and pheochromocytomas increased with age in all groups of rats in this study.

Snell and Stewart[155] observed adenomas or carcinomas of the adrenal cortex in 53 of 59 Osborne-Mendel rats older than 18 months. In this current study, similar age-related findings occurred because the highest incidence of adenomas was found in the oldest rats (Figure 4.18). The female WAG/Rij had the highest peak incidence of 60% in those rats older than 37 months.

Morphologically, there appeared to be a transition of cortical adenomas into carcinomas in this study, and similar observations have been suggested by others.[84] When Tables 4.3 and 4.4 are compared, however, it is impossible to show a trend for groups of rats with a high incidence of adenomas to also have a high incidence of cortical carcinomas. For example, female WAG/Rij rats had a 40% incidence of cortical adenomas, but one of the lowest incidences of cortical carcinomas (only 2%). On the other hand, female BN/Bi and F_1 rats had nearly identical incidences of adenomas (19 and 22%, respectively) and nearly identical incidences of cortical carcinomas (9 and 7%, respectively).

The adrenal cortex of the rats has many capillaries and sinusoids.[84] These vascular channels became distended and cystic in a few of the BN/Bi, WAG/Rij, and (WAG × BN)F_1 rats. Similar lesions have been described as telangiectasis in Fischer rats[42] and in a stock of Wistar rats,[102] as blood lakes,[84] and as dilated vessels.[69] The actual incidence in the different strains and stocks of rats is unknown. This lesion is apparently common in some Sprague-Dawley rats. For example, they were seen frequently in a partially inbred stock of Sprague-Dawley rats at the REP Institutes in Rijswijk, despite the low background in the BN/Bi, WAG/Rij and (WAG × BN)F_1 rats at the same facilities.[83] Kociba et al.[99] have also observed such lesions in Sprague-Dawley (Spartan substrain) rats that they have used in chronic toxicity studies.

Similar variability in the background incidence of lesions is seen in the adrenal medulla. Some strains have a high incidence of pheochromocytomas, namely the Wistar rats reported by Gillman et al.,[69] where the incidence was greater than 76% in male and female rats older than 24 months. The Wistar rats (WAG/Rij) reported in this study (Table 4.6) had only 8% in the females and 1% in the males. Therefore, it is clear that the occurrence of these lesions can differ tremendously among rats, even those supposedly derived from common ancestors, i.e., Wistar rats.

D. Pituitary Gland

1. Nonneoplastic Lesions

Nonneoplastic lesions of the pituitary gland were observed in some rats, but were relatively uncommon or insignificant and were not specifically tabulated. Such lesions included the presence of "castration cells" in the anterior pituitary of a few old rats, cystic structures, and variable amounts of melanin pigment in the BN/Bi and F_1 rats.

2. Neoplastic Lesions

a. Tumors of the Anterior Pituitary Gland

Tumors of the anterior pituitary gland (adenohypophysis) were common. They ranged in size from microscopic lesions causing compression of the normal glandular tissue to large tumors greater than 10 mm in diameter that caused compression of the brain and hydrocephalus. The microscopic appearance varied. Some tumors were composed of well-differentiated cells that formed sheets or nests. Others had variable degrees of anaplasia such as cellular pleomorphism, tinctorial variability, mitoses, giant cells, or, in a few cases, invasion into the meninges and perivascular spaces at the base of the brain. Most were highly vascular with many dilated vascular channels. The microscopic appearance ranged, therefore, from apparently benign lesions to histologically malignant tumors.

Several large pituitary tumors appeared to be composed of two or more nodules, each with a different cellular morphology (Figure 4.22). This was also true of a few microscopic lesions in which multiple microscopic tumors were recognized within one gland (Figure 4.23). Such findings suggest a multicentric origin for at least some of the pituitary tumors.

The incidence in the population is summarized in Table 4.7 Male and female WAG/Rij had the highest incidence at 96 and 95%, respectively. F_1 rats had fewer tumors with 83% in females and 64% in males. BN/Bi had a rel-

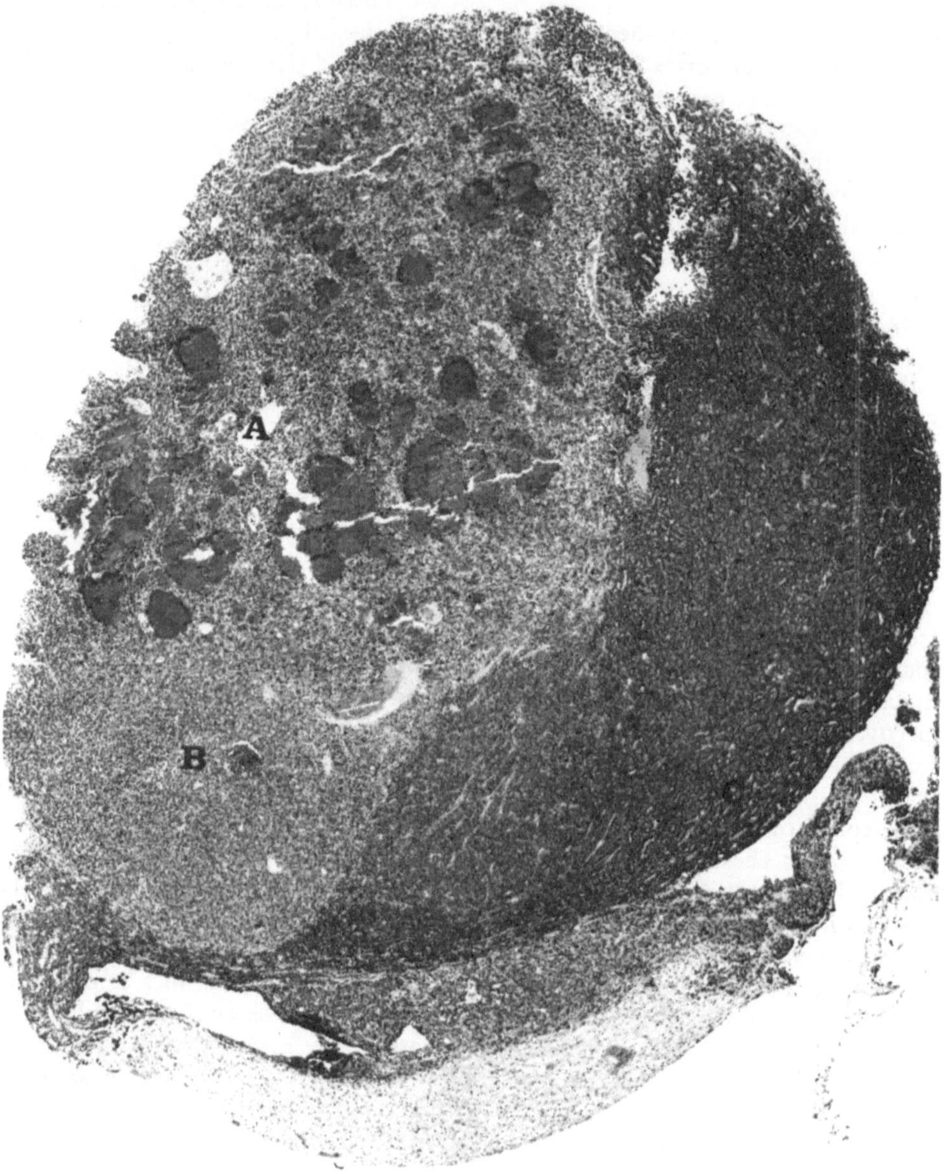

FIGURE 4.22. Three (A, B, and C) morphologically different patterns in a pituitary tumor arising in the anterior pituitary gland of a female WAG/Rij rat. (HPS; magnification ×35.)

atively low incidence with 26% in females and 14% in males.

The age-associated incidence is illustrated in Figure 4.24. Male and female BN/Bi, male and female F_1, and female WAG/Rij all had similar patterns. The tumors began to appear in the age periods between 13 and 24 months. After 24 months, the incidence increased dramatically until the peak incidence was as follows: female WAG/Rij, 97%; male F_1, 85%; female F_1, 79%; female BN/Bi, 33%; and male BN/Bi, 11%. The male WAG/Rij, however, had an earlier onset of pituitary tumors with a rapid increase in incidence already evident in the age

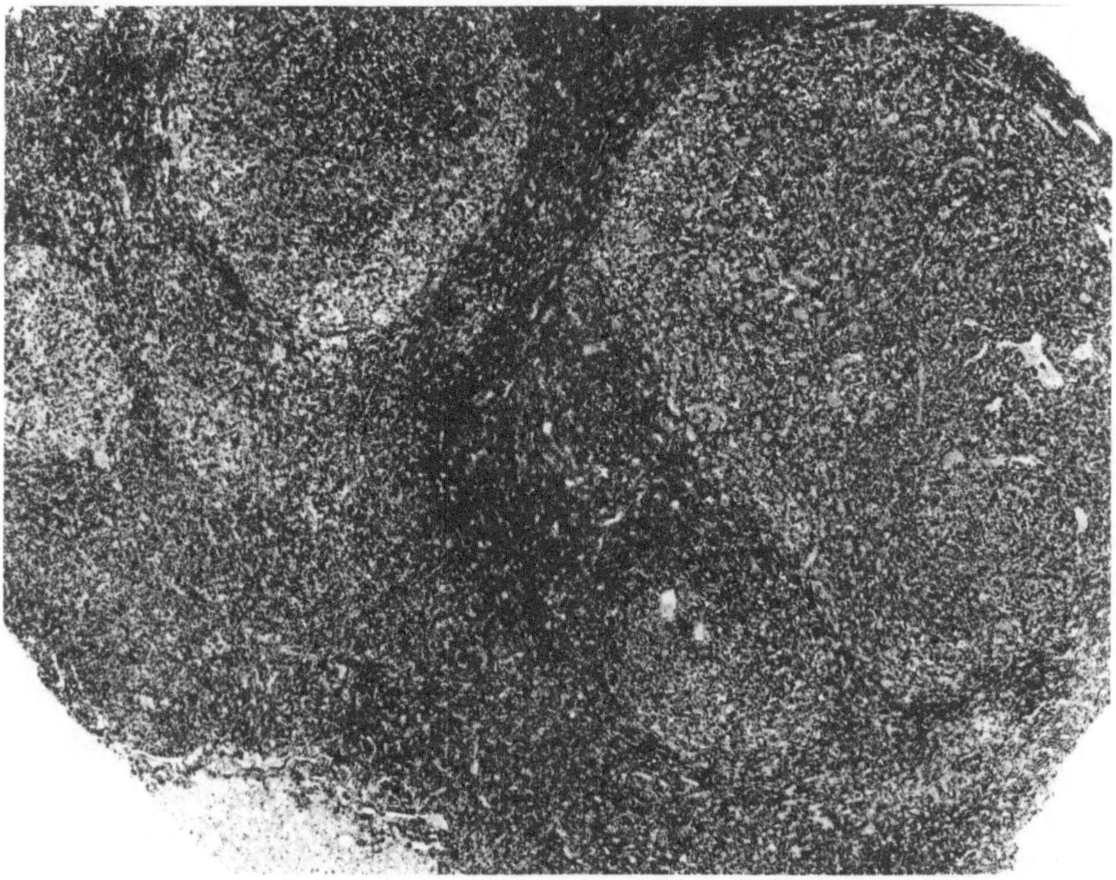

FIGURE 4.23. Multiple microscopic tumors in the anterior pituitary gland of a female WAG/Rij rat. (HPS; magnification × 13.)

TABLE 4.7

Incidence of Tumors in the Anterior Pituitary of Aging BN/Bi, WAG/Rij, and (WAG × BN)F_1 Rats

Strain	Sex	No. examined	No. with pituitary tumor	%	Mean age (range) in months
BN/Bi	Female	236	61	26	31 (17—39)
	Male	74	10	14	28 (15—42)
WAG/Rij	Female	101	96	95	36 (19—46)
	Male	124	118	96	22 (9—31)
F_1	Female	68	57	83	28 (15—40)
	Male	67	43	64	33 (22—42)

periods between 7 and 18 months. The peak incidence occurred in the oldest group (25 to 30 months) and reached 90%.

b. Other Pituitary Tumors

A 25-month-old female F_1 rat had, in addition to an epithelial tumor of the anterior pituitary, a malignant melanoma. The tumor cells were filled with melanin and had infiltrated the posterior pituitary, meninges, pars nervosa, and into the epithelial tumor of the anterior pituitary resulting in a collision of the two tumors (Figure 4.25).

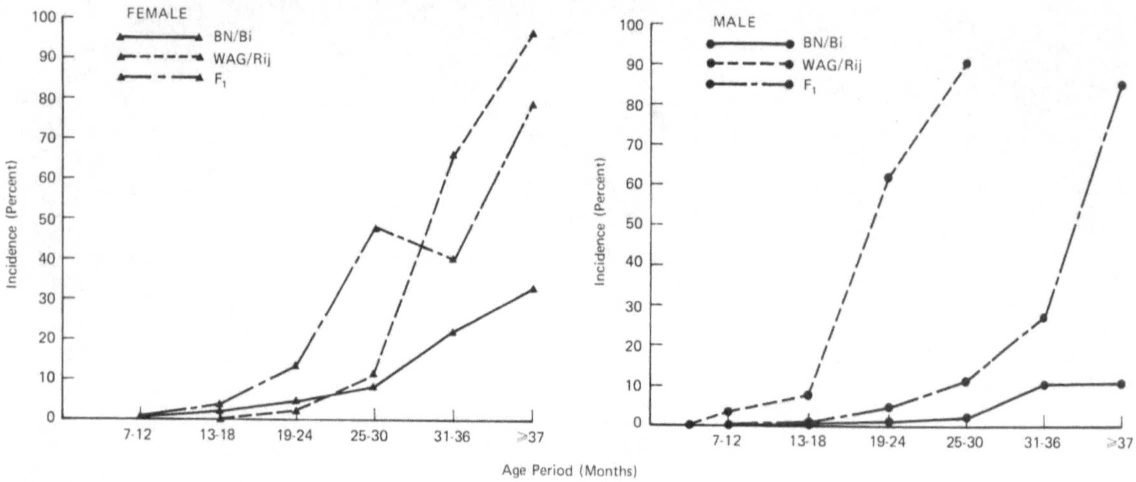

FIGURE 4.24. Percent incidence with age of pituitary tumors in 236 female and 74 male BN/Bi, 101 female and 124 male WAG/Rij, and 68 female and 67 male (WAG × BN)F_1 rats.

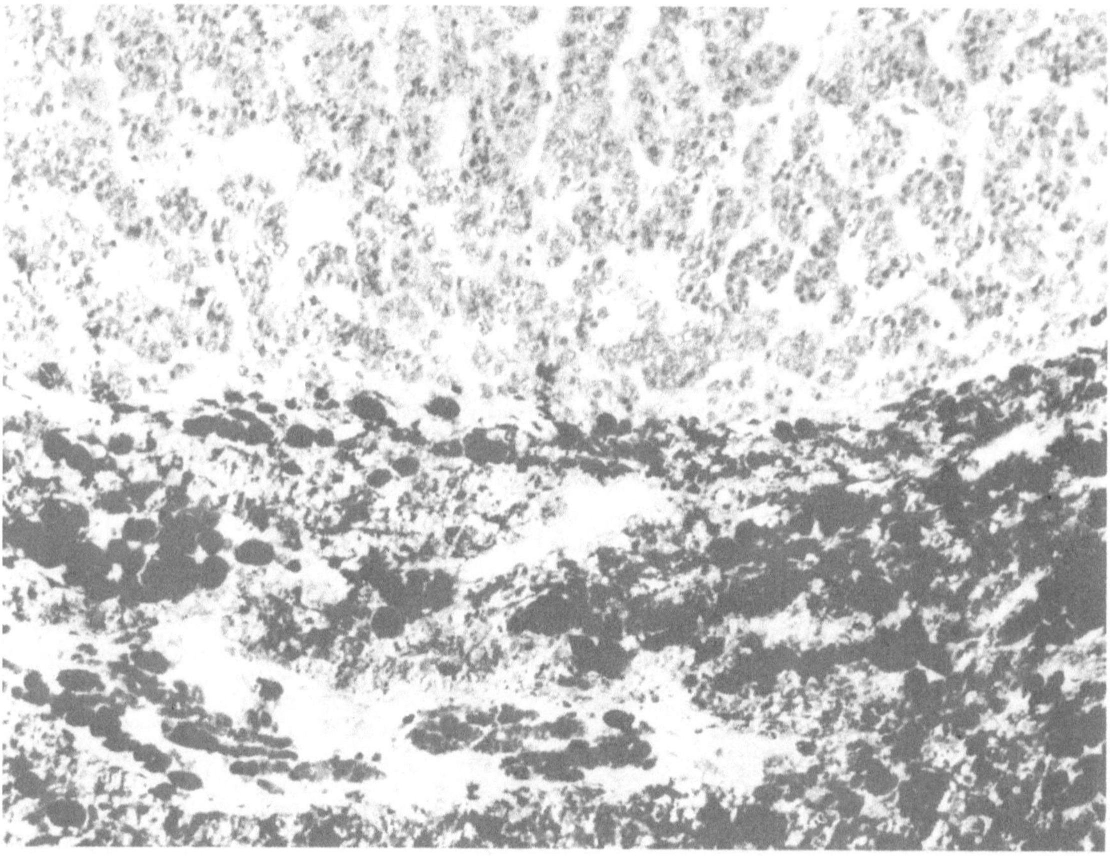

FIGURE 4.25. Pituitary gland from a female (WAG × BN)F_1 rat having a tumor of the anterior pituitary gland (upper half) and a malignant melanoma (lower half). (HPS; magnification × 210.)

3. Discussion

Tumors arising in the adenohypophysis were common in this study, with 385 neoplasms diagnosed. The incidence varied, with 96% in male and 95% in female WAG/Rij rats, 64% in male and 83% in female (WAG × BN)F_1 rats, and 14% in male and 26% in female BN/Bi rats. Nearly all were recognized grossly as enlarged pituitary glands, some greater than 1 cm in diameter. The older the rat, the greater was its risk of dying with a pituitary tumor (Figure 4.24), but the WAG/Rij males appeared to have an earlier onset of this lesion than the other groups studied.

The population of male WAG/Rij rats died about 6 to 9 months earlier than the populations of female WAG/Rij, female and male BN/Bi, or female and male (WAG × BN)F_1 rats. Since the incidence of pituitary tumors was 96% in the WAG/Rij males and since they occurred 6 to 9 months earlier, it would seem that the earlier deaths were caused by pituitary tumors. The problem is much more complex, however, because many neoplastic and nonneoplastic lesions seemed to occur earlier in male WAG/Rij rats (see other organ systems in this chapter).

Spontaneous pituitary tumors are common and have been reported in most strains and stocks of rats as summarized in several review articles.[63,64,147,156,163] They occur in Sprague-Dawley, Wistar, Fischer, Osborne-Mendel, and BN derived rats. In fact, no references were found that described a 0% background in any strain or stock of rats. The incidence varies from less than 20% to over 90%. The true age-associated incidence for different strains and stocks is difficult to assess, however, since only a relatively few reports have been based upon life span studies.

It has been stated that in most strains, the incidence of pituitary tumors was higher in females than males.[163] In this study, the incidence in female BN/Bi and (WAG × BN)F_1 was greater than in males of the same strain. However, the incidence in male and female WAG/Rij rats was identical.

For the purpose of this study, the "typical" rat pituitary tumors were classified simply as anterior pituitary gland tumors. Similar lesions are often designated as chromophobe adenomas by others, but based on the excellent review articles by Furth et al.[63,64] and Ito,[92] it is clearly impossible to determine the cell of origin and possible hormone production using routine histological material. Furthermore, it is equally difficult to classify the tumors as benign or malignant. Approximately one third of the tumors in this study could be malignant as judged by histological criteria. Only three tumors showed local invasion, however, and none had distant metastases. Only one pituitary tumor from a WAG/Rij rat was transplanted[83] and it grew rapidly. Others have transplanted rat pituitary tumors and they generally grew well in the recipient host.[64] Only a limited number of histochemical studies have been done on the WAG/Rij pituitary tumors and all were inconclusive.[83] For these reasons only a simple working diagnosis was given to these tumors until more specific data are available.

Rarely, tumors may occur in other portions of the pituitary than the adenohypophysis. For example, a malignant melanoma had invaded the pars nervosa and pars intermedia in a female (WAG × BN)F_1 rat. Kroes et al.[102] described tumors of the anterior pituitary in their Wistar rat. In addition, 15 to 25% of their rats had hyperplasia of the pars intermedia and some had tumors arising from that area. Therefore, although most rat pituitary tumors are epithelial in origin and arise in the adenohypophysis, other tumors do occur.

E. Pancreas

1. Nonneoplastic Lesions

a. Atrophy

Focal or multifocal atrophy of pancreatic lobules was the most common age-associated lesion in the pancreas. The atrophy occurred in acini and lobules often with complete loss of acinar epithelial cells in such areas (Figure 4.26). Accompanying the loss of acinar cells, a mild inflammatory response was often present that consisted mainly of lymphocytes and increased connective tissue. Fat cells rarely appeared to replace the acini. The lesion was seldom severe and the majority of pancreas appeared to be normal. Despite the loss of acini and accompanying fibrosis, the islets in these areas seemed normal. At times, dilated ductules were observed.

Pancreatic atrophy was recognized in 34% of female and 40% of male BN/Bi, 16 and 11%

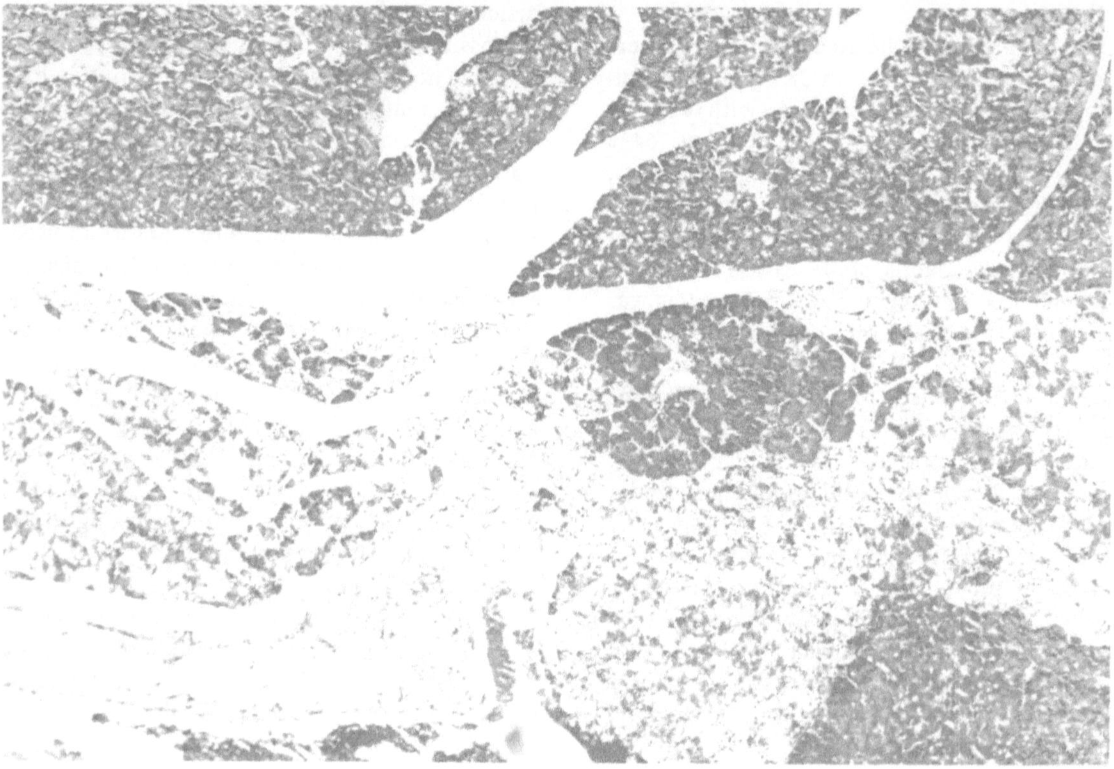

FIGURE 4.26. Focal atrophy of pancreatic acinar tissue. (HPS; magnification × 75.)

TABLE 4.8

Incidence of Pancreatic Atrophy in Aging BN/Bi, WAG/Rij, and (WAG × BN)F_1 Rats

Strain	Sex	No. examined	No. with pancreatic atrophy	%	Mean age (range) in months
BN/Bi	Female	236	81	34	32 (22—54)
	Male	74	30	40	32 (19—44)
WAG/Rij	Female	101	16	16	35 (27—38)
	Male	124	14	11	24 (18—29)
F_1	Female	68	4	6	34 (29—38)
	Male	67	31	46	35 (22—44)

female and male WAG/Rij, and 6 and 46% in female and male F_1 (Table 4.8) rats.

The age-associated incidence is shown in Figure 4.27, with most cases occurring in rats older than 24 months. The male BN/Bi had the highest peak incidence (78%), followed by the F_1 males (56%), BN/Bi females (52%), and WAG/Rij females (28%) and WAG/Rij males (17%). The female F_1 rats had a relatively low incidence of this lesion and were not plotted in Figure 4.27.

2. Neoplastic Lesions

a. Islet Cell Tumors

Adenomas arising in the islets of Langerhans were common. Most were solitary lesions, but a few rats had more than one. Histologically, the cells were polyhedral with distinct cell

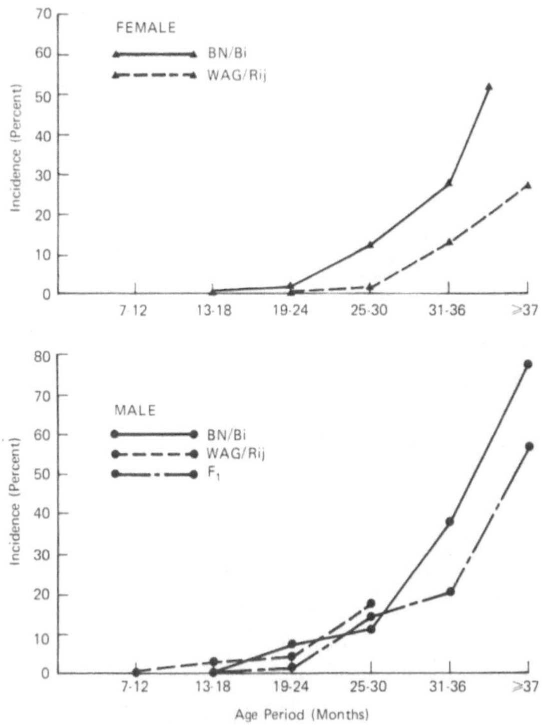

FIGURE 4.27. Percent incidence with age of pancreatic atrophy in 236 female and 74 male BN/Bi, 101 female and 124 male WAG/Rij, and 67 male (WAG × BN)F_1 rats.

boundaries and pink, often finely granular, cytoplasm. The nucleus was round, lightly vesicular, and located in the center of the cell. Cells were arranged in nests that were separated by thin connective tissue trabeculae. Some tumors were surrounded by a connective tissue capsule (Figure 4.28); however, most were not encapsulated and had an irregular zone of contact between the acinar tissue and the tumor (Figure 4.29). Normal and atrophic acinar tissues were often present within the lesion.

Two islet cell tumors were diagnosed as carcinomas. They occurred in male F_1 rats of 22 months and 37 months old. They were histologically similar to the adenomas, but extensive local invasion and destruction of pancreatic acinar tissue were present. However, no metastases were observed from either of these tumors.

The incidence in male and female WAG/Rij and female F_1 rats was relatively low, while the BN/Bi males and females and F_1 males had the greatest incidence at 15, 11, and 15%, respectively (Table 4.9).

The age-associated incidence is shown in Figure 4.30 for BN/Bi males and females and F_1

FIGURE 4.28. Pancreas containing a large islet cell adenoma that is sharply demarcated from the surrounding pancreatic tissue. (HPS; magnification × 30.)

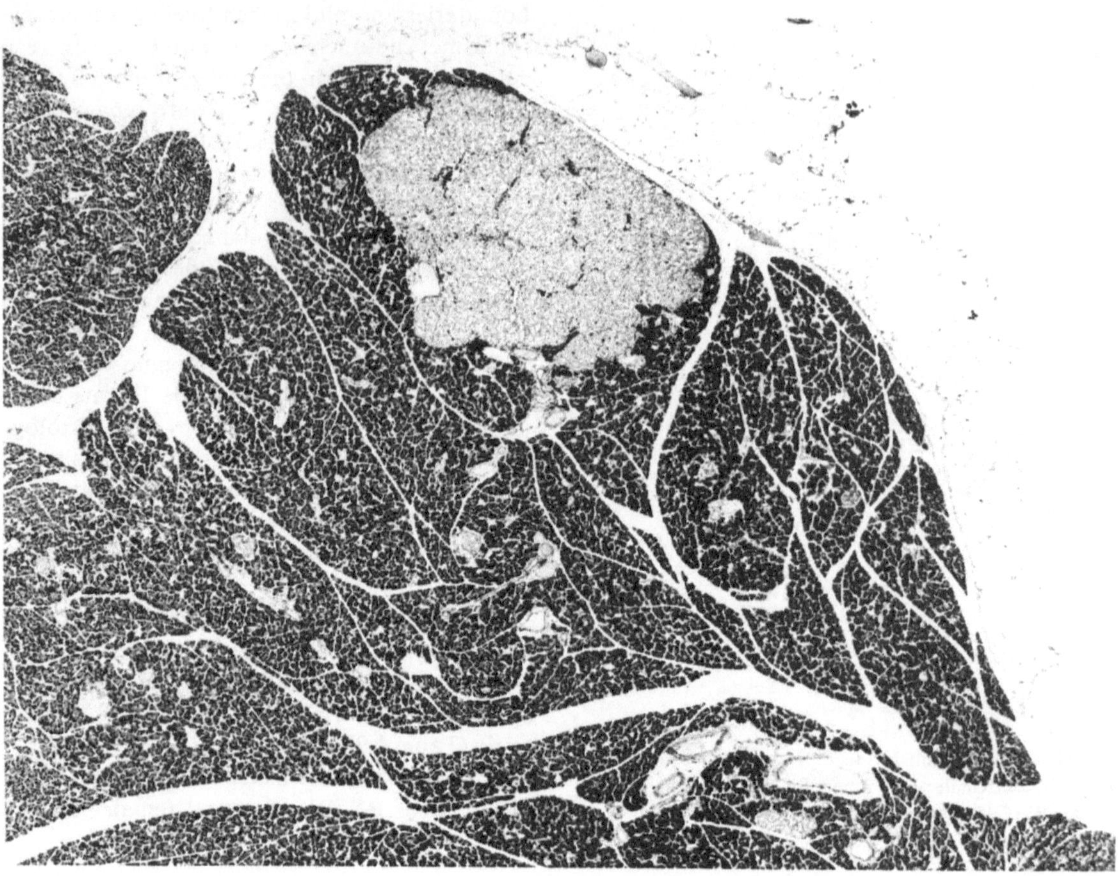

FIGURE 4.29. Pancreas with a small islet cell adenoma that was typical of the pattern found in the aging BN/Bi, WAG/Rij, and (WAG × BN)F_1 rats. Note the irregular contact zone of the adenoma and the surrounding normal acinar tissue. (HPS; magnification × 30.)

TABLE 4.9

Incidence of Islet Cell Tumors in Aging BN/Bi, WAG/Rij, and (WAG × BN)F_1 Rats

Strain	Sex	No. examined	No. with islet tumor	%	Mean age (range) in months
BN/Bi	Female	236	25	11	33 (23—40)
	Male	74	11	15	33 (20—42)
WAG/Rij	Female	101	3	3	40 (35—44)
	Male	124	1	1	23
F_1	Female	68	4	6	32 (26—36)
	Male	67	10	15	32 (22—40)

males. The tumors occurred in the older rats with a peak incidence in BN/Bi males of 32%, BN/Bi females of 24%, and F_1 males of 12%.

b. Exocrine Cell Tumors

Two male F_1 rats, 40 and 36 months old, had a solitary adenoma of the exocrine tissue (Figure 4.31). Both were grossly visible nodules that were sharply demarcated from the surrounding tissue. Microscopically, a thin connective tissue capsule was present. The individual cells in both tumors were similar. They had basophilic cytoplasm filled with numerous zymogen-like granules and tended to form acinar structures similar to the normal pancreatic acini. The nodules compressed the adjacent pancreatic tissue. There were no ducts or islets of Langerhans in either of the adenomas.

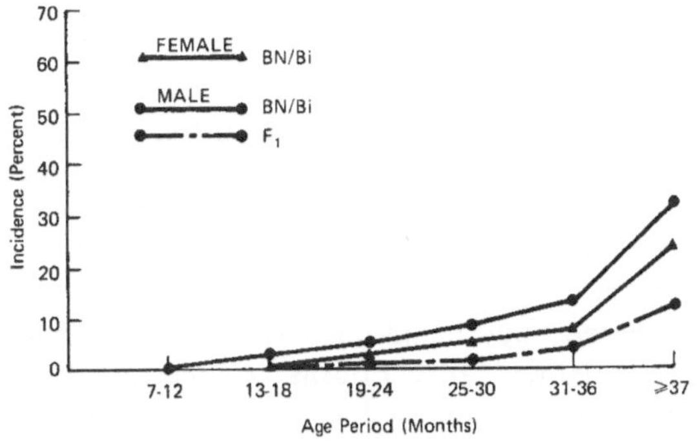

FIGURE 4.30. Percent incidence with age of islet cell tumors in 236 female and 74 male BN/Bi and 67 male (WAG × BN)F_1 rats.

3. Discussion

Atrophy of the pancreatic acini and lobules was the most common age-associated lesion in the pancreas of aging rats. It was present in 78% of male BN/Bi, 56% of male (WAG × BN)F_1, and 52% of female BN/Bi rats older than 37 months (Figure 4.27). Therefore, it can be a significant pancreatic lesion in some aging rats. As it was often difficult, if not impossible, to recognize at the gross necropsy examination, histopathologic evaluation was needed to determine its presence or absence.

As reviewed by others,[1,138,166,163] tumors of the islets are usually adenomas and the incidence can vary depending upon the strain or stock. Similarly, tumors of the pancreatic islets in rats of this study were mostly adenomas, and an apparent strain difference was seen in the incidence of these lesions. Female and male BN/Bi rats had an incidence of 11 to 15%, while male and female WAG/Rij rats had an incidence of only 1 to 3% (Table 4.9). These neo-

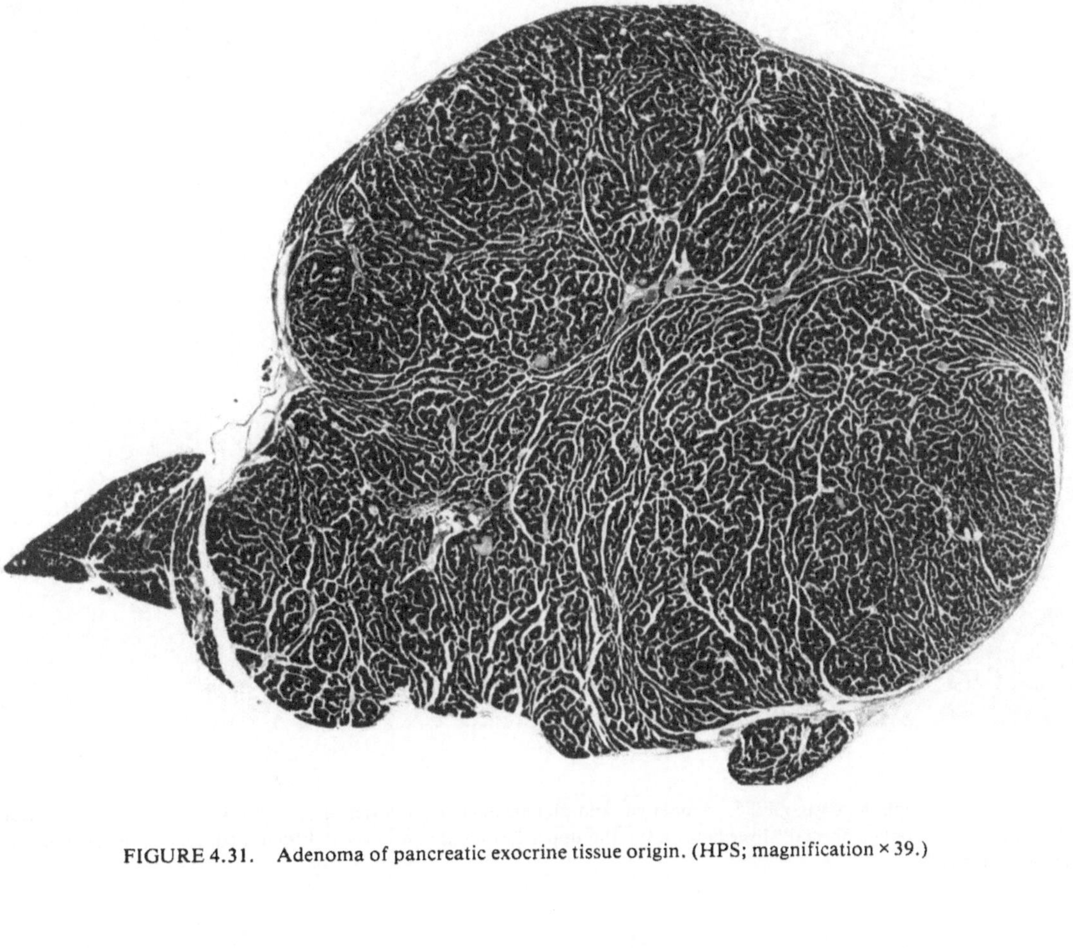

FIGURE 4.31. Adenoma of pancreatic exocrine tissue origin. (HPS; magnification × 39.)

plasms were age-associated and were relatively common (24 to 32%, Figure 4.30) in male and female BN/Bi rats older than 37 months. Likewise, Coleman et al.[42] found islet adenomas in 9 of 144 male Fischer rats with most occurring in the oldest animals (5 of the 9 were in rats between 30 to 33 months of age).

F. Liver

1. Nonneoplastic Lesions

a. Foci and Areas of Cellular Alterations

A common age-associated lesion in the liver was an alteration in groups of hepatocytes. The alteration was seen as tinctorial or morphological changes in hepatocytes that set them apart from the normal cells (Figures 4.32 and 4.33). Such changes, when smaller than a lobule in size, were called foci. When they were larger than a lobule, they were considered to be areas. Liver cells within these foci or areas were either larger or smaller than normal (Figures 4.34 and 4.35). Similarly, the cell nuclei were larger or smaller than normal and either had normal chromatin pattern or were vesicular or hyperchromatic. The cytoplasm was either basophilic or pink, finely granular or homogeneous, or finely or coarsely vacuolated. The margins of the foci and areas were usually sharp without clear zones of compressed normal liver parenchyma.

In the foci and areas it was not uncommon to find a few mitoses, focal extramedullary hematopoiesis, and dilated or telangiectatic vascular spaces. Such findings were also seen in the normal portions of the liver, but they were more common in the regions with cellular change.

Dilated endothelial-lined spaces were often associated with areas of cellular alteration. The smaller spaces usually had varying numbers of red blood cells in the lumina, while the larger spaces contained only a pink proteinaceous ma-

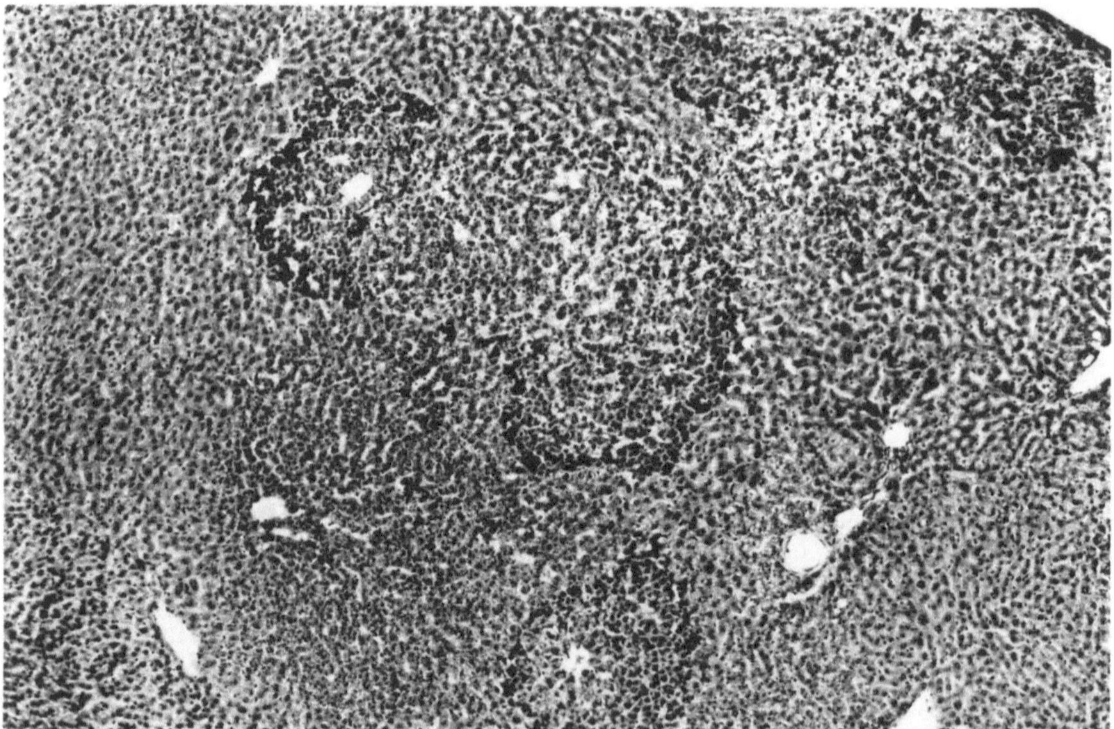

FIGURE 4.32. Basophilic, eosinophilic, vacuolated, and pleomorphic hepatocytes are present within an area of hepatocellular alteration. The only "normal" liver is found in the upper left-hand and lower right-hand corners of the figure. (HPS; magnification × 75.)

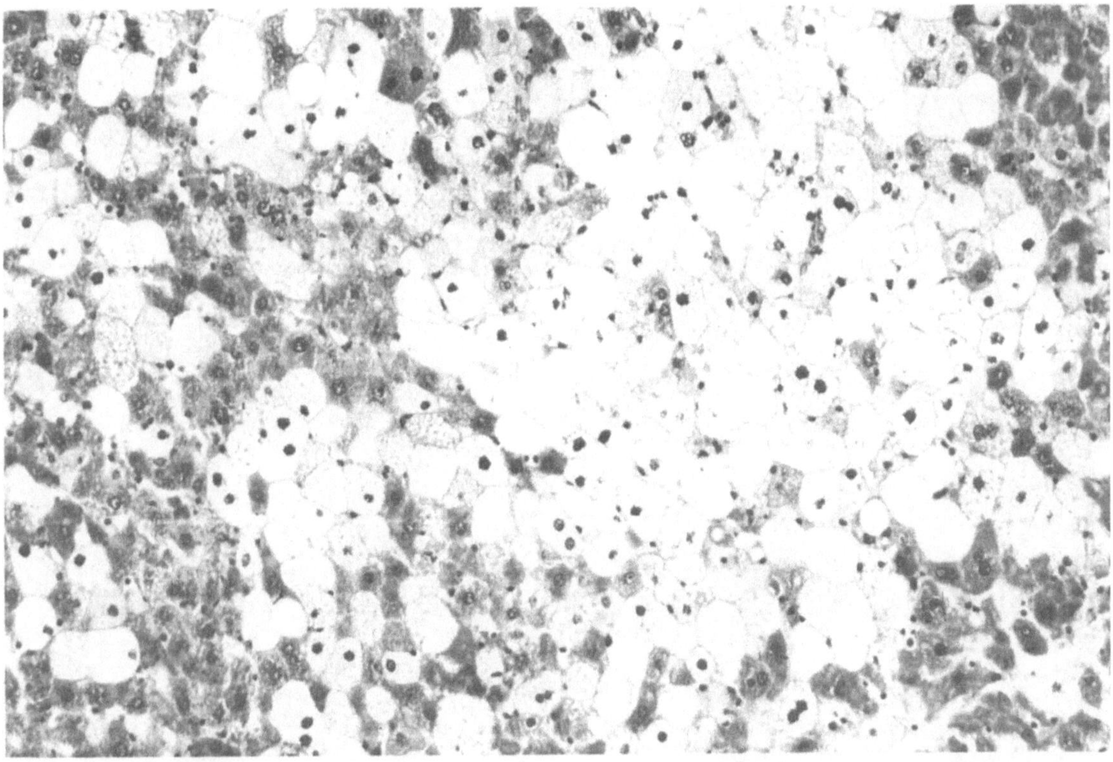

FIGURE 4.33. Area of altered hepatocytes. Many of the cells tend to be larger than normal and have vacuolated cytoplasm and small dark nuclei. (HPS; magnification ×465.)

terial (Figure 4.36). These areas varied somewhat in size, but most were relatively small areas that occupied only a few hepatic lobules in diameter. Along the margins, individual hepatocytes were present that were vacuolated and degenerative. Necrosis was not evident, however, and the pathogenesis of this change is unknown.

The BN/Bi rat strain seldom had foci or areas of cellular alterations in the liver. Fewer than 5% of the males and females died with such changes and, when present, the lesions were focal and relatively insignificant. This is in sharp contrast to the findings in the WAG/Rij and F_1 rats, where such changes were common (Table 4.10); 36% of male WAG/Rij, 37% of male F_1, and 47% of female F_1 rats died with these lesions. In contrast, 84% of all WAG/Rij females died with these lesions. In addition, the changes in the WAG/Rij females were usually very numerous and, at times, more area was occupied by the altered cells than by the normal cells. As Figure 4.37 shows, the peak age-associated incidence was in the oldest rats and reached 76% for the female WAG/Rij, 45% for male WAG/Rij, and 60 and 35% for female and male F_1 rats, respectively.

b. Biliary Cysts

Many rats died with grossly visible cysts in the liver. They ranged in size from a few millimeters to several centimeters, were solitary or multiple, multiloculated, and contained a clear amber-colored fluid (Figure 4.38). Microscopically, all cysts were similar irrespective of their size. The lesions consisted of many large and small spaces lined by simple cuboidal or flat endothelial-like cells (Figures 4.39 and 4.40). The trabecular tissue between the spaces was formed by a thin layer of connective tissue. Occasional nests of hepatocytes were present between the cysts and were usually compressed. A few lymphocytes and macrophages were seen in the trabeculae, but such cells were relatively in-

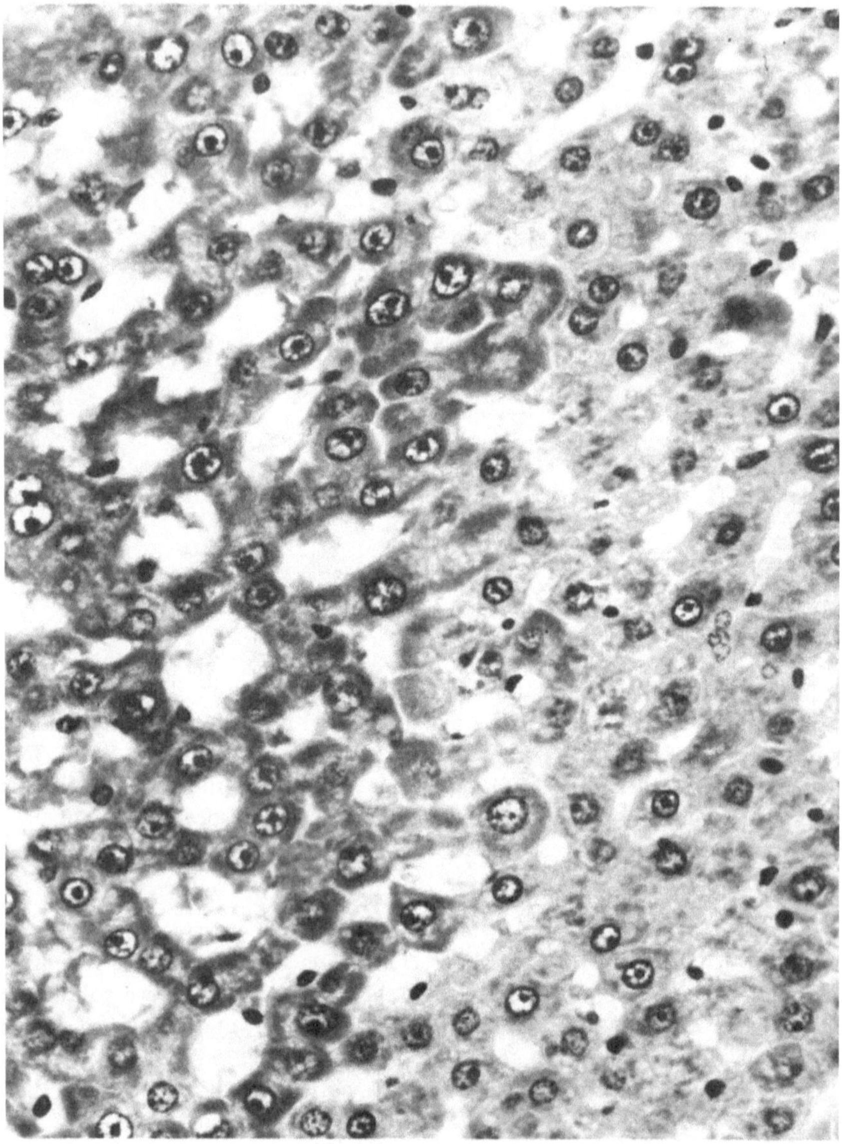

FIGURE 4.34. The junction between an area of altered hepatocytes (upper half) and normal liver (lower half). In this example, the altered hepatocytes were darker red in color, larger than normal, and contained nuclear pleomorphism. (HPS; magnification × 500.)

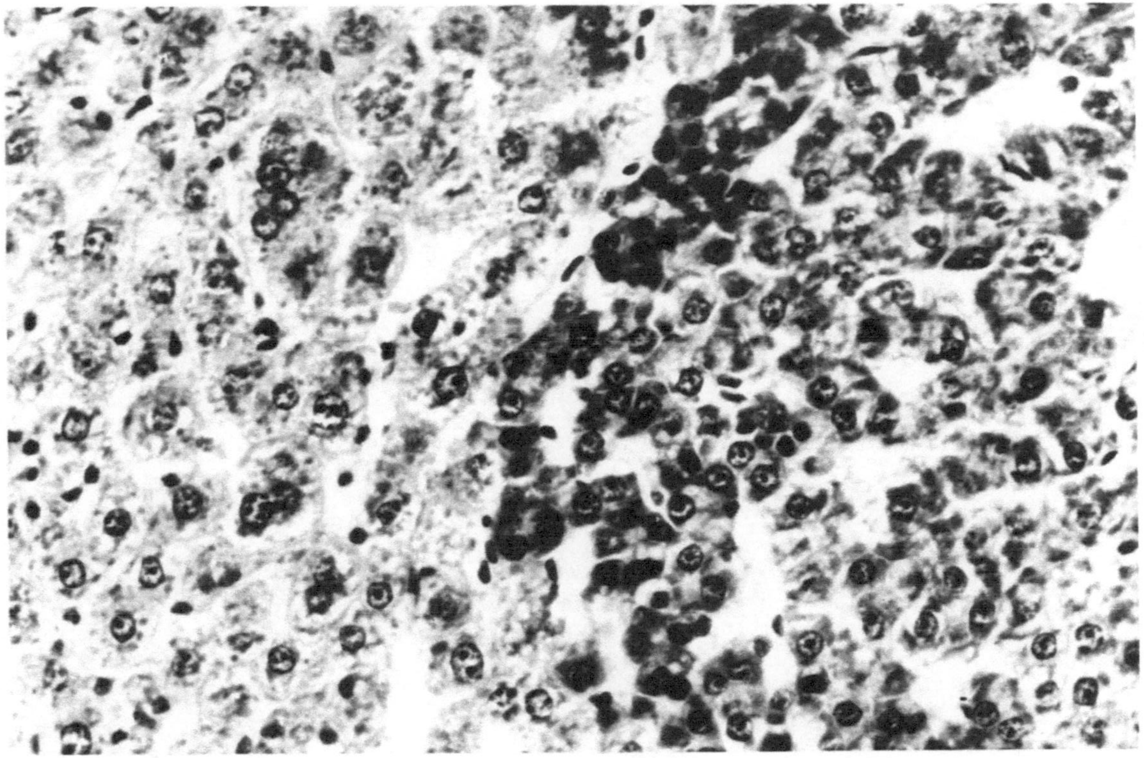

FIGURE 4.35. The junction between normal hepatocytes (left) and a focus of smaller, basophilic hepatocytes (right). (HPS; magnification × 465.)

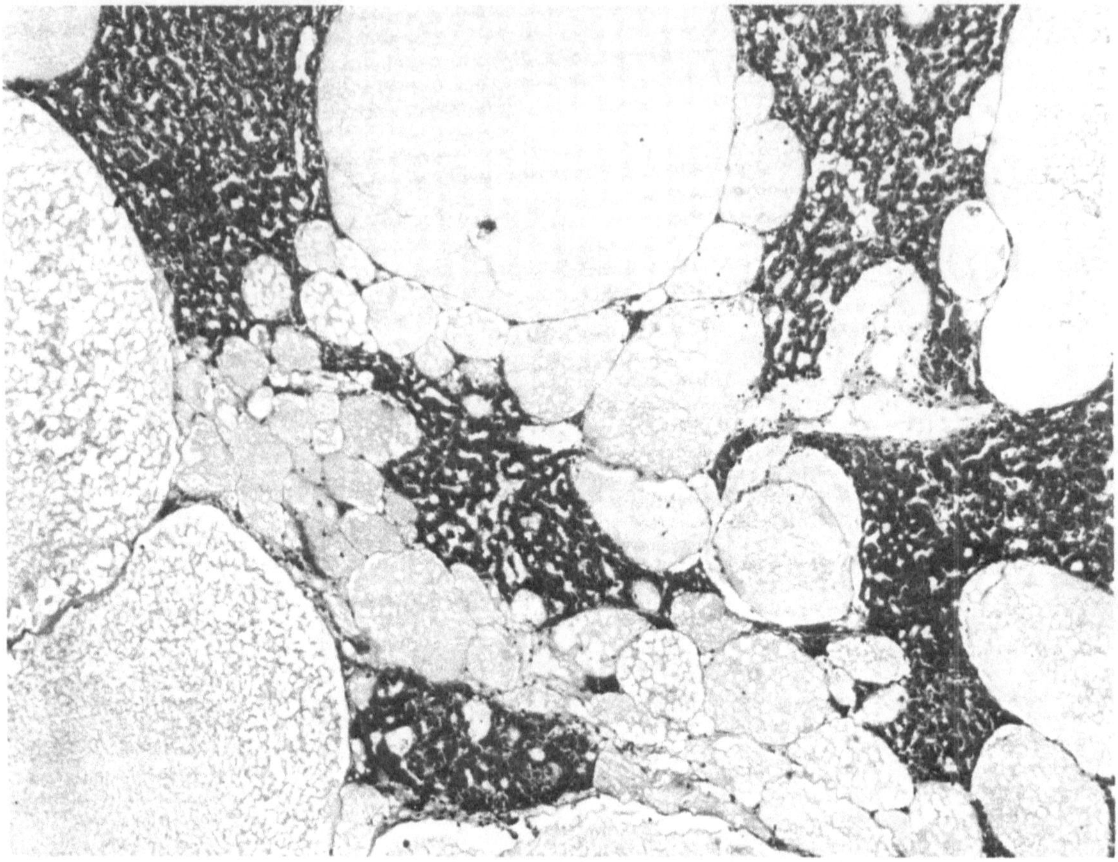

FIGURE 4.36. Many endothelial-lined spaces within an area of basophilic altered hepatocytes. The spaces contain pink, protein-like material that is granular in some locations and homogeneous in others. (HPS; magnification × 80.)

TABLE 4.10

Incidence of Foci and Areas of Cellular Alterations in the Livers of Aging WAG/Rij and (WAG × BN)F_1 Rats

Strain	Sex	No. examined	No. with lesions	%	Mean age (range) in months
WAG/Rij	Female	101	85	84	34 (19—46)
	Male	124	45	36	24 (18—31)
F_1	Female	68	32	47	30 (24—40)
	Male	67	25	37	35 (22—40)

frequent. Compressed hepatic tissue was seen along the margin of some but not all of the cysts.

The origin of these cysts could not be determined. The smallest were identical to the largest. None contained blood or bile and none showed an unequivocal connection to an identifiable bile duct or vascular channel. Since the lining appeared most like bile duct epithelium, such cysts were classified as bilary cysts and appeared to represent a benign, proliferative, cystic lesion in these aging rats.

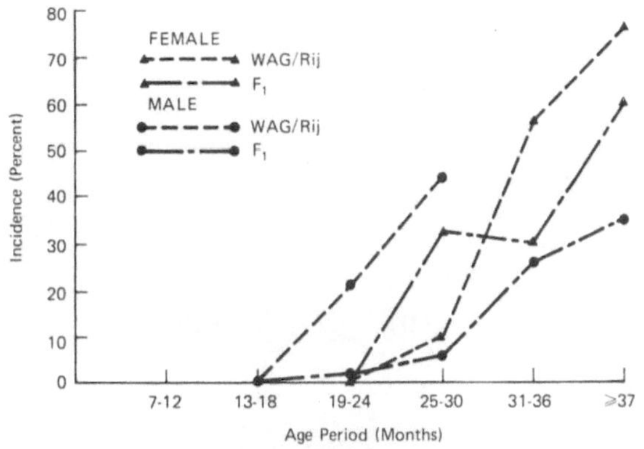

FIGURE 4.37. Percent incidence with age of foci and areas of altered hepatocytes in 101 female and 124 male WAG/Rij and 68 female and 67 male (WAG × BN)F_1 rats.

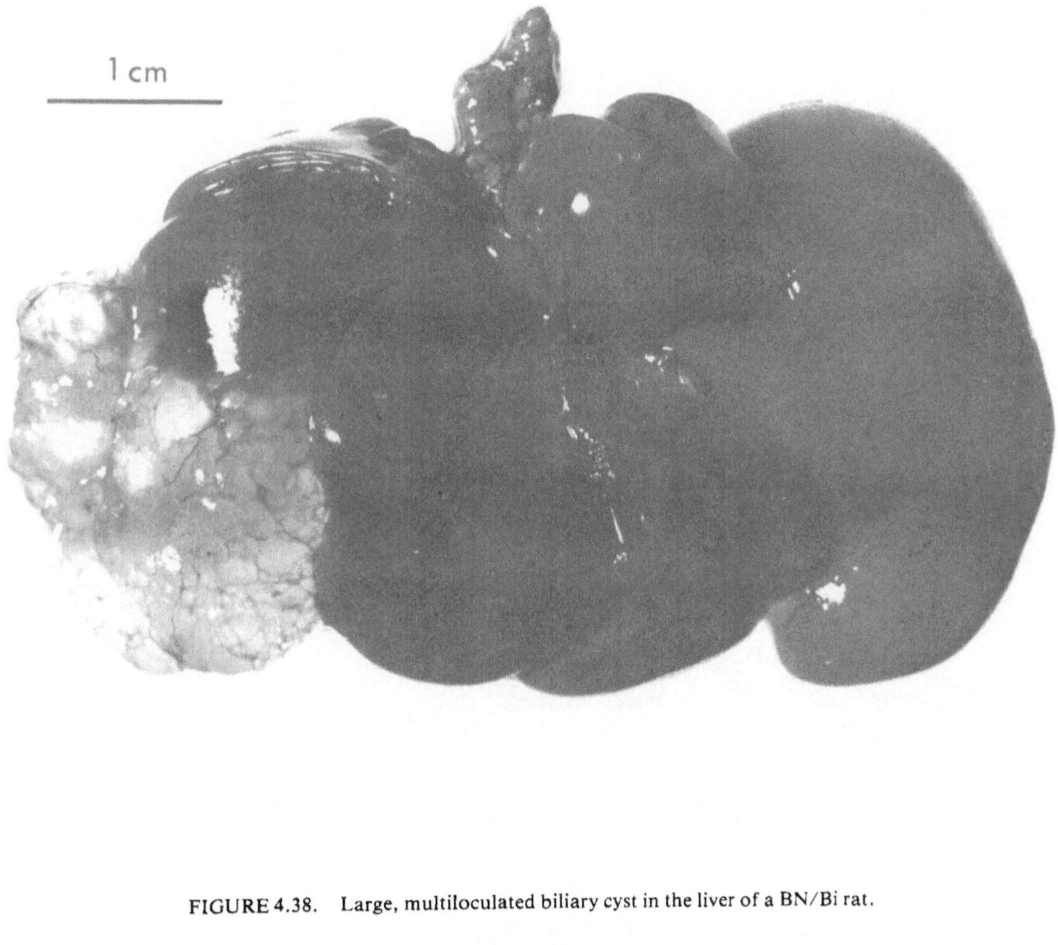

FIGURE 4.38. Large, multiloculated biliary cyst in the liver of a BN/Bi rat.

FIGURE 4.39. Biliary cyst in the liver of a BN/Bi rat. (HPS; magnification × 30.)

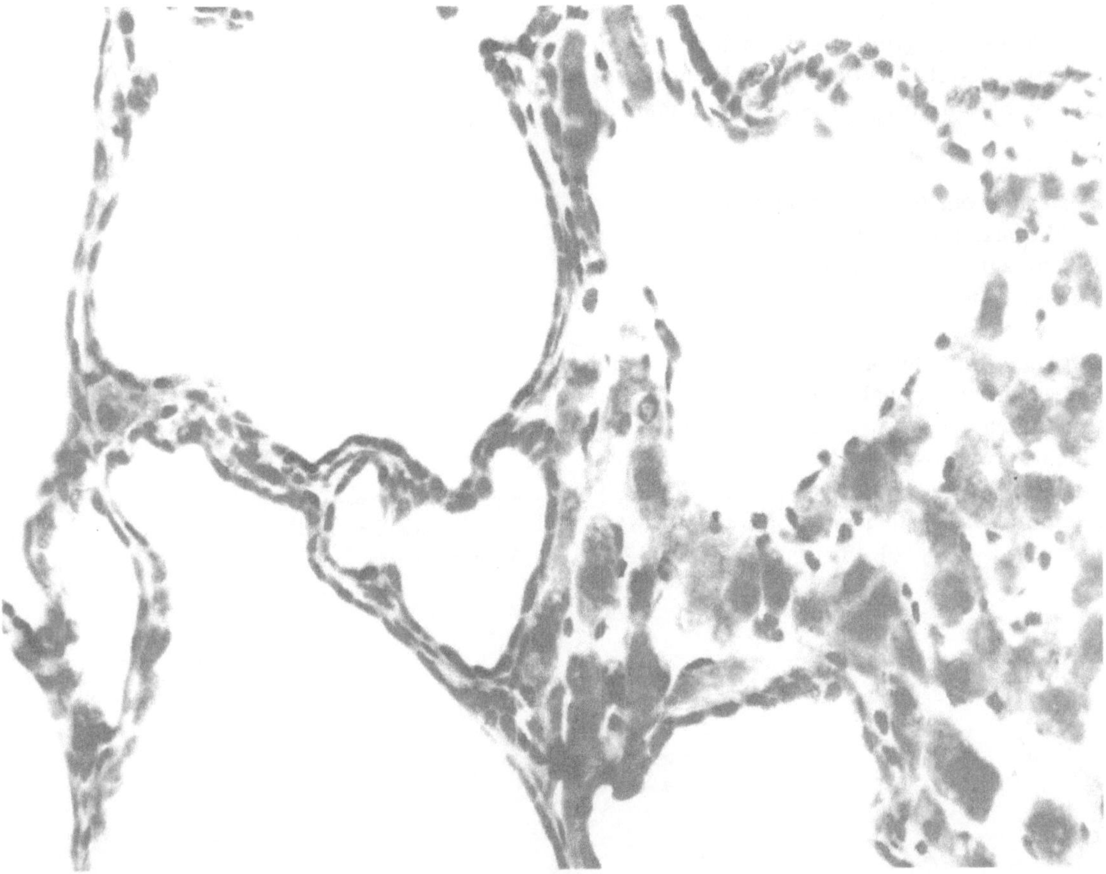

FIGURE 4.40. Higher magnification of the biliary cyst shown in Figure 4.39. (HPS; magnification × 500.)

TABLE 4.11

Incidence of Cysts in the Liver of Aging BN/Bi, WAG/Rij, and (WAG × BN)F_1 Rats

Strain	Sex	No. examined	No. with cysts	%	Mean age (range) in months
BN/Bi	Female	236	129	55	32 (13—54)
	Male	74	19	26	32 (20—42)
WAG/Rij	Female	101	10	10	36 (30—43)
	Male	124	9	4	24 (18—29)
F_1	Female	68	13	19	31 (26—38)
	Male	67	6	9	38 (29—44)

The incidence of such biliary cysts is shown in Table 4.11. Females had a higher incidence than males and the BN/Bi strain had a higher incidence than the WAG/Rij strain. For example, 55% of BN/Bi females had cysts in contrast to 26% of the males. The WAG/Rij females had an incidence of 10% while the males had only 4%. The incidence in F_1 rats was intermediate between the BN/Bi and WAG/Rij strains.

The age-associated incidence was plotted for female BN/Bi and F_1 rats and for male BN/Bi rats (Figure 4.41). The peak incidence occurred in the oldest age groups and reached 75% of the female BN/Bi, 50% of female F_1, and 33% of male BN/Bi rats that were older than 37 months.

c. Miscellaneous Nonneoplastic Lesions

Additional lesions were also recognized in the livers of aging rats, but were not specifically tabulated. Among these was the presence of lipofuscin pigment in hepatocytes. This pigment was common in the older rats. Nearly all rats may have it, but special stains, such as PAS and Ziehl-Nelsen acid fast, were not used routinely to identify the pigment. Variations in the size of hepatocyte nuclei and polyploidy, also very common in these aging rats (Figure 4.42), were not studied in detail. Similarly, hemosiderin

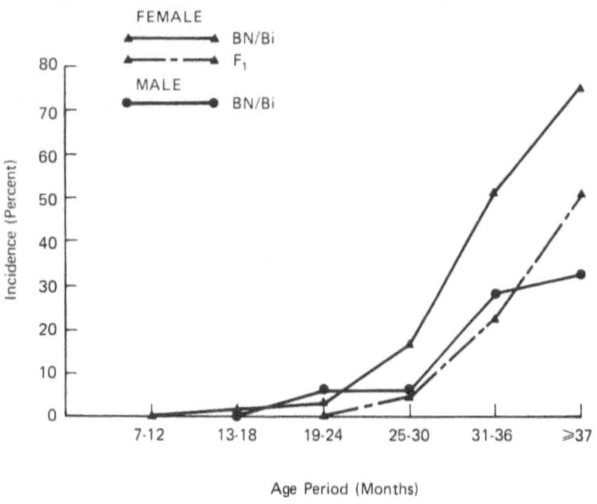

FIGURE 4.41. Percent incidence with age of biliary cysts in 236 female and 74 male BN/Bi and 68 female (WAG × BN)F_1 rats.

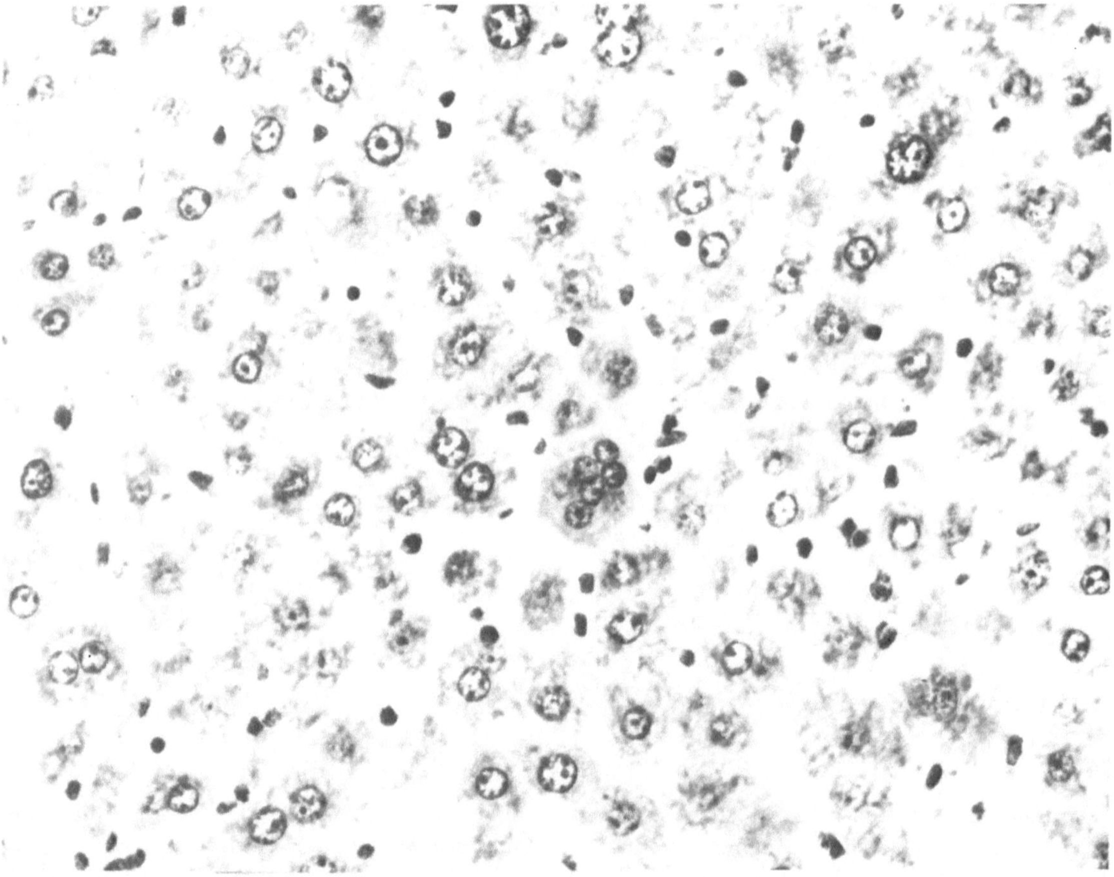

FIGURE 4.42. "Normal" appearance of the liver of an aged BN/Bi rat. Note the multinucleated hepatocytes, the variation in the size of nuclei, and the dark granular material in the cytoplasm which is lipofuscin-like pigment. (HPS; magnification × 500.)

pigment was present in hepatocytes and Kupffer's cells of some aged rats, but Prussian blue reaction for iron was not done routinely. Many of the rats that died with large tumors elsewhere in the body had extramedullary hematopoiesis randomly distributed in the liver. Focal areas of necrosis, microabscesses, or focal aggregations of mononuclear inflammatory cells were also seen in a few livers, but only sporadically. Bile duct hyperplasia was recognized in the livers of some rats, but appeared to be a relatively insignificant lesion in these rats. Finally, pink intranuclear inclusion bodies of varying size were seen in hepatocytes of a few rats. They did not appear to be of viral origin and probably increased with age, but more studies would be needed to clearly show an age-associated pattern.

2. Neoplastic Lesions

a. Neoplastic Nodules

Neoplastic nodules were morphologically benign nodules that could not be differentiated from nodular hyperplasias or adenomas. However, they appeared to be more than areas. The nodules were often morphologically similar to the areas of cellular alteration described previously, with the cells and architecture showing many of the same tinctorial and cellular variations. In general, there was more cellular pleomorphism and more mitoses and the hepatic architecture was more disorganized within the nodules. The major criterion for diagnosing nodules was, however, the presence of distinct zones of compression of normal liver along the contact regions between the normal liver and the nodule (Figure 4.43) and a disorganization

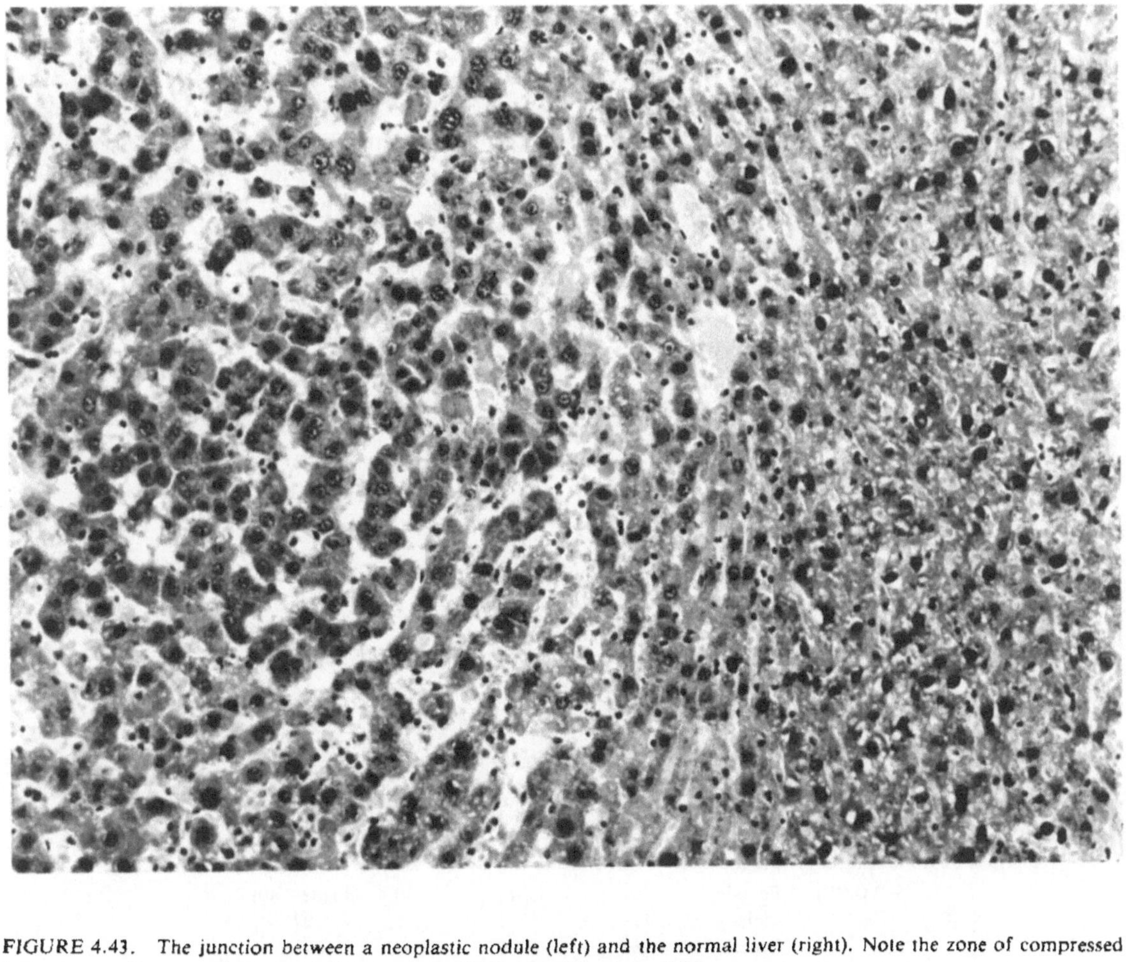

FIGURE 4.43. The junction between a neoplastic nodule (left) and the normal liver (right). Note the zone of compressed hepatocytes adjacent to the hepatocellular nodule. (HPS; magnification × 210.)

of the normal pattern of cords and sinusoids.

As tabulated in Table 4.12, the female WAG/Rij rats had the highest incidence (10%) of neoplastic nodules. The other groups of rats had a relatively low incidence with 1% in male WAG/Rij, 3% in male F_1,and none in female F_1 or male and female BN/Bi rats. They were only seen in the oldest groups of rats. For example, the youngest rat with a nodule was the male WAG/Rij rat and it was 27 months old. The other 12 nodules were found in rats 33 months or older.

b. Hepatocellular Carcinoma

A 44-month-old male F_1 rat died with an hepatocellular carcinoma. The tumor was confined to one liver lobe and was composed of irregular sheets, cords, and trabeculae of cells and areas of cellular atypia (Figures 4.44 and 4.45). Few mitoses were found and no vascular invasion or distant metastases was seen.

c. Malignant Hemangioendothelioma

A 31-month-old female WAG/Rij rat died with a large malignant hemangioendothelioma that appeared to be arising in the liver. The tumor was typical of a endothelial cell tumor with many blood-filled vascular spaces and some with fibrin thrombi. The spaces were lined by anaplastic endothelial cells with mitoses and localized destruction of the liver. The tumor was confined to one lobe of the liver and no distant metastases were observed.

3. Discussion

Squire and Levitt[162] defined liver foci, areas, and neoplastic nodules in their report of a workshop on the classification of hepatocellular lesions in rats. Squire and Goodman[163] reiterated these views in a review article on the tumors of laboratory animals. Some disagreement still exists, however, about the correct terminology for such lesions. A detailed discussion of the various arguments is not within the scope of this chapter. Briefly, some investigators have classified lesions similar to neoplastic nodules as hyperplastic nodules,[54,133] while others considered them to be minute-deviation hepatomas.[128] The terminology suggested by Squire and Levitt[162] was used to classify these liver lesions in the BN/Bi, WAG/Rij, and (WAG × BN)F_1 rats.

The observations on the BN/Bi, WAG/Rij, and (WAG × BN)F_1 rats do not solve the controversy of terminology, but certain trends were seen that may help to clarify the biological behavior of these lesions. For example, the BN/Bi males and females had a very low background of foci and areas of cellular alteration (<5%). They also had a 0% incidence of hepatocellular neoplastic nodules. On the other hand, WAG/Rij females had the highest background incidence of foci and areas (84% incidence in those older than 37 months) as well as the highest incidence of nodules (10%), with half in animals 36 months or older. It would appear then that foci, areas, and nodules are all age-associated, with the percentage increasing as the population ages. Also, strains with a high background of foci and areas may be more likely to develop liver cell nodules than strains with a low background incidence. The implication here suggests that foci and areas progress to nodules. This hypothesis has not been tested, however, and additional comparisons of differ-

TABLE 4.12

Incidence of Neoplastic Nodules in the Liver of Aging BN/Bi, WAG/Rij, and (WAG × BN)F_1 Rats

Strain	Sex	No. examined	No. with nodules	%	Mean age (range) in months
BN/Bi	Female	236	0	—	—
	Male	74	0	—	—
WAG/Rij	Female	101	10	10	36 (33—40)
	Male	124	1	1	27
F_1	Female	68	0	—	—
	Male	67	2	3	39, 40

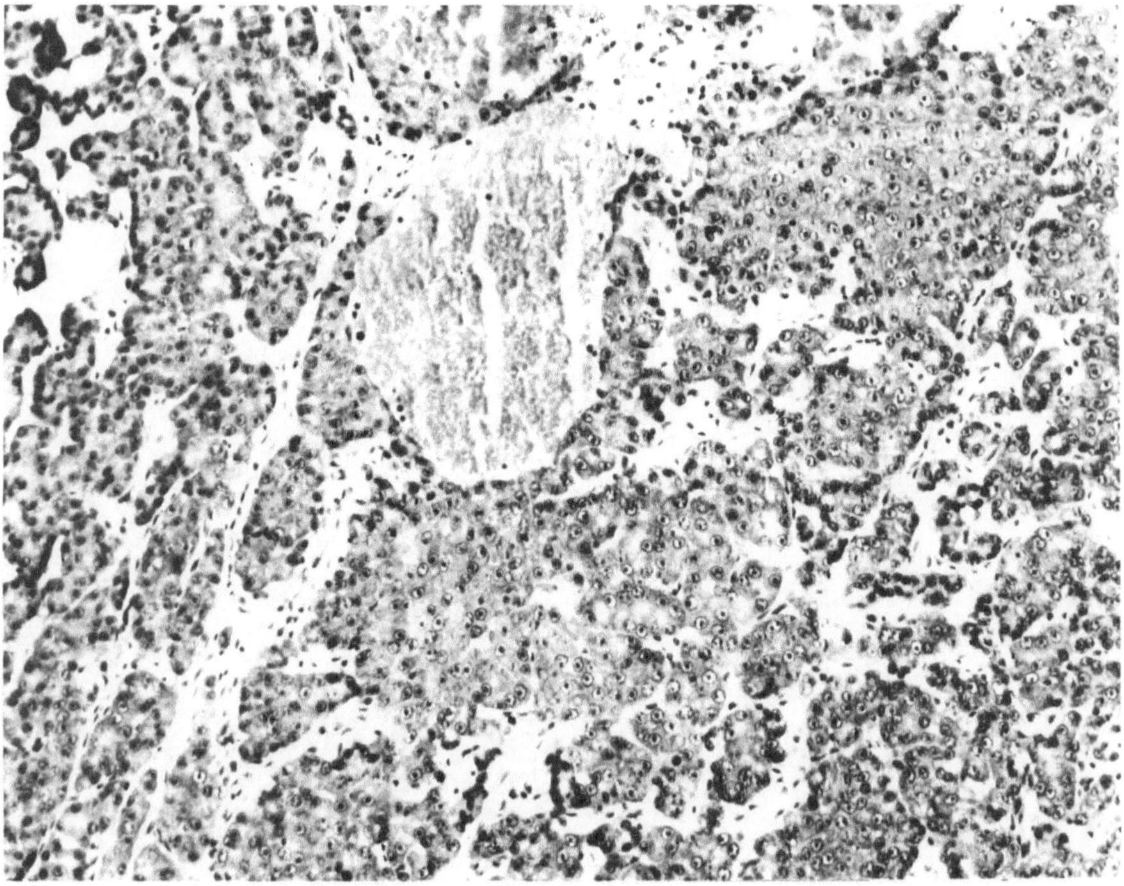

FIGURE 4.44. Hepatocellular carcinoma in the liver of a 44-month-old male (WAG × BN)F_1 rat. (HPS; magnification × 210.)

ent strains of rats are needed before such conclusions can be made.

Only one rat died with an hepatocellular carcinoma, a 44-month-old F_1 male. Too few rats developed carcinomas to determine if there was a trend for rats with neoplastic nodules to develop carcinomas. Hepatocellular carcinomas have been seen in a few WAG/Rij females that were part of other studies at REP Institutes,[83] but the background incidence is 1% or less. This low number of carcinomas likewise makes it difficult to suggest that neoplastic nodules give rise to hepatocellular carcinomas.

Biliary cysts increased in incidence, number per rat, and size with age and they were benign proliferative lesions. Therefore, some controversy exists about the nomenclature of these lesions. For example, Schäuer and Kunze[151] described cystic cholangiomas which closely resembled the biliary cysts reported in the BN/Bi, WAG/Rij, and (WAG × BN)F_1 rats. Therefore, this is yet another area where additional studies are clearly needed.

The contrasting pattern of lesions in the livers of the BN/Bi vs. the WAG/Rij rat is worth stressing because these lesions may influence an investigator's choice of an animal for aging liver studies. BN/Bi rats have a high background of bile duct lesions, but the hepatocytes were morphologically normal. On the other hand, the WAG/Rij rats had morphologically altered hepatocytes. Therefore, if one must select a rat in which to study aging in liver cells, one should do so only after considering the "normal" morphological alterations occurring in the livers of the rats to be used.

Aging BN/Bi, WAG/Rij, and (WAG × BN)F_1 rats also had several additional lesions

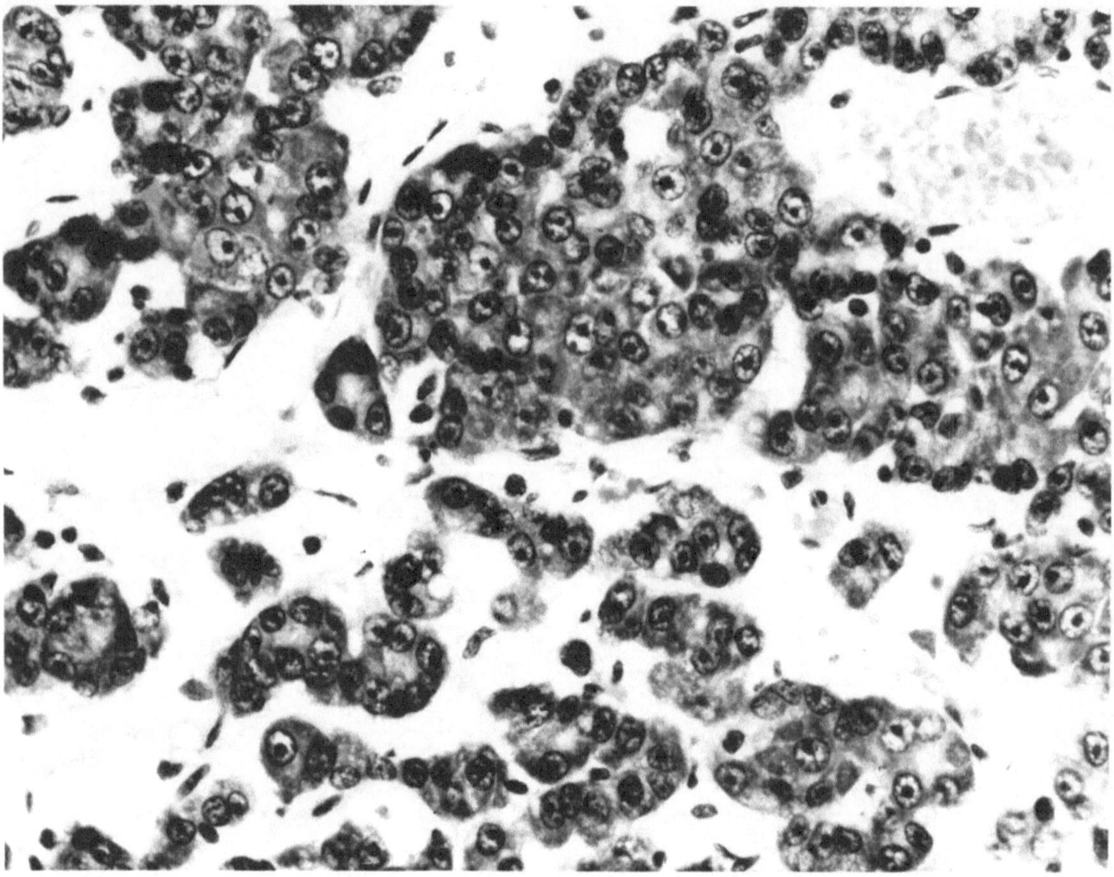

FIGURE 4.45. Higher magnification of the hepatocellular carcinoma in Figure 4.44. (HPS; magnification × 500.)

in the liver. Such lesions were often difficult to assess and to establish a clear age-associated incidence, however. For example, lipofuscin pigment in hepatocytes was clearly age related and the incidence may approach 100% in rats older than 30 months. Hemosiderin pigment and extramedullary hematopoiesis were also observed in the older rats, but they appeared to occur secondary to large neoplasms elsewhere in the body.

Polyploidy also appeared to be age related because it could be found in most of the aged rats. de Leeuw-Israel[51] has shown, however, that liver cell polyploidy increased up to 6 months of age and stabilized between 6 and 12 months in RU rats. Beyond 12 months, the amount of polyploidy remained constant. Additional studies by van Bezooijen et al.[176] indicated that only minor changes in the polyploid state were found in WAG/Rij rats after 4 months of age. Therefore, polyploidy increases with age, but only up to a point (6 to 12 months). Afterwards, no further age-associated increase can be shown.

G. Alimentary Tract and Abdominal Cavity

1. Nonneoplastic Lesions

The buccal cavity and its associated structures, such as the tongue, teeth, and gingiva, were not prepared for histopathology unless a lesion was observed at the time of necropsy. Therefore, lesions of the mouth were limited to gross findings such as overgrown incisor teeth and grossly observed masses. Specific age-associated nonneoplastic lesions were not recognized.

a. Gastric Ulcers

The most significant nonneoplastic lesion was ulceration of the forestomach (nonglandular). Ulcers occurred in all groups of rats (Table 4.13) and appeared as focal or multiple lesions that varied from microscopic to greater than 1 cm in diameter. Blood was found in the lumen of the stomach and intestine of some rats. Histologically, a few were shallow ulcers with little associated inflammation. However, most were deeply ulcerated areas extending through the submucosa into the muscular tunics. Associated with the larger ulcers was a severe inflammatory response with edema, arteritis, and many neutrophils. Occasionally, foreign bodies and accompanying granulomatous inflammation were present. Acanthosis, hyperkeratosis, and parakeratosis of the mucosa occurred along the margins of the ulcers and at times diffusely over the mucosa of the entire forestomach. Approximately one third (Table 4.13) had ulcers and inflammation that penetrated through the entire stomach wall, resulting in associated peritonitis. Despite the severe ulcerations in the forestomach, the glandular stomach and intestine were uninvolved.

The age-associated incidence is shown in Figure 4.46. Male WAG/Rij and F_1 rats had peak incidences of 28 and 25%, respectively, while WAG/Rij and F_1 females had peak incidences of 16 and 10%.

b. Miscellaneous Nonneoplastic Lesions

In addition to ulcers and associated gastritis of the forestomach, a few other gastric lesions were recognized. Included were findings such as dilated crypts and crypt abscesses of the mucosa, increase in collagen in the lamina propria and submucosa of both the squamous and glandular stomach, and mild nonspecific inflammation in the submucosa. Such lesions were not common. Some may have been age associated, but as specific calculations were not done, no conclusions were made.

Similarly, nonspecific changes were recog-

TABLE 4.13

Incidence of Ulcerative Gastritis and Associated Peritonitis of the Forestomach in Aging BN/Bi, WAG/Rij, and (WAG × BN)F_1 Rats

Strain	Sex	No. examined	No. with ulcers	%	Mean age (range) in months	No. with peritonitis	%	Mean age (range) in months
BN/Bi	Female	236	2	1	28, 32	0	—	—
	Male	74	1	1	34	0	—	—
WAG/Rij	Female	101	17	17	32 (19—38)	3	18	35 (33—36)
	Male	124	31	25	23 (18—28)	12	39	23 (18—28)
F_1	Female	68	6	9	32 (26—37)	2	33	33, 37
	Male	67	15	22	32 (22—39)	4	27	28 (22—36)

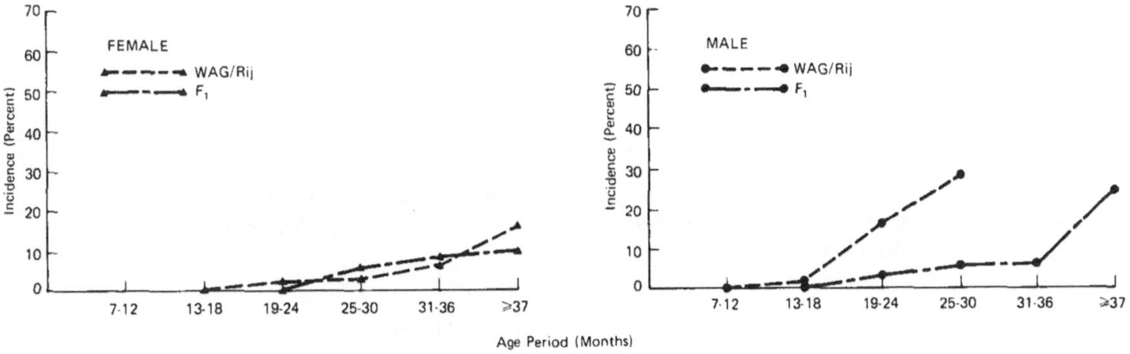

FIGURE 4.46. Percent incidence with age of ulcers and gastritis of the nonglandular stomach in 101 female and 124 male WAG/Rij and 68 female and 67 male (WAG × BN)F_1 aging rats.

nized in the intestinal tract of a few rats, but it was usually impossible to evaluate subtle age-associated changes on the routine sections of the intestinal tract. Sporadic chronic, nonulcerative colitis, and typhlitis occurred in less than 5% of the rats. It consisted of a mild increase in lymphocytes, plasma cells, and neutrophils in the lamina propria and submucosa, usually with some edema. Two BN/Bi rats had chronic colitis, with numerous eosinophils. In these rats the inflammation extended into the muscular wall of the colon. Specific organisms were not recognized using sections stained with HPS, Giemsa, PAS, or Gomori's silver stain. Mild enteritis or enterocolitis was found in approximately 2% of the aging rats of this study.

Pinworms were present in the cecum, colon, or both of some rats, but they were not numerous and were not associated with any recognized histological changes.

2. *Neoplastic Lesions*

a. Head, Neck, and Mouth

Squamous cell carcinomas appeared to arise within the buccal cavity of four female rats (Table 4.14). All were highly infiltrative tumors that invaded and destroyed normal tissues and occupied the mouth, head, and neck with ulcerations. Two had distant metastases to lymph nodes or lungs. Histologically the tumors were composed of squamous epithelial cells, epithelial pearls, cornifying cells, keratin, and many mitoses. Along the margins, the cells were less differentiated and were often in irregular nests.

Five malignant mesenchymal tumors involved the head, neck, and mouth (Table 4.15) and, like the squamous cell carcinomas, were highly infiltrative and destructive. The exact site of origin could not be determined. One had widespread metastases to several organs. They were similar histologically, with a uniform to slightly pleomorphic population of undifferentiated spindle cells that were further characterized by the lack of giant cells, failure to produce much collagen, and lack of palisading or Verocay bodies.

b. Stomach

Solitary squamous cell papillomas of the forestomach occurred in one male BN/Bi, two female WAG/Rij, and one female F_1. The lesions were benign wart-like elevations on the mucosa and consisted of a fibrovascular supporting stroma lined by keratinized squamous epithelium.

One 26-month-old female BN/Bi rat had a large, undifferentiated spindle cell sarcoma of the stomach. Most areas resembled a leiomy-

TABLE 4.14

Occurrence of Squamous Cell Carcinoma in the Mouth of Aging Female BN/Bi and (WAG × BN)F_1 Rats

Strain	No. examined	No. with tumor	Mean age (range)	No. with distant metastases
BN/Bi	236	3	35 (31—38)	2
F_1	68	1	37	0

TABLE 4.15

Occurrence of Highly Invasive, Undifferentiated Sarcomas Involving the Region of the Mouth, Head, and Neck in Aging Female BN/Bi and Male (WAG × BN)F_1 Rats

Strain	Sex	No. examined	No. with tumor	Mean age (range)	No. with distant metastases
BN/Bi	Female	236	3	22 (14—34)	1
F_1	Male	67	2	22, 27	0

osarcoma, with fusiform cells containing pale pink cytoplasm with distinct cell boundaries and elongated nuclei with blunt ends. Other areas were less differentiated, however, and resembled a neurofibrosarcoma with small cells having indistinct cell boundaries and spindle- to fusiform-shaped nuclei. Mitoses were frequent and the tumor was locally invasive but did not metastasize.

c. Intestine

An apparently benign nodular lesion in the wall of the small intestine was recognized in 12 of the 236 (5%) female BN/Bi rats and 2 of the 68 (3%) female F_1 rats (Table 4.16). They were diagnosed as fibroleiomyomas. The lesions were grossly visible as solitary, firm white nodules. They were about 5 mm in diameter and were located in the distal duodenum or proximal jejunum. Microscopically they arose in the muscle wall (Figure 4.47) and consisted of variable amounts of muscle and fibrous connective tissue. The cell boundaries were indistinct, the cytoplasm pink to yellow with HPS stain, and the nuclei round to elongated (Figure 4.48). The tumor cells appeared to spread between and

TABLE 4.16

Incidence of Fibroleiomyoma in the Small Intestine of Aging Female BN/Bi and (WAG × BN)F_1 Rats

Strain	No. examined	No. with tumor	%	Mean age (range) in months
BN/Bi	236	12	5	32 (19—37)
F_1	68	2	3	36, 40

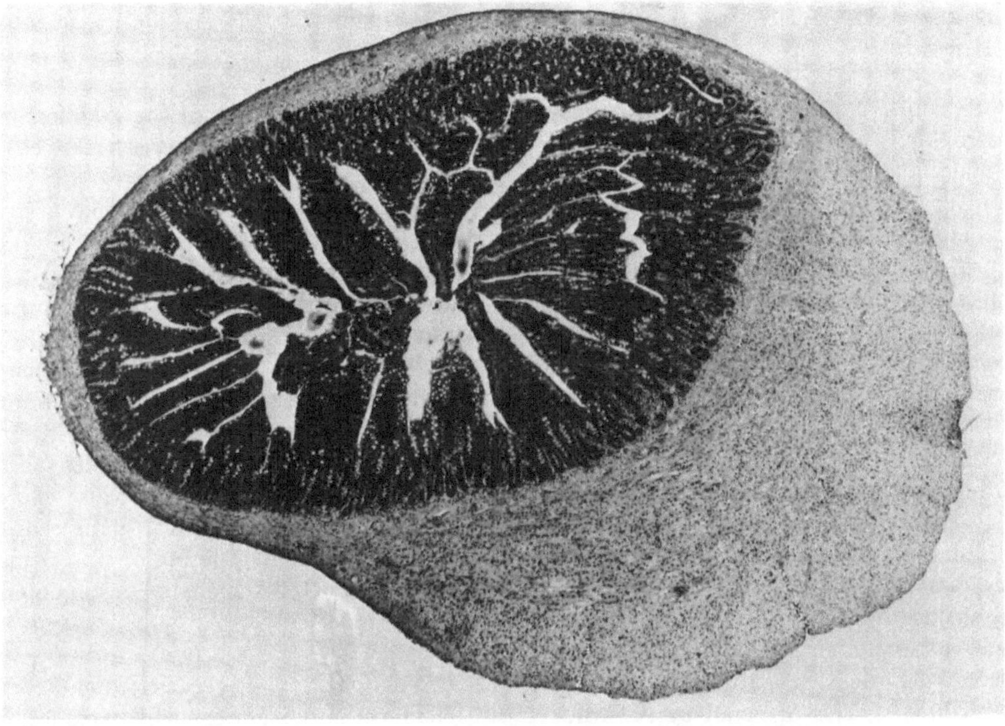

FIGURE 4.47. Fibroleiomyoma of the wall of the small intestine in a female BN/Bi rat. (HPS; magnification × 30.)

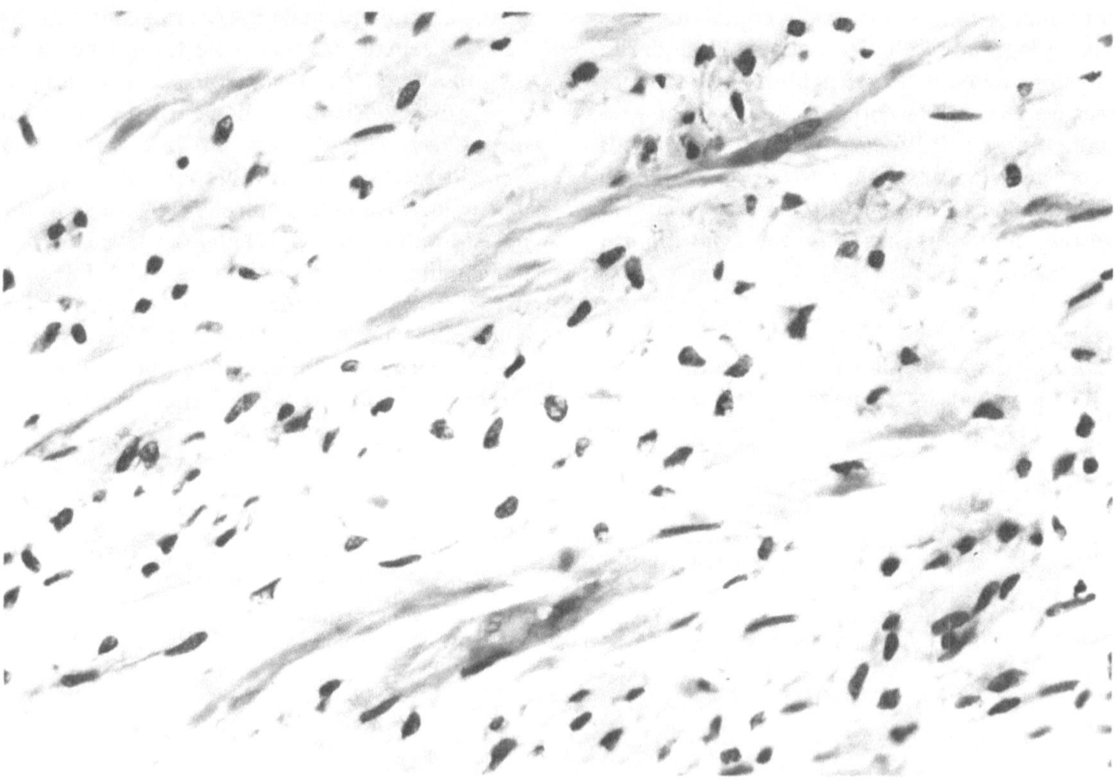

FIGURE 4.48. Higher magnification of the fibroleiomyoma shown in Figure 4.47. (HPS; magnification × 465.)

around the normal muscle, which often appeared atrophic.

Two male BN/Bi rats, both 33 months old, had leiomyosarcomas originating in the wall of the small intestine. The tumors were composed of fusiform cells with abundant pink cytoplasm and elongated nuclei with blunt ends. Both tumors had areas of anaplastic, poorly differentiated cells with many mitoses, and one had metastasized to the lungs.

d. Abdominal Cavity

Mesotheliomas, arising from the serosal lining cells, were present in two female BN/Bi rats 28 and 35 months of age. Grossly, nodules and papillary masses of varying sizes were found over the surface of the small intestine, spleen, pancreas, and omentum. Microscopically the lesions consisted of several patterns such as tubular structures, spaces or small cystic structures lined by papillary mesothelium, and areas of spindle cells. The stroma of the tumors was highly vascular and contained variable amounts of collagen. The lining of the nodules was by cuboidal to pleomorphic mesothelial cells. Mitoses were infrequent.

One 28-month-old female F_1 had a large lipoma in the omental fat. The tumor was composed of well-differentiated fat cells and was sharply demarcated from the normal surrounding fat tissue.

A 39-month-old female BN/Bi rat died with the abdominal cavity filled with a pleomorphic sarcoma. The tumor had invaded the liver, pancreas, intestine, uterus, lymph nodes, abdominal wall, and diaphragm. The cells were pleomorphic with many mitoses, giant cells, and multinucleated cells. The tumor was of mesenchymal origin, but the site of origin could not be determined.

3. Discussion

The most important age-associated lesions in the rats of this study were ulceration and accompanying gastritis of the forestomach. The ulcers were apparently strain dependent because less than 1% occurred in the BN/Bi rats while 9 to 25% of the WAG/Rij and (WAG × BN)F_1 rats had these lesions. Between 18 and 39% of the ulcers had perforated the entire thickness of the stomach wall and resulted in peritonitis. Such lesions were severe and must have contributed to the death of these animals.

The etiology of the ulcers is unknown. The ulcers in rats were usually multifocal, randomly distributed in the squamous portion of the stomach, and were seen as acute, chronic, and apparently healed ulcers. Stress by a number of causes may produce ulcers in the stomach of rats.[3] Such lesions were usually in the glandular stomach, however.

As reviewed by others,[119,130,146,163] spontaneous tumors of the rat gastrointestinal tract are not common. In this series of BN/Bi, WAG/Rij, and (WAG × BN)F_1 rats, neoplastic lesions were not rare, but only a few were clearly invasive and destructive tumors that probably contributed to the death of some rats. Lesions such as the fibroleiomyomas seemed to have a strain and sex predilection since they were seen only in females and occurred in 5% of the BN/Bi rats, but in 0% of the WAG/Rij.

Pinworms are common in most rat colonies and it is often difficult to eliminate them or to keep them out of even SPF colonies. In this series, a low but constant infection by pinworms was present in the aging colony. Histologically, no lesions could be attributed to these nematodes, indicating that low numbers of these worms do not appear to have any significant effect on the longevity or the morphological features of the cecum and colon of aging rats.

H. Salivary Gland

1. Nonneoplastic Lesions

Nonneoplastic changes in the salivary glands were not common. The only recognized change was atrophy of the parotid salivary gland. It consisted of atrophy and loss of acinar tissue, dilated ducts with flattened epithelium, a mild mononuclear inflammatory response, and increased connective tissue between acini. Occasional calcium deposits were recognized within ducts. Atrophy was recognized in less than 5% of any of the groups studied and therfore was not tabulated.

2. Neoplastic Lesions

Three tumors were found in the salivary glands of the 670 aging rats. The first tumor consisted of a well-differentiated adenoma arising in the salivary gland of a 34-month-old BN/Bi female. The other two were adenocarcinomas, one in a 28-month-old female BN/Bi rat, the other in a 22-month-old male WAG/Rij. Both tumors formed cystic spaces lined by epithelial cells. The epithelial cells were often anaplastic and mitoses were numerous. They tended to form sheets of cells in some areas and papillary projections in other regions. Both tumors were surrounded by a connective tissue capsule. The tumor in the BN/Bi female had not metastasized, but the carcinoma in the WAG/Rij male had multiple metastases to the lungs.

3. Discussion

Neoplastic and nonneoplastic age-associated lesions appear to be rare in rats as judged by the relative lack of references in the literature. Glucksmann and Cherry[70] discussed several induced tumors of the salivary gland in rats, but did not describe spontaneous lesions. Similarly, Squire and Goodman[163] reported spontaneous salivary gland tumors to be rare in rats. Likewise, only three salivary gland tumors were observed in the 670 rats of this study. Equally uncommon were age-associated nonneoplastic changes.

I. Cardiovascular System

1. Nonneoplastic Lesions

a. Myocardial Degeneration and Fibrosis of the Ventricular Wall

Fibrosis with loss and degeneration of cardiac muscle fibers was the most common age-associated lesion of the heart. It occurred most frequently on the left side with involvement of the left ventricular wall (especially the inner one third), the apex, the interventricular septum, and papillary muscles. Involvement of the right ventricle was infrequent. The lesion usually consisted of solitary or multiple areas of fibrosis (Figure 4.49) that were often cellular with fibroblasts and occasionally with numerous Anitschkow's cells (Figure 4.50). Lymphocytes and granulocytes were not numerous, but some were found in most hearts. In the areas of fibrosis, cardiac muscle was either degenerative or absent, often being fragmented, vacuolated, or showing loss of cross striations. Fragmented

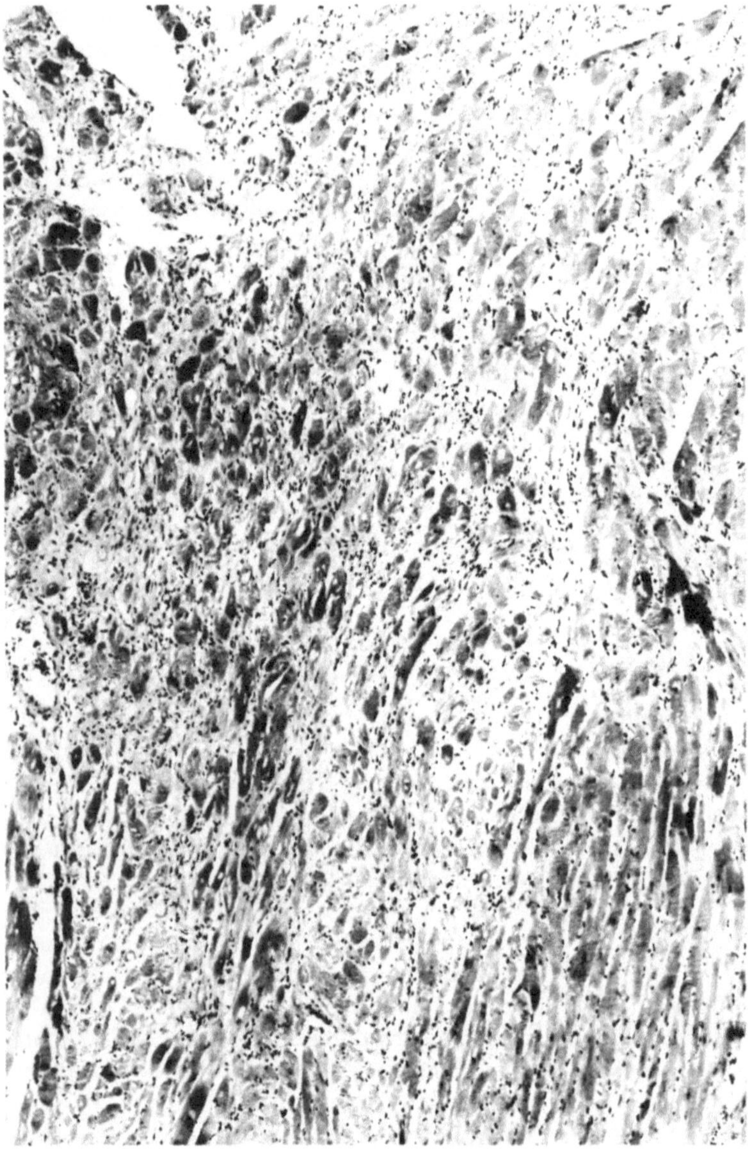

FIGURE 4.49. Myocardial degeneration and fibrosis in the left ventricular wall of the heart. (HPS; magnification × 75.)

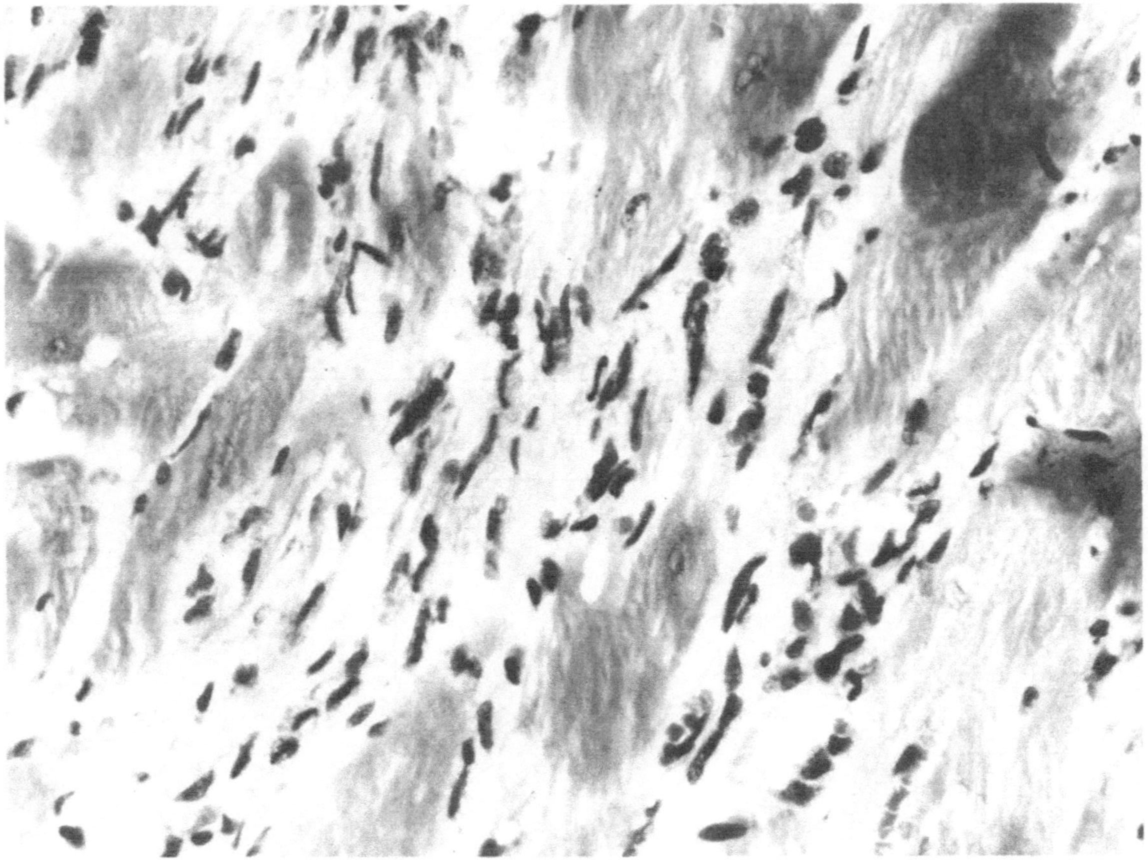

FIGURE 4.50. Higher magnification of Figure 4.49 showing the interstitial fibrosis, Anitschkow's cells, and degeneration of myocardial cells. (HPS; magnification × 500.)

and degenerative cardiac muscle cells were also observed along the margins.

The peak age-associated incidence was between 80 and 90% in male and female BN/Bi and female WAG/Rij that were older than 37 months (Figure 4.51). The age of onset was also similar with fibrosis, beginning between the ages of 13 to 24 months. The percent incidence remained below 3% until 24 months and then increased sharply. The male WAG/Rij had an earlier onset of this lesion with 8 and 43% in the age period of 13 to 18 and 19 to 24 months, respectively. The peak incidence reached 65%, which was below the 80 to 90% peak incidence for the other groups.

The hearts from the male and female BN/Bi and WAG/Rij rats were selected to grade the severity of the myocardial fibrosis. This was done in order to determine if one sex or strain had more severe lesions than the other. The hearts were graded twice using routine sections stained with HPS, without a knowledge of the sex or age of the rat. With the HPS, collagen stained yellow, which contrasted sharply with the red staining of the cardiac muscle. The grading system ranged from no fibrosis, to slight (+), to moderate (+ +), to severe (+ + +). The results of the grading are summarized in Figure 4.52 for BN/Bi and Figure 4.53 for WAG/Rij rats. Grade + + + severity was not recognized in female WAG/Rij and in only two male WAG/Rij, while the peak incidence of + + + fibrosis was 14 and 20% in female and male BN/Bi, respectively. The BN/Bi rats, therefore, had more severe fibrosis than the WAG/Rij. In the male BN/Bi, the age of onset of grade + + + fibrosis appeared to be about 6 months earlier than in females, while the age of onset of grade + and + + fibrosis was similar in males and females. In addition,

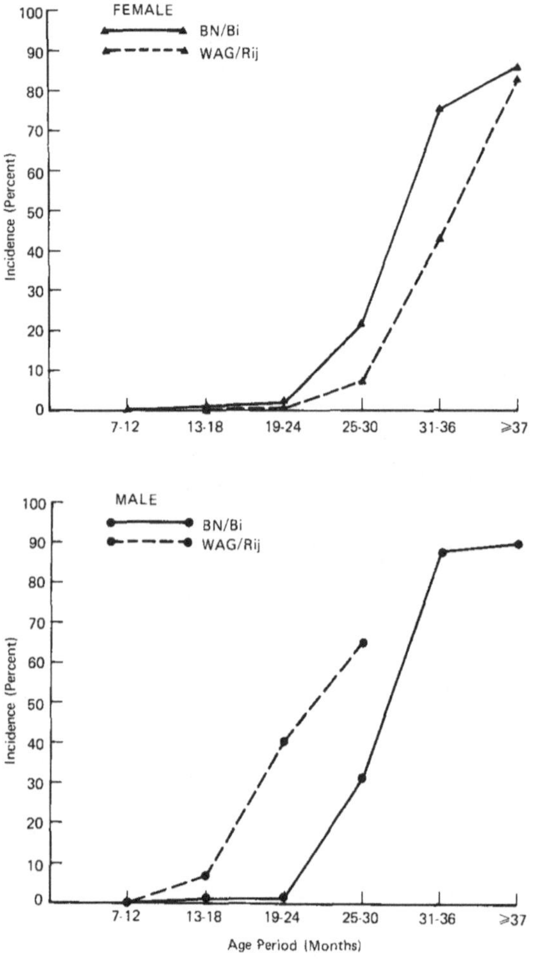

FIGURE 4.51. Percent incidence with age of myocardial fibrosis in 236 female and 74 male BN/Bi and 101 female and 124 male WAG/Rij rats.

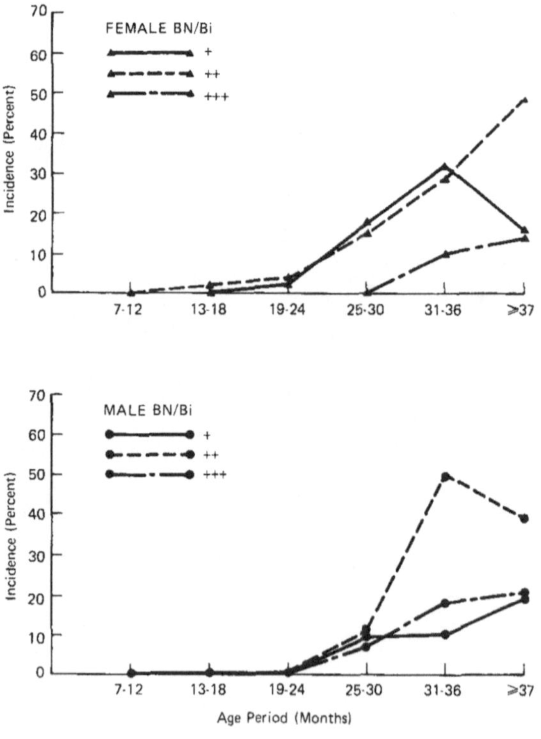

FIGURE 4.52. Severity of myocardial fibrosis plotted as percent incidence with age from 236 female and 74 male BN/Bi rats. The range of severity is as follows: + = mild, + + = moderate, and + + + = severe.

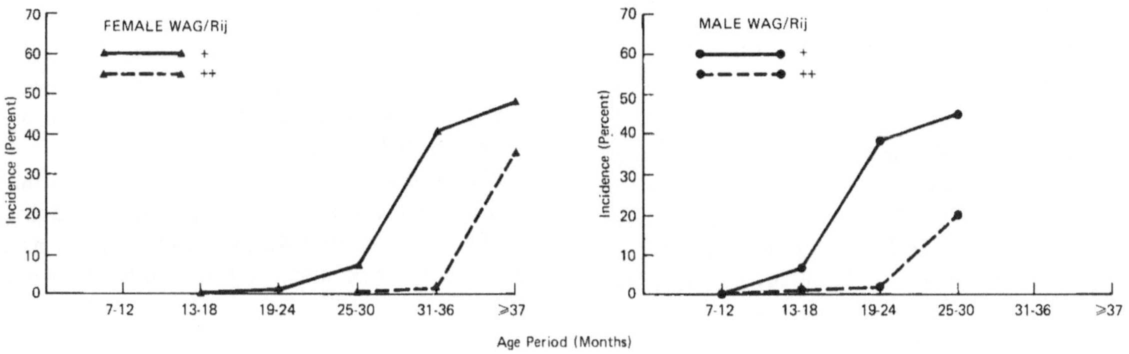

FIGURE 4.53. Severity of myocardial fibrosis plotted as a percent incidence with age from 101 female and 124 male WAG/Rij rats. The range of severity is as follows: + = mild and + + = moderate.

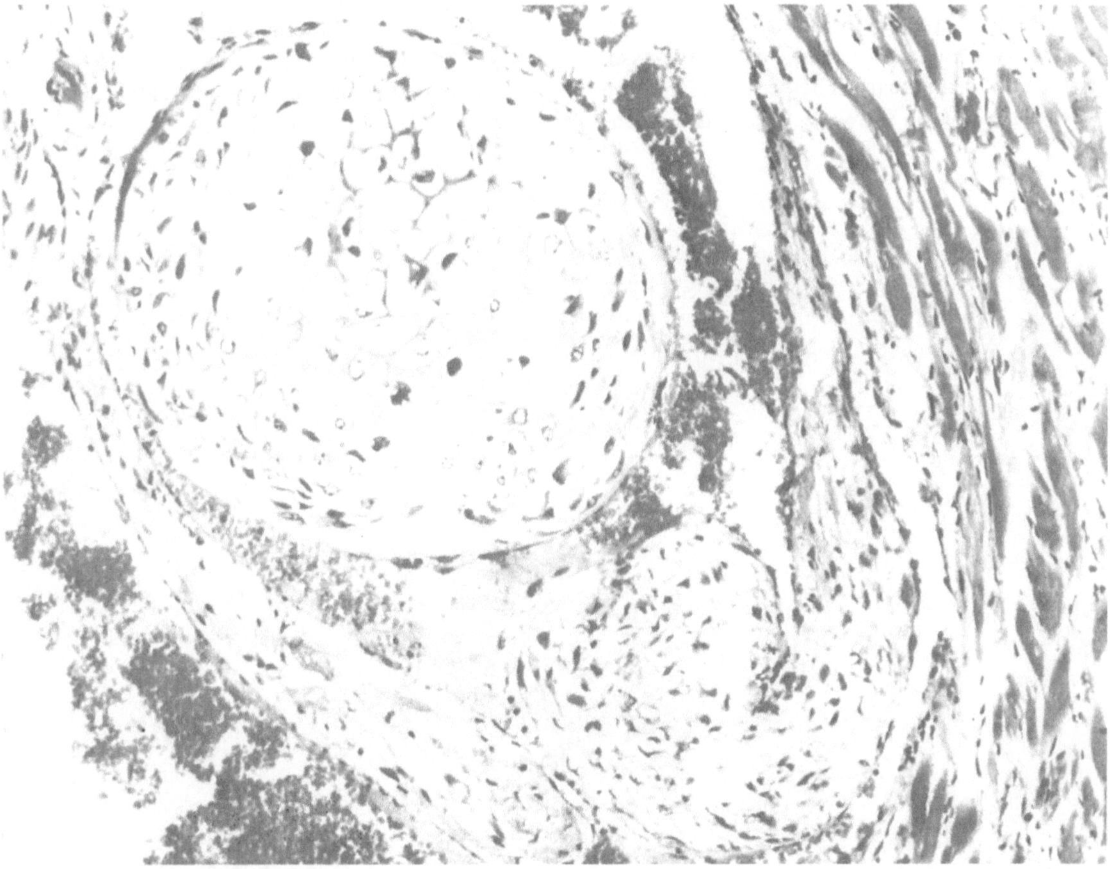

FIGURE 4.54. Focal cartilaginous metaplasia in a papillary muscle of the heart. (HPS; magnification × 210.)

there was an apparent earlier onset of fibrosis in WAG/Rij males based on the shift to the left of the curves compared to female WAG/Rij or male and female BN/Bi rats. Time did not permit a grading of the male and female F_1 rats.

b. Cartilaginous and Osseous Metaplasia

In some of the hearts with fibrosis, metaplastic bone or cartilage was found in the left heart. It was either within papillary muscles or located at the apex of the left ventricle (Figure 4.54).

Such metaplastic changes were observed in 9% of male and 5% of female BN/Bi, but not in any of the WAG/Rij or F_1 rats.

c. Cartilaginous Focus at the Base of the Aorta

Several rats had a focus of cartilage in the wall of the aorta adjacent to the aortic valve (Figure 4.55). These foci were not routinely recorded when evaluating the histologic findings of the heart. Therefore, the actual percentages among the different groups of rats were not determined.

d. Heart Valves

Lesions of the heart valves occurred in more than 75% of all rats examined. They were second only to myocardial fibrosis as an age-associated lesion in the heart. They could not be accurately tabulated, however, because some hearts had no valves in the section and others had only partially sectioned valves. The commonly observed lesions were defined, however, and a general assessment of the valves most often affected was made.

The most common lesion was a myxomatous degenerative change that resulted in thickening of the valve (Figure 4.56). Severe lesions had areas of fibrosis and, at times, inflammatory cells (Figure 4.57). Some had small fibrin thrombi and others had organizing thrombi on the valve surface. A few had mineral deposits in the valve. The most frequently affected appeared to be the mitral valve alone, followed by involvement of both the mitral and the aortic valves. Less affected was the aortic valve alone or the tricuspid valves. No lesions were observed in the pulmonary valves, but due to the plane of sectioning, pulmonary valves were seldom observed in the sections.

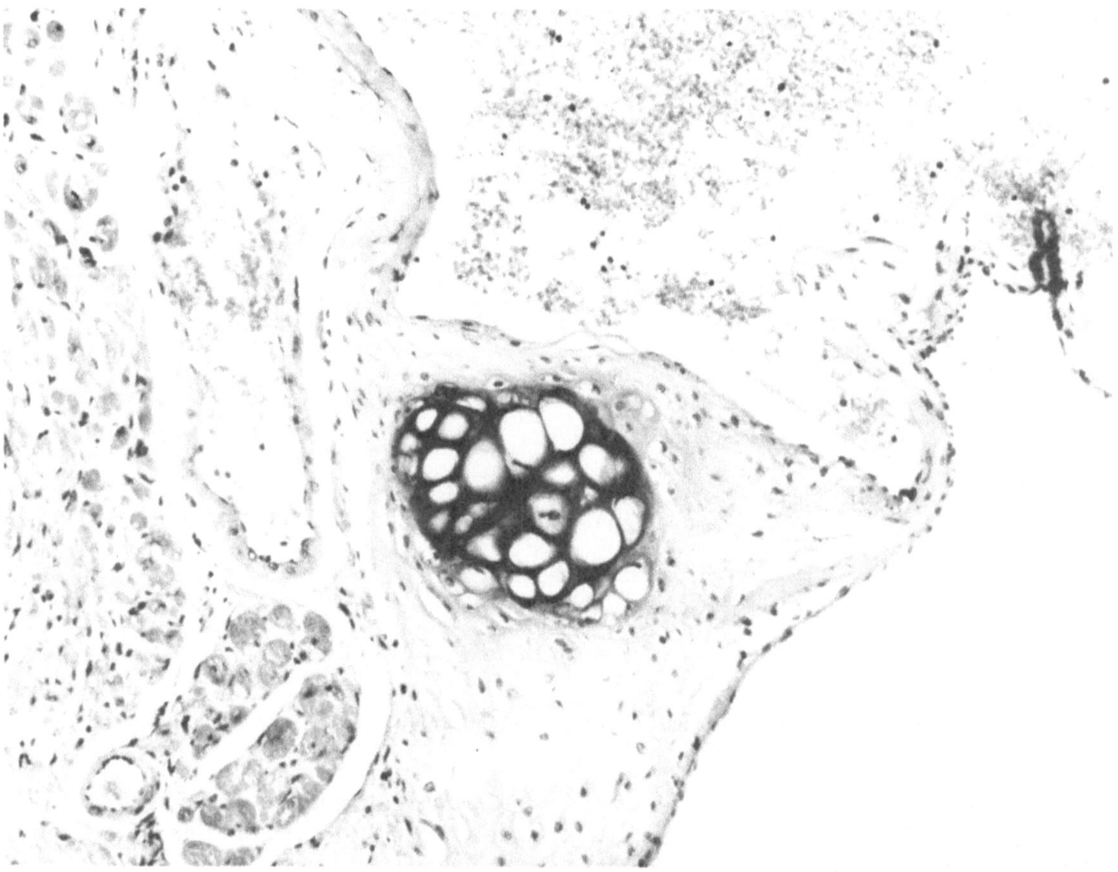

FIGURE 4.55. Carilaginous focus at the base of the aorta and aortic valve. (HPS; magnification × 210.)

FIGURE 4.56. Myxomatous degeneration of the mitral valve of the heart. (HPS; magnification × 240.)

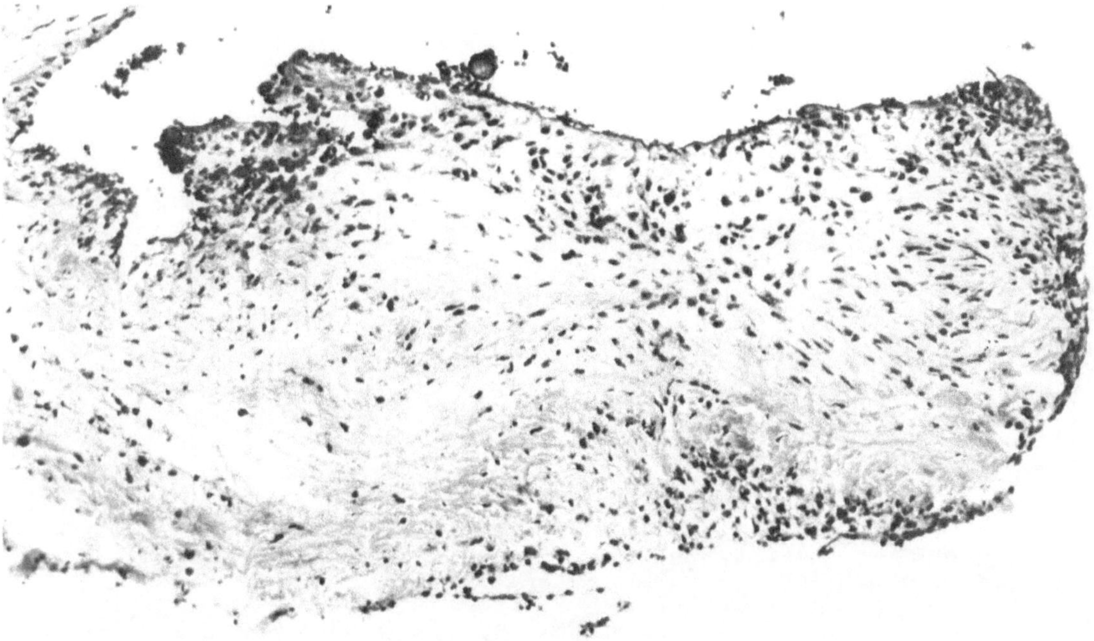

FIGURE 4.57. Mitral valve showing myxomatous degeneration, inflammation, focal loss of endothelial cells, and focal fibrin deposits. (HPS; magnification × 195.)

e. Endocardial Disease

Endocardial disease was a lesion in the heart characterized by a proliferation of undifferentiated mesenchymal cells in the subendocardium. It was usually confined to the left ventricle, but in a few cases mild involvement of the right ventricle, mitral valve, and left atrium was infrequently observed. The severity varied from mild, with a thickness of only a few cell layers (Figure 4.58), to severe, with several layers of cells under the endothelium (Figures 4.59 and 4.60). The individual cells had indistinct boundaries with pale pink and often vacuolated cytoplasm. Their nuclei were often fusiform in the layers nearest the myocardium, but tended to be round or pleomorphic in the layers closest to the lumen of the ventricle. The lesion was usually sharply demarcated from the myocardium, but in some the subendocardial proliferation extended irregularly around myocardial fibers, resulting in focal degeneration of cardiac muscle.

Inflammation was not a prominent feature of endocardial disease, but a few lymphocytes and even necrotic cellular debris were present in some cases, especially along the margin of the lesion and the myocardium. Two female BN/Bi rats showed a focal loss of endothelial cells over the lesion with inflammation and thrombus formation. In one, the thrombus was large and organizing, while the other consisted only of fibrin deposits on the luminal surface.

A 22-month-old-male WAG/Rij rat (Table 4.17) had endocardial disease and a large undifferentiated sarcoma (see Section II.I.2). The tumor appeared to arise in the wall of the left ventricle either in or adjacent to the endomyocardial disease lesion.

As shown in Table 4.17, 34 cases of endocardial disease were observed in this study. The incidence ranged from 0% in female WAG/Rij to 9% in female BN/Bi. The male and female BN/Bi strain had the highest incidence (8 to 9%), followed by the male and female F_1 (3 to 6%). The male and female WAG/Rij had the lowest incidence, with none found in females and only one in males. This lesion was clearly age associated, with only two cases found in rats less than 24 months old and 17 in rats older than 30 months.

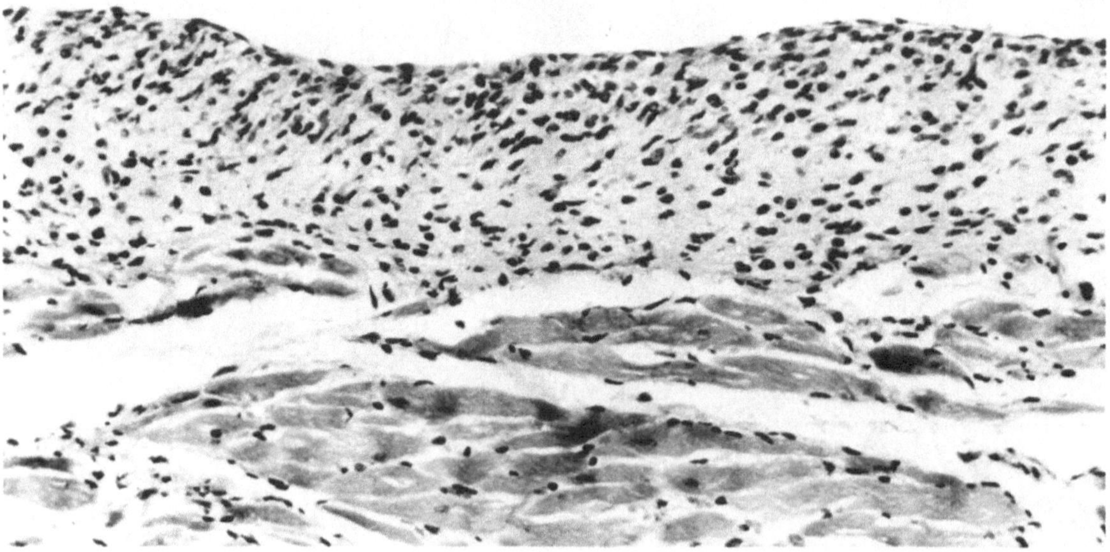

FIGURE 4.58. Mild endomyocardial disease of the subendothelial region of the left ventricle. (HPS; magnification × 240.)

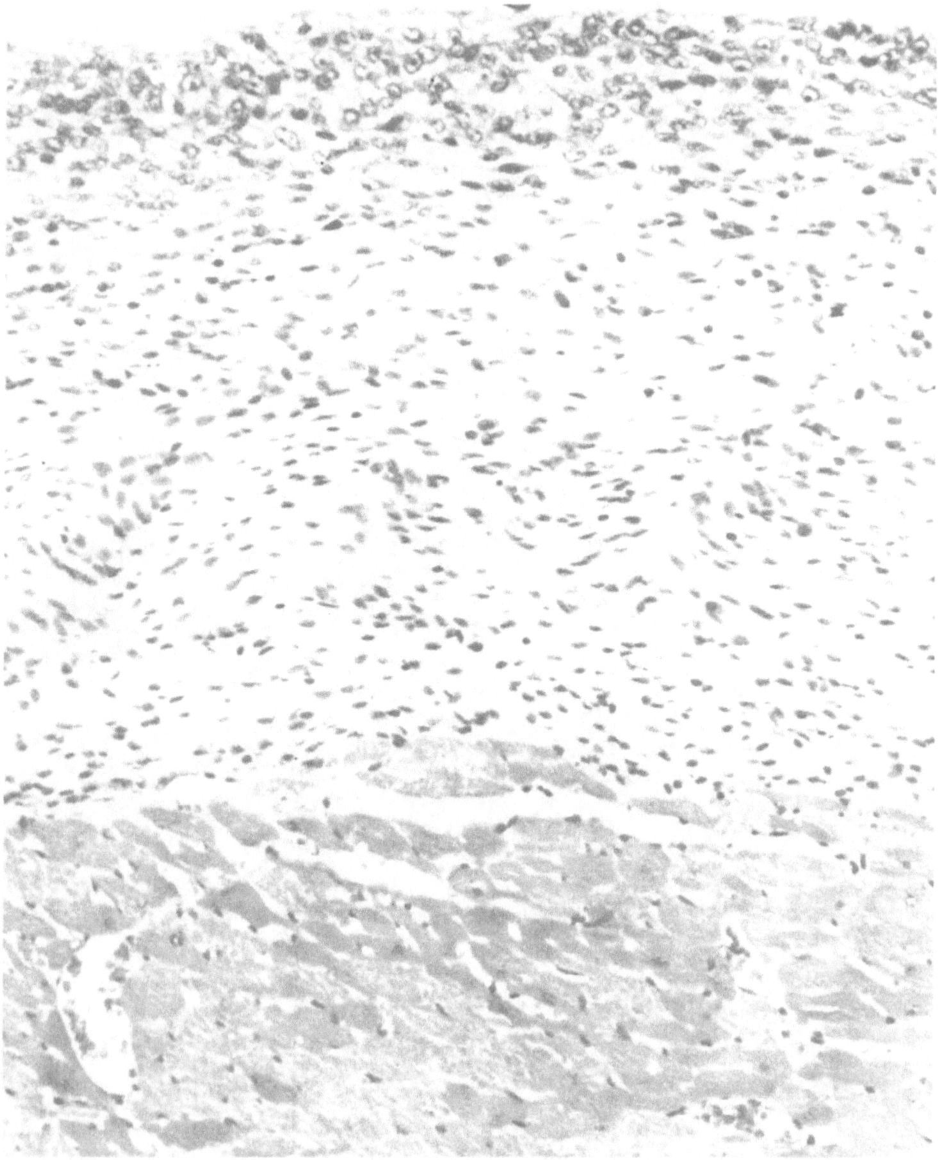

FIGURE 4.59. Severe endomyocardial disease of the left ventricle. Note the round nuclei in the cells nearest the ventricular chamber, the spindle-shaped cells nearest the myocardium, and the sharp demarcation between the myocardium and the subendothelial proliferation. (HPS; magnification × 260.)

The rats with the most severe endocardial disease also had lesions suggesting heart failure. These included pulmonary edema and congestion, hemosiderin-filled macrophages in the lung, and chronic passive congestion of the liver with degeneration and loss of hepatocytes around central veins.

f. Lesions of the Atrium and Auricle

Several lesions occurred in the atria and auricles of the heart. The most common was a thickening of the subendothelium of the left atrium and less commonly the left auricle. The lesion was found in over half of all hearts examined and in all strains and sexes of rats. The mildest form was a multifocal, relatively acellular accumulation of a homogeneous material that stained pale yellow with the HPS stain. The accumulation was diffuse and more cellular in the more severe cases than in the mild

FIGURE 4.60. Papillary muscle of the left heart showing severe endomyocardial disease. (HPS; magnification ×210.)

TABLE 4.17

Incidence of Endocardial Disease in Aging BN/Bi, WAG/Rij, and (WAG × BN)F_1 Rats

Strain	Sex	No. examined	No. with endocardial disease	%	Mean age (range) in months
BN/Bi	Female	236	21	9	32 (24—40)
	Male	74	6	8	33 (31—37)
WAG/Rij	Female	101	0	0	—
	Male	124	1	1	22
F_1	Female	68	4	6	31 (22—37)
	Male	67	2	3	40, 42

cases (Figure 4.61). The etiology of this lesion is unknown, but it appeared to be age associated.

Degeneration of the myocardium of the atria in the absence of myocarditis was observed in a few hearts. The lesion was similar to that observed in the ventricles, with muscle cells usually vacuolated, some fragmented, and some with a loss of cross striations. The degeneration was usually accompanied by a mild increase in fibrous tissue between muscle fibers.

Chronic myocarditis and endocarditis, espe-

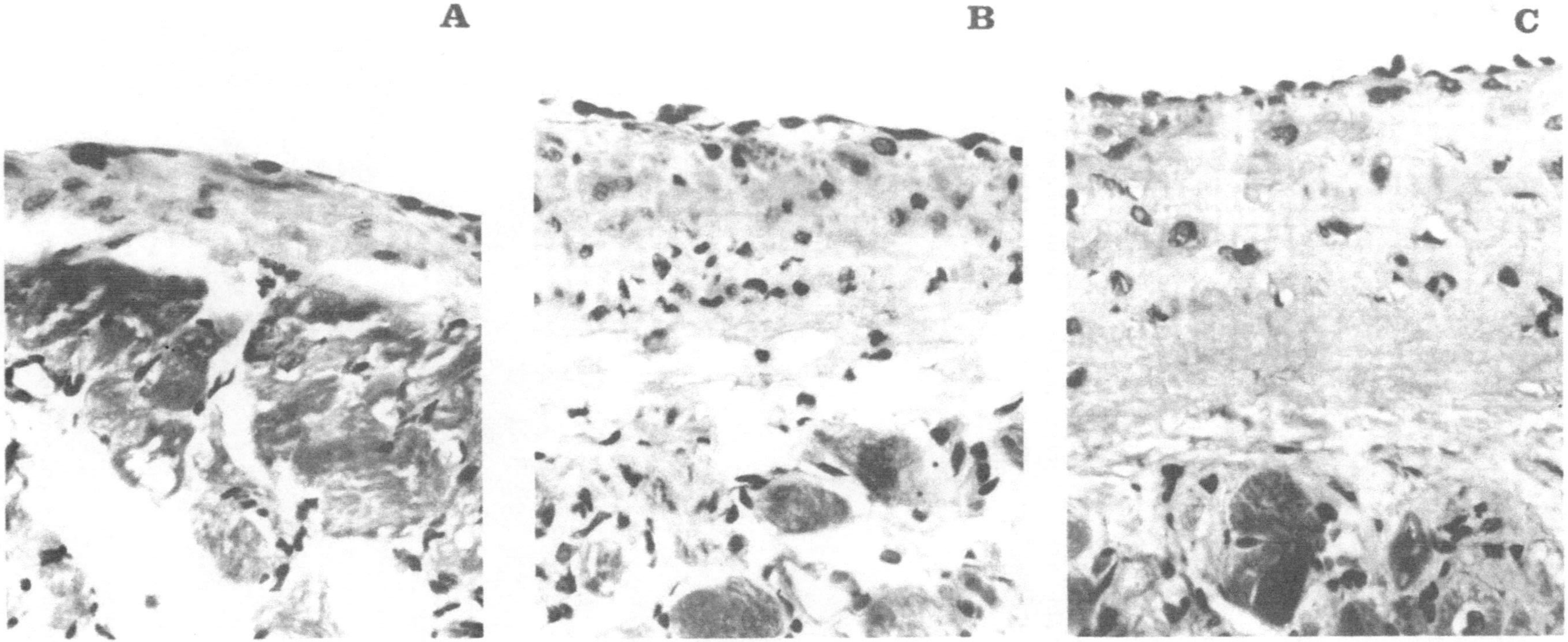

FIGURE 4.61. Three ranges of severity of subendothelial cell proliferation in the left atrium of the heart. A is mild, B is moderate, and C represents severe proliferation. (HPS; magnification × 310.)

cially of the left atrium and auricle, was characterized by a mild, diffuse infiltration by lymphocytes and occasionally plasma cells and granulocytes. There was also an apparent increase in the number of sarcolemmal nuclei. At times, an endocarditis was also present. Occasionally, focal loss or denudation of the endothelium was present, accompanied by a few granulocytes and fibrin thrombi overlaying the endocardium. Such lesions were seen in 15 and 8% of the male and female BN/Bi rats, respectively. This was in contrast to less than 5% in male and female WAG/Rij and F_1 rats.

Large, organizing thrombi were found in the left atrium or auricle of some hearts (Figure 4.62). In such cases, there was also a myocarditis of the atrium. Such large thrombi were present in the left atrium or auricle in 3% of the female and 12% of the male BN/Bi rats, but none were observed in any of the WAG/Rij or F_1 rats.

g. Periarteritis Nodosa

Periarteritis (polyarteritis) nodosa was an acute, subacute, or chronic disease of the medium and small arteries. It was characterized by subendothelial edema, damage to the elastic membrane, fibrinoid or hyaline necrosis of the media, and inflammatory cell infiltration into all layers of the arterial wall (Figures 4.63 and 4.64). The inflammatory cells were most abundant in the adventitia and were composed of numerous neutrophils, lymphocytes, and plasma cells and, to a lesser degree, macrophages and eosinophils. It occurred as a focal or segmental lesion with some areas of an artery severely involved and other areas normal. Slight aneurysmal dilatation occurred in a few

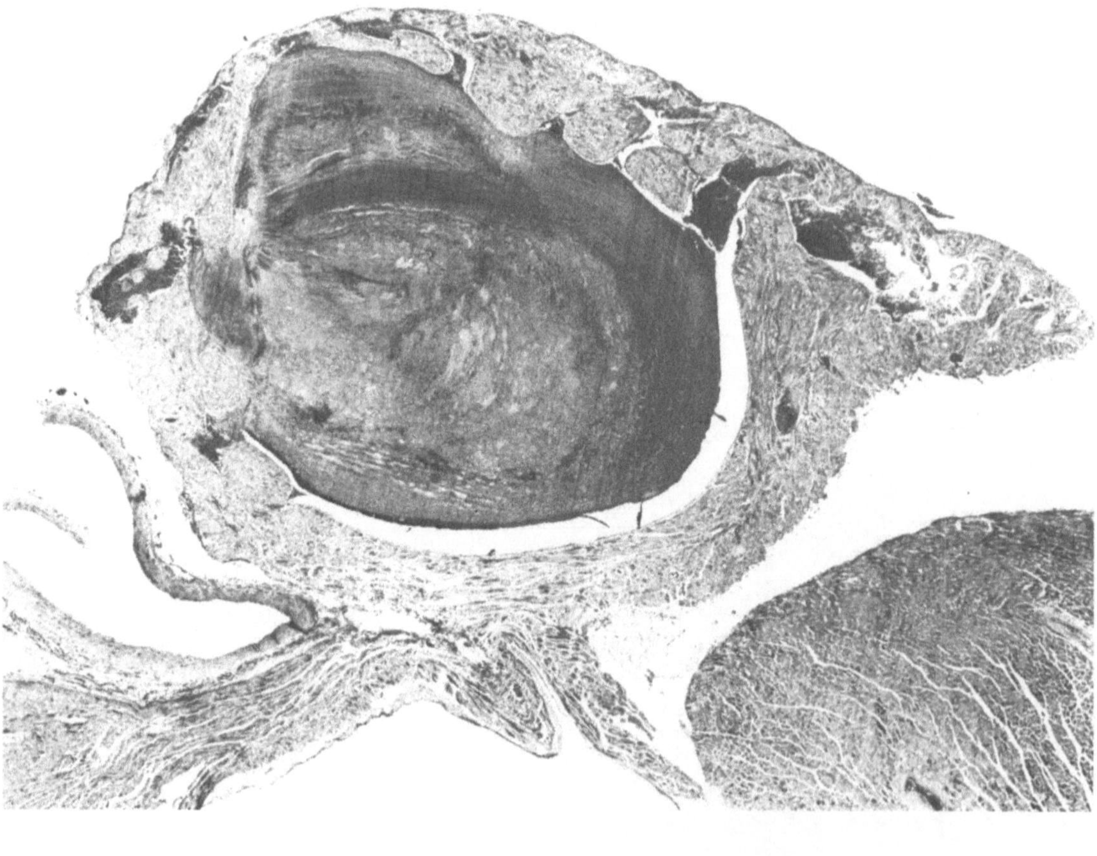

FIGURE 4.62. Thrombus in the atrium of the heart. (HPS; magnification × 20.)

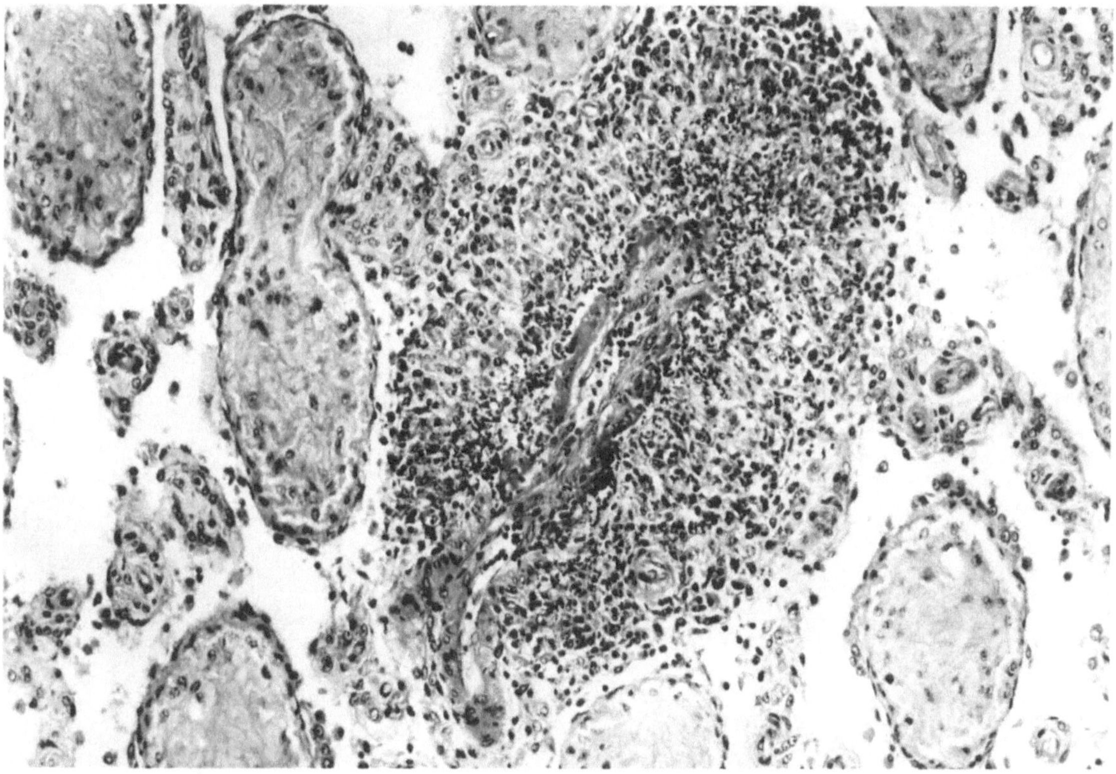

FIGURE 4.63. Periarteritis nodosa of a testicular artery in a male rat. (HPS; magnification × 195.)

cases but was not common. Fibrin thrombi were sometimes present in the lumen of involved arteries. Periarteritis was usually focal and limited to one organ, but a few rats had multifocal involvement of the arteries of several organs. The most common sites in males were the arteries of the testicle and to a lesser extent the arteries in the spermatic cord. In females the mesenteric arteries were the most commonly affected.

The incidence of periarteritis nodosa was variable, as shown in Table 4.18. Female BN/Bi and male F_1 rats had the highest incidence, 20 and 16%, respectively. The male BN/Bi, male and female WAG/Rij, and female F_1 rats had a much lower incidence, ranging between 0 and 5%. The age-associated incidence is illustrated in Figure 4.65 for female BN/Bi and male F_1 rats. For these rats, the age of onset was similar. The peak incidence was 30% in male F_1 and 18% in female BN/Bi rats.

h. Incidental Vascular Lesions

A few rats had focal fibromuscular dysplasia or degeneration of the wall of the coronary arteries. The lesion consisted of a loss of the normal muscle in the media with replacement by a hyaline- to collagen-like material (Figure 4.66). This lesion was not common and was not specifically tabulated.

Medial calcification and degeneration occurred in the wall of the ascending aorta in a 24- and a 31-month-old female BN/Bi and in a 35-month-old female WAG/Rij rat. The 31-month-old BN/Bi had a large squamous cell carcinoma of the left ureter that had metastasized to several organs. It resulted in obstruction of both ureters with secondary, bilateral, severe hydronephrosis. This rat apparently died of uremia and had calcification in other organs in addition to the aorta. The 24-month-old BN/Bi rat and the WAG/Rij, however, lacked histological evidence of uremia or renal failure.

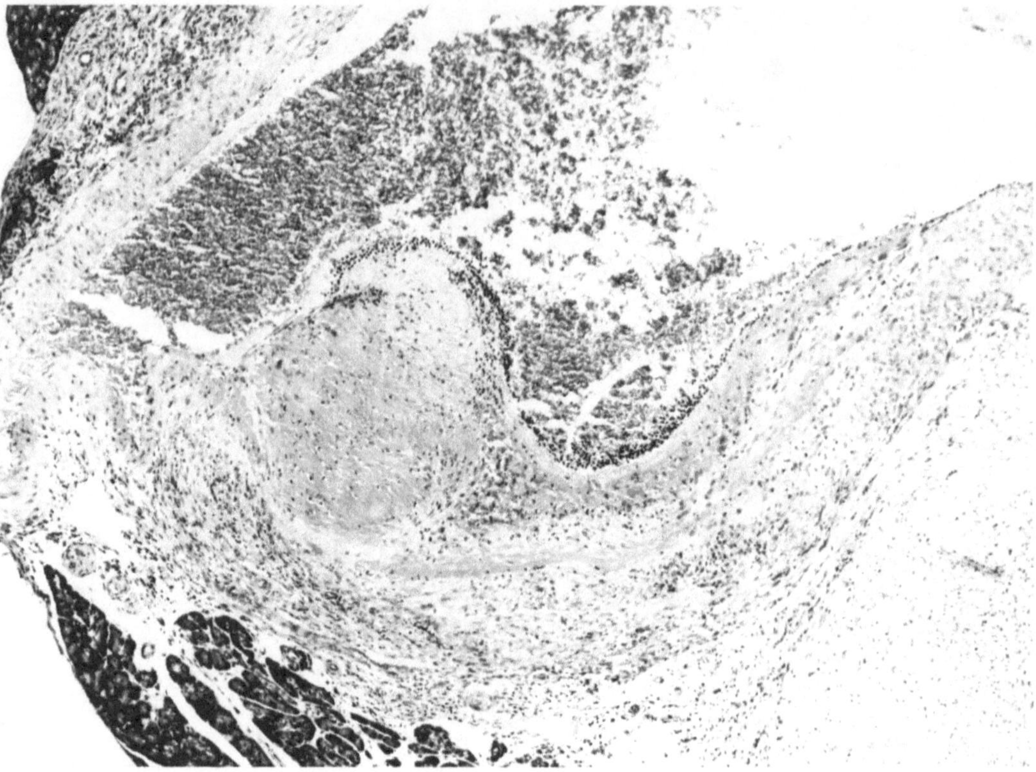

FIGURE 4.64. Periarteritis nodosa of a pancreatic artery in a female rat. (HPS; magnification × 75.)

TABLE 4.18

Periarteritis Nodosa in Aging BN/Bi, WAG/Rij, and (WAG × BN)F_1 Rats

Strain	Sex	No. examined	No. with periarteritis	%	Mean age (range) in months
BN/Bi	Female	236	15	20	33 (22—40)
	Male	74	4	5	32 (22—43)
WAG/Rij	Female	101	2	2	25, 34
	Male	124	0	0	—
F_1	Female	68	1	1	30
	Male	67	11	16	36 (22—40)

Focal calcium deposition in the media of arteries, without associated degenerative changes, was recognized. It consisted of basophilic, amorphous, spherical deposits of calcium and was found frequently in the larger arteries of the lung (Figure 4.67) in males and females and in arteries of the testicle in males. It was not associated with any identifiable etiology and appeared to be relatively insignificant. It was common, but was not always recorded. Therefore, the age-associated incidence in each strain and sex was not evaluated.

2. Neoplastic Lesions

A 13-month-old male BN/Bi rat died with a fibrosarcoma of the heart. The tumor involved most of the left and right atria, heart valves, and left and right ventricles. The cells were surrounded by a collagenous stroma, had indistinct cell boundaries, pale pink cytoplasm, and fusiform or spindle-shaped nuclei that were lightly vesicular. Mitoses were present but infrequent, and no distant metastases were found.

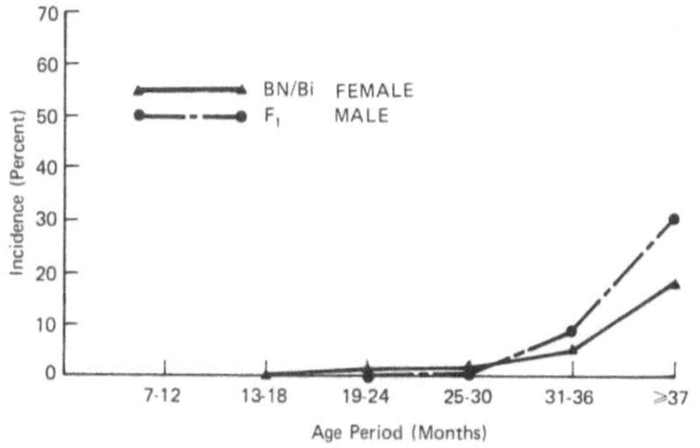

FIGURE 4.65. Percent incidence with age of periarteritis nodosa in 236 female BN/Bi and 67 male (WAG × BN)F_1 rats.

A 22-month-old male WAG/Rij rat died with an undifferentiated sarcoma arising in the wall of the left ventricle. The tumor consisted of undifferentiated mesenchymal cells that infiltrated through the ventricle wall to the epicardium and formed a large solid mass in the chamber of the left heart. It was highly infiltrative and contained many mitoses, but metastases were not observed. Endocardial disease was also present in the left and right ventricle and the left atrium.

a. Chemoreceptor (Aortic Body) Tumors

Twelve female WAG/Rij rats (mean age of 37 months with a range of 32 to 46 months), two male WAG/Rij rats (23 and 26 months), and one male F_1 rat (37 months) had microscopic, cellular nodules at the base of the heart near the aorta. The nodules were composed of nests or clusters of epithelioid cells surrounded by thin reticulum and fibrous trabeculae that formed an organoid pattern. The cells were

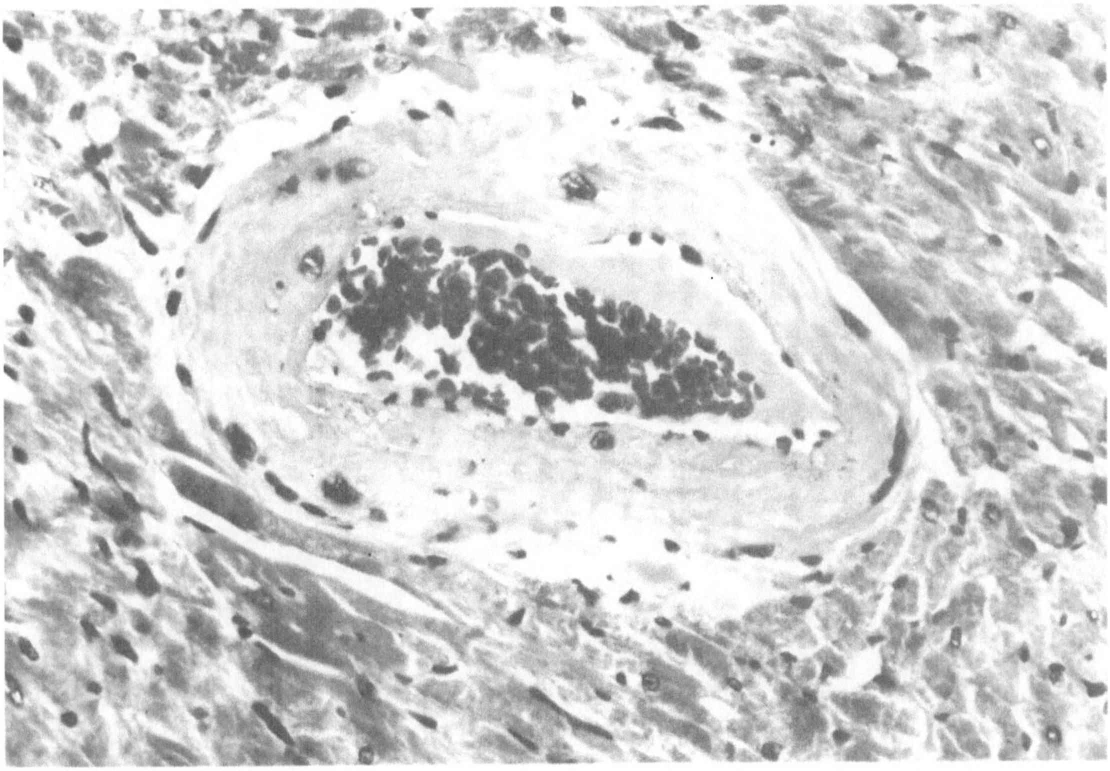

FIGURE 4.66. Artery in the heart showing focal fibromuscular degeneration or dysplasia of the wall.

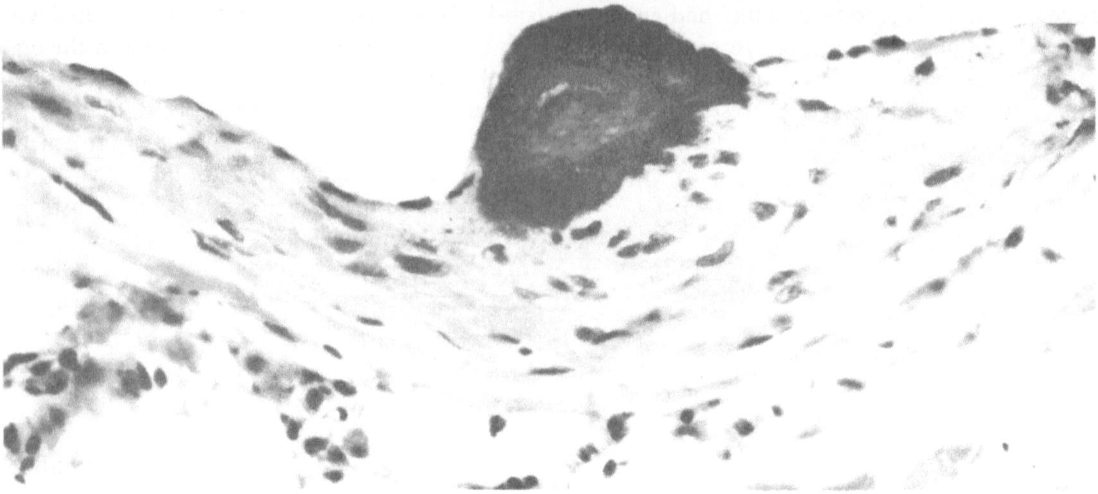

FIGURE 4.67. Pulmonary artery with a focal, subendothelial mineral deposit. (HPS; magnification ×465.)

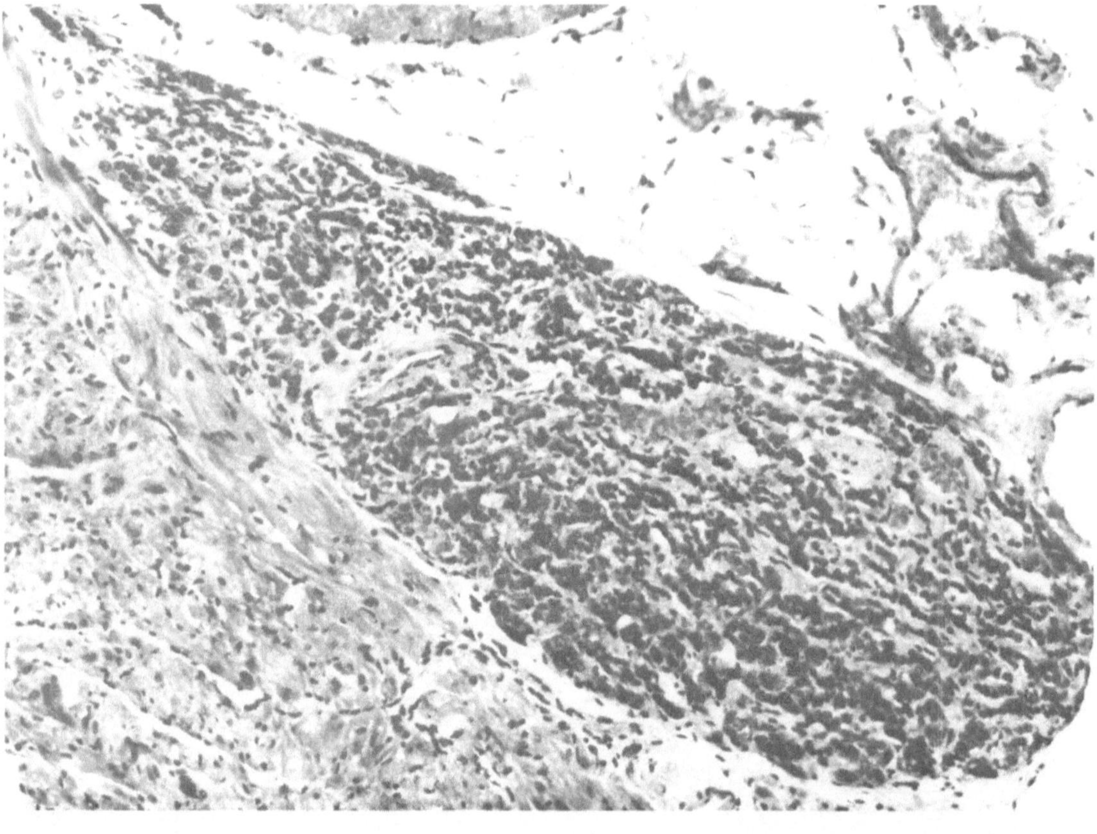

FIGURE 4.68. Nodular proliferation of the aortic body of the heart. (HPS; magnification ×210.)

round with clear to pale pink vacuolated cytoplasm. Their nuclei were round and contained clumped chromatin. In three nodules, the nuclei tended to be ellipsoidal and the chromatin was arranged similar to that in Anitschkow's cells. Most lesions were sharply demarcated from the surrounding tissue (Figure 4.68), but three had infiltrated into the heart between myocardial cells and along blood vessels (Figures 4.69 and 4.70). One had subendothelial growth and appeared to form projections into the lumen of the vessel (Figure 4.71). In two of the large nodules, cellular pleomorphism and atypia were observed.

The nodules were all compatible with enlarged chemoreceptor organs (i.e., aortic bodies). It is possible that most or even all of these lesions were chemoreceptor tumors, but additional studies would be needed to rule out the possibility that the smaller lesions were hyperplasias.

b. Blood Vessels

Tumors of blood vessels were not common and are discussed in the sections dealing with the organ or organ system in which they were found.

3. Discussion

Myocardial degeneration and fibrosis in the hearts of aging rats have been documented in several reviews.[4,41,53,107] They are clearly age associated, but the etiology is unknown.

Several mechanisms have been proposed to explain the lesion, but none of the explanations can account for the high incidence in all rat strains examined. For example, Lehr[107] considered the mechanism to be secondary to severe renal injury in albino rats. This cannot explain the fibrosis in the hearts of BN/Bi, WAG/Rij, and (WAG × BN)F_1 rats because severe renal disease was uncommon (see Section II.K.2.a). Wilens and Sproul[185,186] suggested that sclerosis of coronary arteries produced myocardial damage because of ischemia. Lesions of the coronary arteries were rare in Sprague-Dawley rats studied by Berg,[14] however. Similarly, the inci-

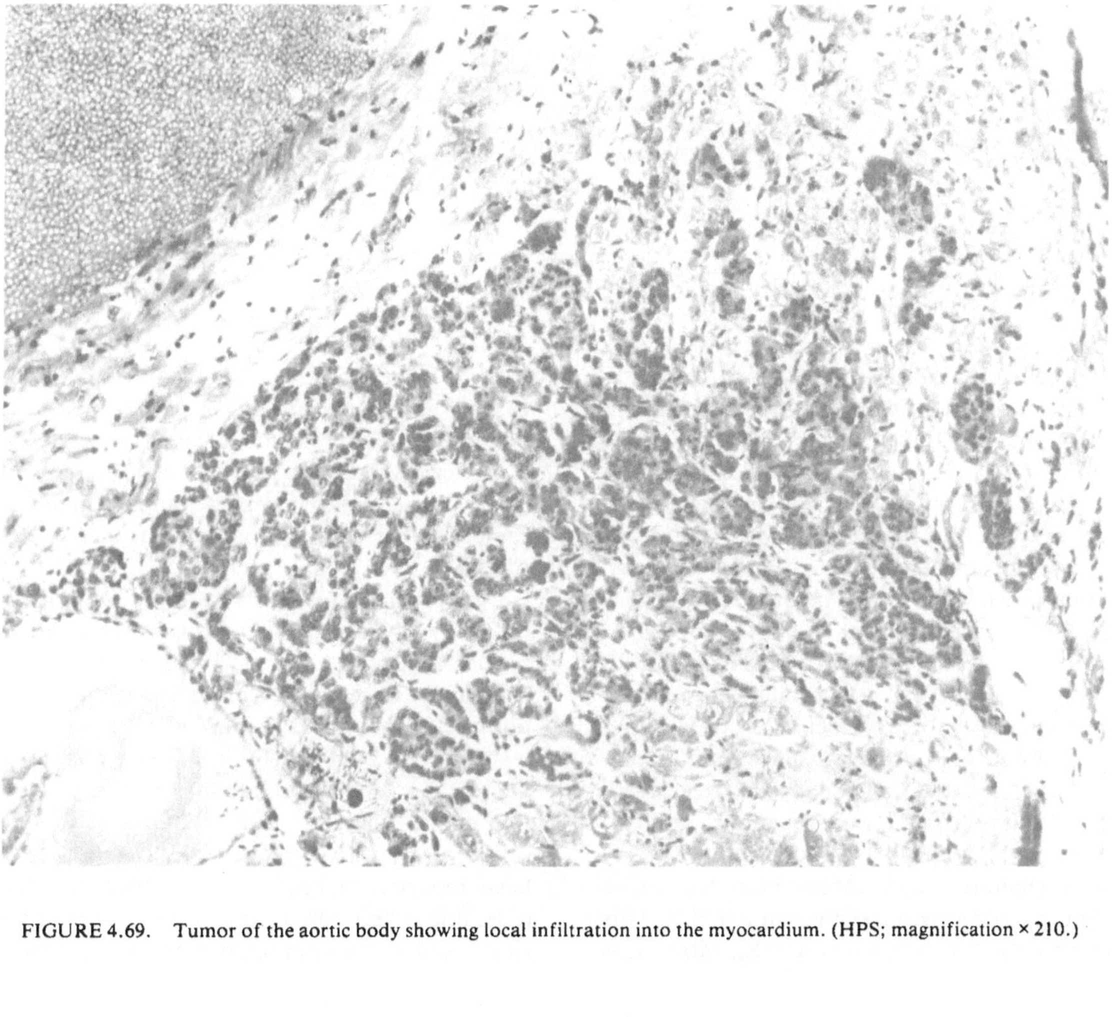

FIGURE 4.69. Tumor of the aortic body showing local infiltration into the myocardium. (HPS; magnification × 210.)

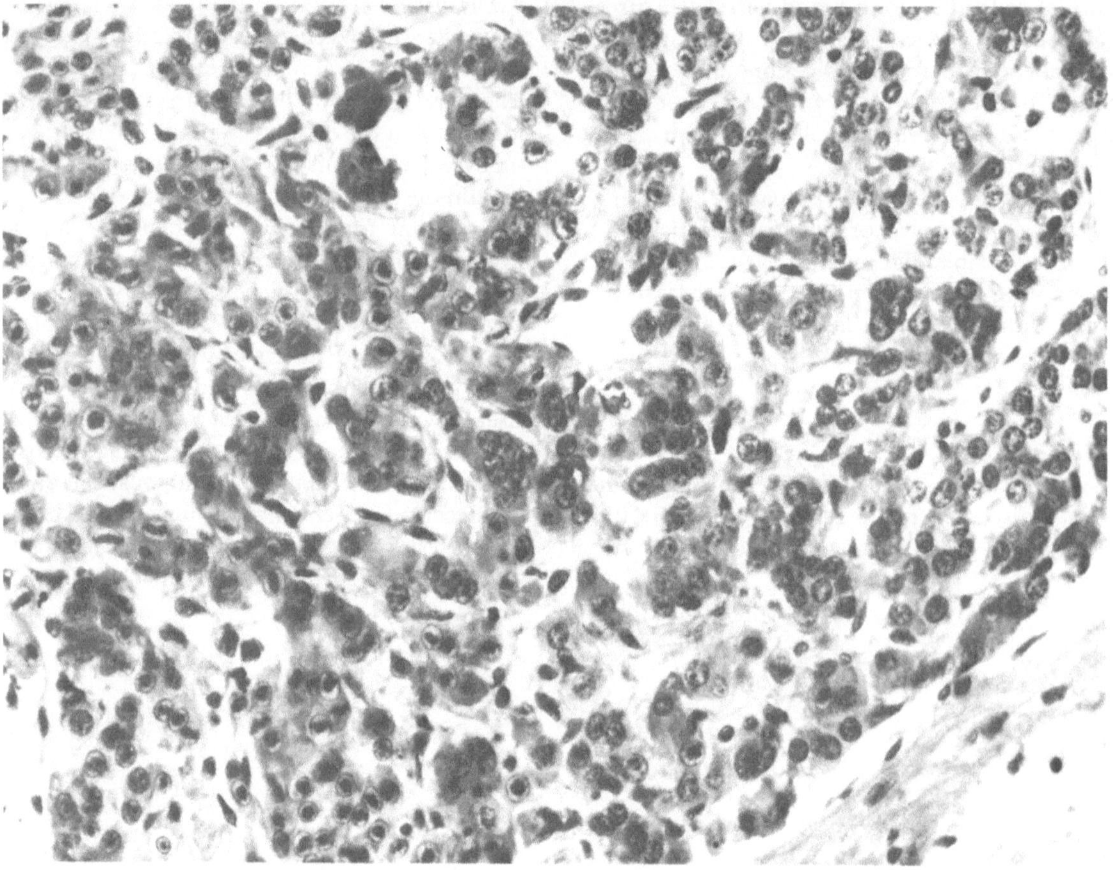

FIGURE 4.70. Higher magnification of the aortic body tumor seen in Figure 4.69, showing some cellular pleomorphism and variation in patterns of nuclear chromatin. (HPS; magnification × 500.)

dence of coronary artery lesions was too low to explain the high incidence of myocardial damage in the BN/Bi, WAG/Rij, and (WAG × BN)F_1 rats of this study. Finally, Coleman et al.[42] suggested that myocardial fibrosis may be secondary to myocarditis in male Fischer 344 rats because the incidence of chronic inflammation was highest in rats below 6 months old. Large numbers of young BN/Bi, WAG/Rij, and (WAG × BN)F_1 rats (between 2 and 10 months old) have also been evaluated as part of other studies,[83] and no evidence has been found to suggest that myocarditis, necrosis, or infarcts are precursors of the myocardial degeneration and necrosis in these rat strains.

Berg and Simms[14] have shown that myocardial degeneration and fibrosis occurred earlier in Sprague-Dawley males than in females. This was also true for the male WAG/Rij rats of this study (Figure 4.51). The male and female BN/Bi rats, however, had similar ages of onset and nearly identical peak incidences of 80 to 90%. It is clear, therefore, that the age of onset of this lesion is earlier in males than in females of some, but certainly not all, strains and stocks of rats.

Endocardial disease is a distinct morphological entity of the rat heart that was originally described by Boorman et al.[20] It is not related to myocardial degeneration and fibrosis; however, both lesions may be observed in the same heart. It appears to be a slowly progressive condition and can result in severe chronic passive congestion of the lung and liver, which seems to be a result of heart failure. Three sarcomas have been reported in hearts with endocardial disease, one in this study, one in the Boorman series,[20] and a fibromatous tumor-like prolifera-

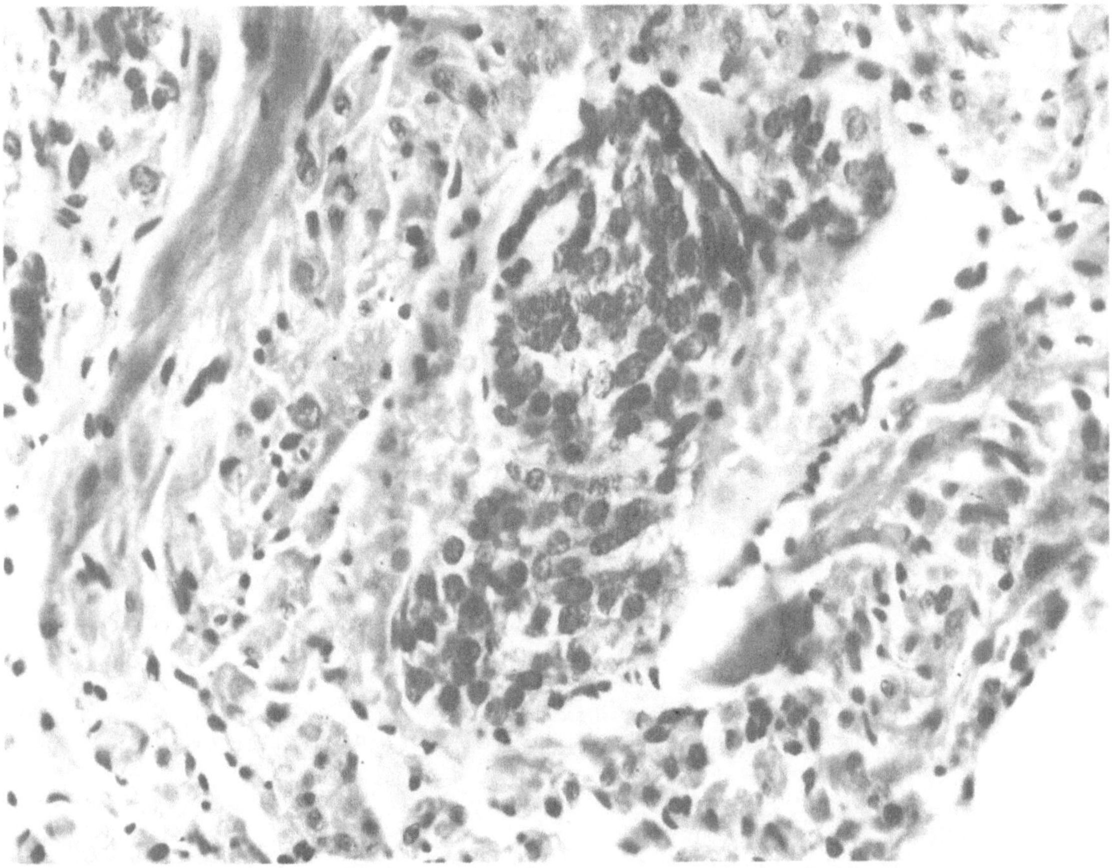

FIGURE 4.71. Same aortic body tumor shown in Figure 4.69. Note the intravascular clumps of tumor cells that are lined by endothelial cells. (HPS; magnification × 500.)

tion found by Frith et al.[60] in the heart of a rat. Since primary tumors in the heart of rats are not common, the findings of sarcomas associated with endomyocardial disease suggest that this lesion may occasionally progress from a benign proliferative lesion to invasive cancer in aged rats.

A cartilaginous focus was often found at the base of the aorta of all rat strains examined. The structure occurred in rats of all ages and has been observed in younger rats evaluated as part of other studies.[83] A similar cartilaginous focus has been described by Hueper,[87] who found it in young rats and thought it was a normoplastic transformation caused by mechanical stress. Hollander[78] demonstrated that it can be found in rats as young as 1 week of age. Therefore, this change is apparently a normal physiological structure and is not an age-associated lesion.

Tumors are rare in the hearts of rats, as reviewed by Squire and Goodman,[163] Ivankovic,[93] and Altman and Goodman.[1] In this current series, two sarcomas were found. In addition, 15 microscopic nodules of the aortic body were observed, all but one in WAG/Rij rats. Three aortic body nodules were clearly tumors with infiltration into the myocardium and along blood vessels. The others were probably tumors, but hyperplasia could not be completely excluded since Hebel and Stromberg[76] have documented normal carotid bodies that were nearly as large as some of the aortic body nodules in this series.

Trevino and Nessmith[172] reported a large aortic body tumor in a 3-year-old white rat, but the WAG/Rij rats are the only published strain in which large numbers of aortic body lesions have been observed. The incidence of this lesion

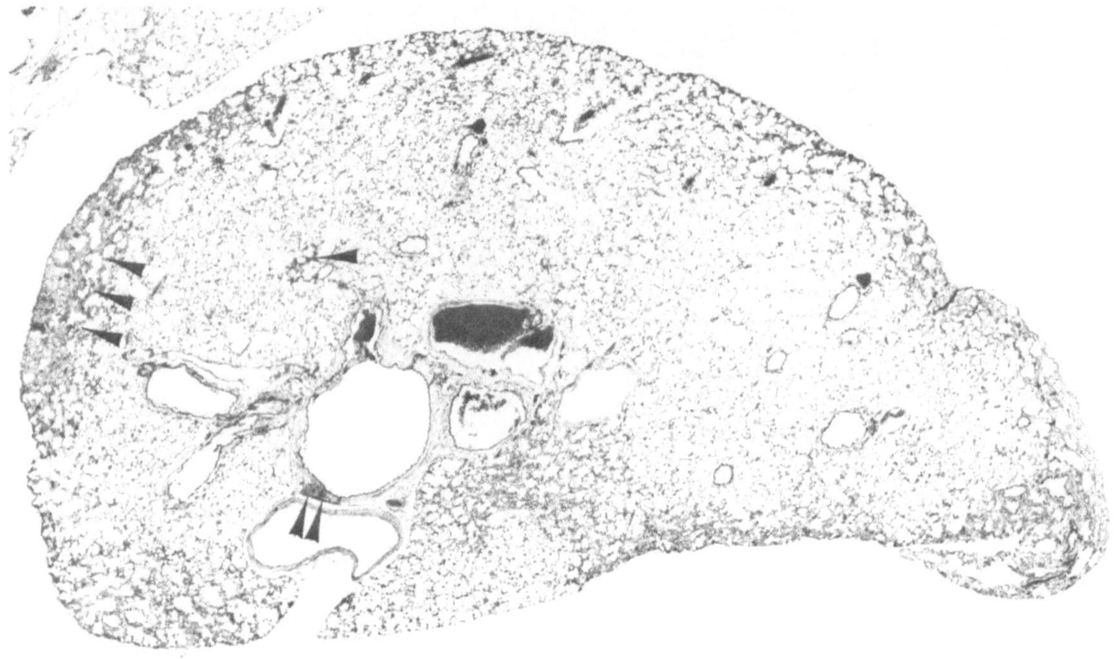

FIGURE 4.72. Lung showing focal aggregation of perivascular lymphoid cells (↑), focal aggregation of peribronchial lymphoid cells (↑↑), and focal subpleural aggregation of alveolar macrophages (↑↑↑). (HPS; magnification × 18.)

in female WAG/Rij of this series was 12%, indicating it is not a rare lesion. Quast and Jersey[131a] have observed three similar chemoreceptor lesions in Sprague-Dawley (Spartan substrain) rats. One was grossly recognized and had metastasized to the lungs. In a separate publication, van Zwieten et al.[180] have reviewed aortic body lesions in the hearts of aging rats and have shown them to be strain and sex dependent and age associated.

Periarteritis nodosa has been described by several investigators.[41,53,107] Briefly, the cases of periarteritis nodosa in the rats of this report were similar to those cited in the previously published reports. The lesion was uncommon in male BN/Bi, male and female WAG/Rij, and female (WAG × BN)F_1 rats. It was most common in BN/Bi female and (WAG × BN)F_1 males where the age-associated peak incidence reached 30 and 18%, respectively.

J. Respiratory System

1. Nonneoplastic Lesions

A variety of nonneoplastic lesions were observed in the lungs, but few were considered to be age associated. Most appeared to be secondary to some process or processes occurring elsewhere in the body of the rat. For example, chronic passive congestion of the lungs with pulmonary edema, hemosiderin deposits, and hydrothorax was seen in rats with severe endocardial disease (see Section II.I). Rats dying with neoplasms that had pulmonary metastases also had secondary lung lesions such as edema.

Figure 4.72 illustrates the typical appearance of the lungs from the aging rats. The most common findings were focal or multifocal perivascular aggregations of lymphocytes (Figure 4.73). Rarely, focal aggregations of lymphocytes were found near bronchi or bronchioles. The next most common lesion was focal or multifocal aggregations of macrophages in the alveolar spaces (Figure 4.74). Neither of these lesions could be shown to be age associated, but both were relatively common in all groups of rats.

Other lesions were observed, but were much less common and included such findings as chronic passive congestion, focal aspiration granulomas, focal hemorrhage, pulmonary edema, macrophages containing hemosiderin, and others. Focal mineral deposits were seen in

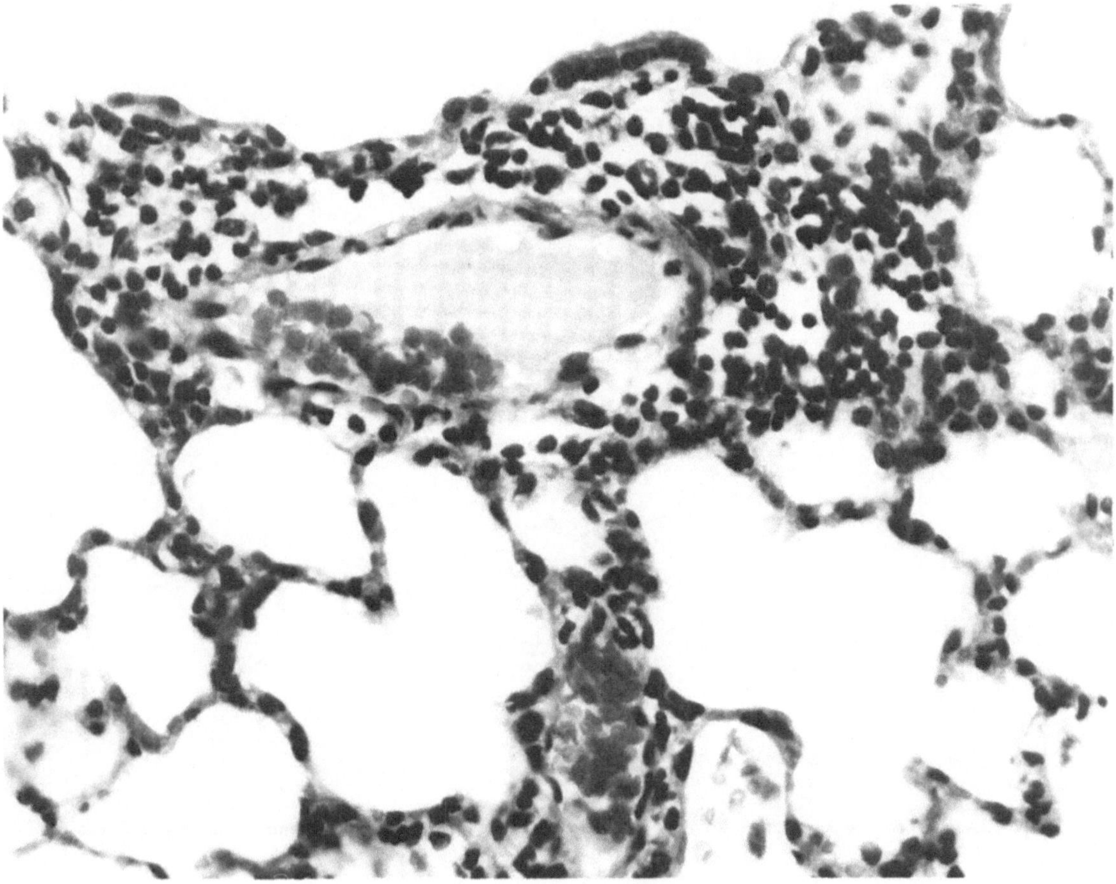

FIGURE 4.73. Higher magnification of perivascular aggregation of lymphoid cells. (HPS; magnification × 500.)

the pulmonary arteries, as described in section II.I.

Focal adenomatous hyperplasia occurred as a proliferative lesion of the alveolar epithelial cells that were cuboidal in shape and formed small gland-like structures. It was observed as a solitary lesion in two (WAG × BN)F_1 males (ages 29 and 36 months), 1 (WAG × BN)F_1 female (31 months old), 1 BN/Bi female (31 months old), and 1 BN/Bi male (35 months old). None were seen in the WAG/Rij males or females.

2. Neoplastic Lesions

Only two primary lung tumors were recognized. One was a squamous cell carcinoma in a 33-month-old (WAG × BN)F_1 female. It was composed of sheets and nests of squamous epithelial cells, formed typical keratin pearls, and had invaded surrounding pulmonary tissue. Distant metastases were not found. The second tumor was an angioleiomyosarcoma in a 35-month-old BN/Bi female. It arose in the pulmonary arteries and extended from the pulmonary trunk into the right and left lungs. The tumor cells appeared to have originated from the smooth muscle wall of the pulmonary arteries. The cells were round to spindle shaped, mitoses were infrequent, and local invasion and destruction of pulmonary tissue were present. No distant metastases were found.

Several tumors had metastasized to the lungs, but these are described under each specific primary tumor.

3. Discussion

Pour et al.,[129] Squire and Goodman,[163] and Altman and Goodman[1] have alluded to an apparent low incidence of spontaneous tumors in the respiratory tract of rats in their literature

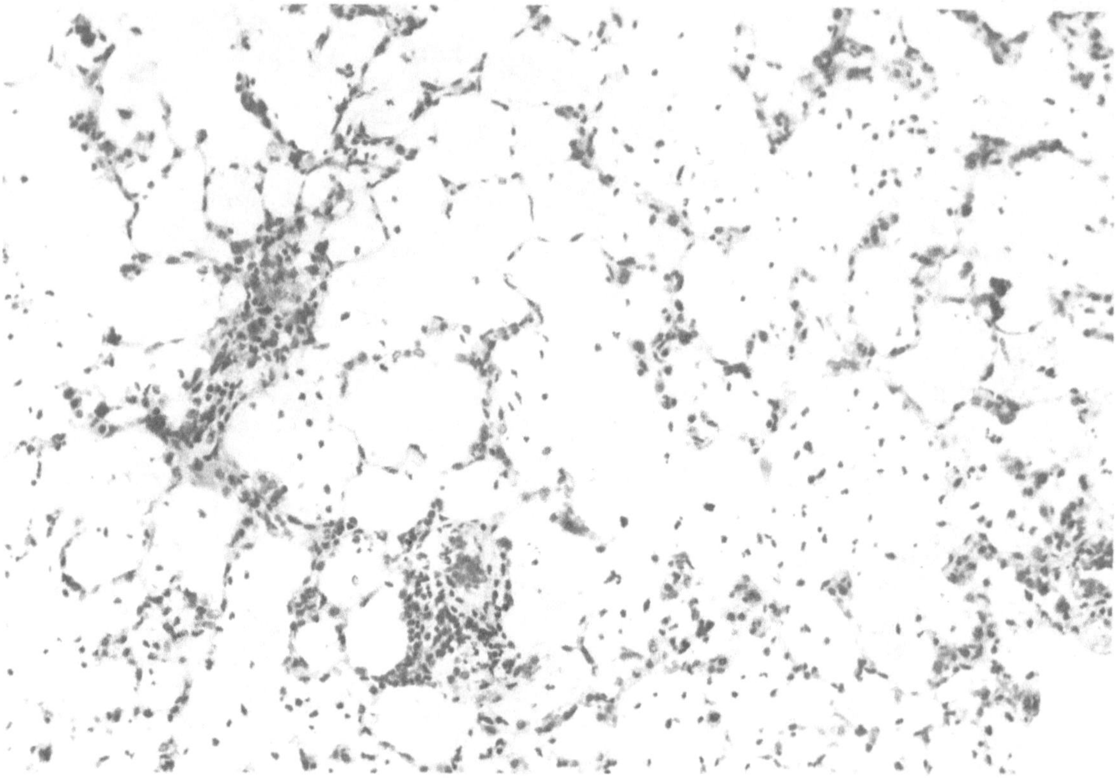

FIGURE 4.74. Focal area of lung with alveolar macrophages and perivascular aggregations of lymphoid cells.

reviews. Similarly, only two primary lung tumors were found in the respiratory tract of the 670 aging rats of this study. Metastatic tumors to the lungs were more frequent, however, and occurred in rats with several different primary tumors. Boorman and Hollander[19] reported a mesothelium of the pleura in a 39-month-old female WAG/Rij, but none were seen in this study.

Nonneoplastic age-associated lesions were also uncommon. In a review, Anver and Cohen[4] described only two conditions in aging rat lungs: pulmonary foam cells (alveolar macrophages) and murine mycoplasmosis. Previously, Innes[89,90] reviewed the literature on these two conditions and also reported that bone spicules, pulmonary edema, hemosiderin, and foreign body granulomas may be seen in rat lungs. Lindsey et al.[110] reviewed the literature on murine mycoplasmosis and described the severe impact the disease can have on rats.

Aggregations of alveolar macrophages were seen in the BN/Bi, WAG/Rij, and (WAG × BN)F_1 rats of this study, but it is unclear if the frequency increased with age. Such lesions have been seen in rats of these strains that were younger than 6 months old.[83] No attempt was made to determine whether this lesion increased in frequency or severity with age. Yang et al.[187] and Giddens and Whitehair[67] have also described this condition in rat lungs, but did not describe an age-associated increase. Recently, Flodh[58] was able to show an increased frequency and severity of pulmonary foam cells in the lungs of Sprague-Dawley rats at 22 months old compared to younger rats of 6 and 9 months old. However, it is not known if other rat strains also show a similar increase.

Aggregations of lymphocytes were observed around some arterioles in the lungs. Like the occurrence of alveolar macrophages, it was common, but relatively insignificant. It occurred in all the lungs of all groups, but an age-associated increase was not apparent. The etiology of this change is unknown. As GF BN/Bi and WAG/Rij rats at the REP Institutes[83] did not have this lesion, one might speculate that it is caused by an infectious agent. Lamb[106] also observed similar lesions in SPF rat lungs and speculated on a viral etiology. The only virus that was latent in the aging rat colony throughout this study was pneumonia virus of mice (PVM) (see Chapter 2), but studies have not shown that PVM can produce such lesions.

In Chapter 2, it was noted that Sendai virus entered the aging rat colony during part of this study. The disease in rats and mice of the REP Institutes has been documented in detail.[31,190] The virus did not produce mortality in rats, but some rats dying after Sendai virus entered the colony had new lung lesions.

Approximately 20% of the 670 rats reported in this study died after Sendai virus entered the colony. The new lesions in the lungs included focal or multifocal interstitial pneumonia, increased perivascular cuffing by lymphocytes and plasma cells, and an increase in the severity and frequency of peribronchial cuffing by lymphoid tissue. Bronchiectasis, lung abscesses, or consolidation were not found in any of these rats, however. None of the lesions associated with Sendai virus were considered age associated.

K. Urinary System

1. Nonneoplastic Lesions

a. Hydronephrosis

Hydronephrosis ranged from mild to severe, as illustrated in Figure 4.75. Mild (+) cases consisted of only slight dilation of the renal pelvis. Severe lesions (+ + + +) consisted of an enlarged, cystic kidney with a thin rim of renal tissue compressed along the outer edge of the cyst. The examples of + + + and + + + + hydronephrosis were invariably accompanied by severe hydroureter.

The occurrence of hydronephrosis, regardless of severity, is seen in Table 4.19 for the BN/Bi, WAG/Rij, and (WAG × BN)F_1 rats. It occurred more frequently in the males of all three strains; 16% of the WAG/Rij males had the lesion in contrast to 0% in the females. Male F_1 rats had a 10% incidence, while the females had only a 2% incidence.

The BN/Bi males and females had the greatest incidence — 43 and 39%, respectively, In

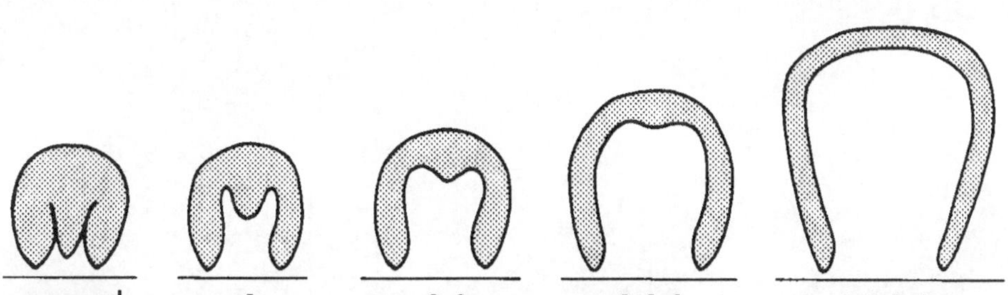

FIGURE 4.75. Schematic illustration of the severity of hydronephrosis seen in aging rats. + and + + = mild hydronephrosis, + + + = moderate, and + + + + = severe.

TABLE 4.19

Prevalence of Hydronephrosis in Kidneys of Aging BN/Bi, WAG/Rij, and (WAG × BN)F_1 Rats

Strain	Sex	No. examined	No. with hydronephrosis	%
BN/Bi	Female	236	93	39
	Male	74	32	43
WAG/Rij	Female	101	0	—
	Male	124	20	16
F_1	Female	68	2	2
	Male	67	7	10

this strain, it was often severe, with about one third of both the males and females having + + + or + + + + severity. Most of the severe cases were associated with urolithiasis, urothelial cancers, or both. Bilateral lesions were found in approximately 65% of the BN/Bi rats with hydronephrosis. The cases of unilateral lesions were approximately equally divided between the right and left kidney.

Most of the examples of hydronephrosis in WAG/Rij or (WAG × BN)F_1 rats, on the other hand, were + or + + severity, were unilateral, usually occurred in the right kidney, and were not associated with a causative agent such as stones or urothelial cancers. For example, 20 cases of hydronephrosis occurred in male WAG/Rij rats (Table 4.19). All were unilateral with 15 in right kidneys, 2 in left kidneys, and 3 in which the exact kidney was not recorded. One of the 20 was + + + and it was in a left kidney. The seven cases in male F_1 rats were all of + or + + severity and all were in right kidneys.

b. Urolithiasis

Stones or calculi were observed in every group, but the size and location of the uroliths varied. BN/Bi rats had the largest stones that were usually composed of calcium oxalates. The stones were usually in the lower third of the ureter, bladder, or both and were hard, had rough spiny surfaces, were dark to light brown in color, and often had concentric laminations (Figure 4.76). On occasion, a stone was white or was found in the renal pelvis.

The calculi in WAG/Rij and (WAG × BN)F_1 rats, on the other hand, were microcalculi appearing as small, birefringent crystals attached to the surface of the renal pelvis. Calculi in the ureter or bladder were not common.

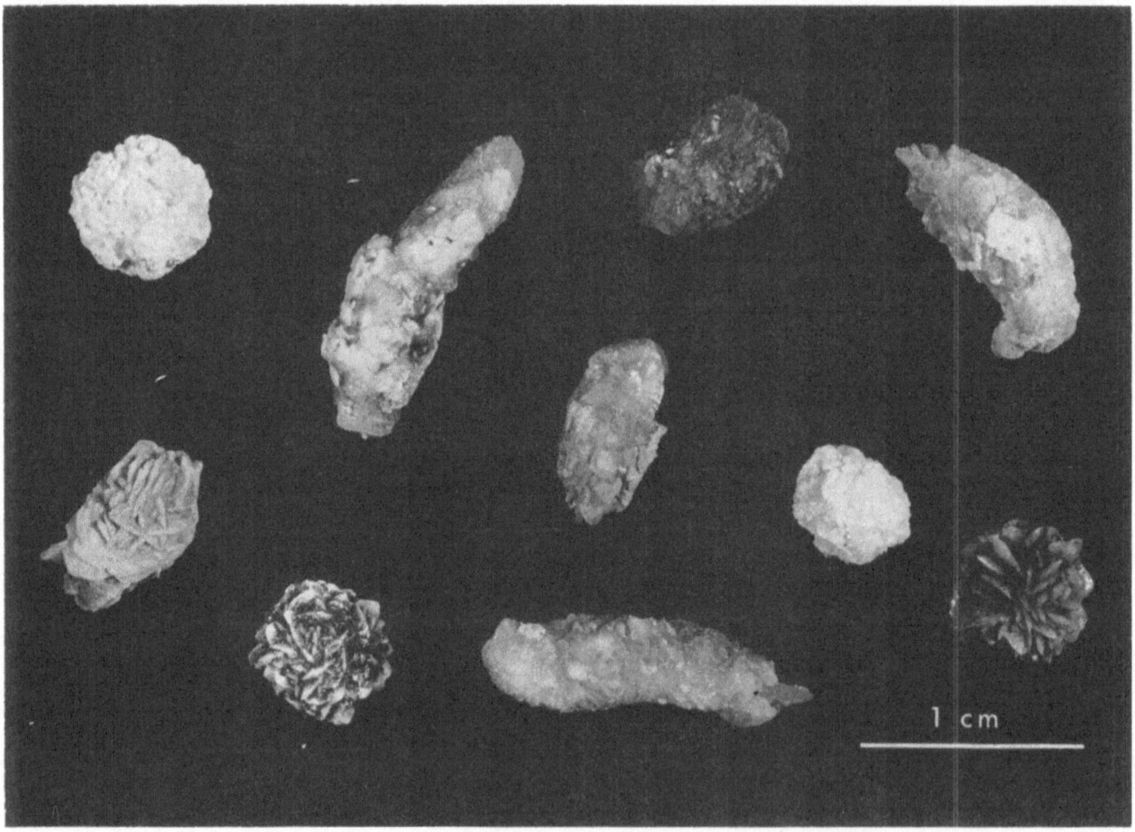

FIGURE 4.76. Uroliths removed from the ureter and bladder of aging BN/Bi rats.

c. Chronic Progressive Glomeruloneophropathy

Severe, progressive glomerulonephropathy was not a serious problem in the aging rats of this study. Some lesions were observed in older rats, but most were mild and focal. Glomerular lesions consisted of mild sclerosis of Bowman's capsule and the mesangium of the glomerular tufts. Lesions in the cortex were focal, dilated renal tubules with proteinaceous casts; occasional, focal lymphocytic infiltration and fibrosis in the interstium; focal, perivascular, lymphocytic cuffing; and focal scarred areas with atrophied tubules, fibrosis and scattered inflammatory cells. Moderate to severe renal disease was not common. Figure 4.77 illustrates

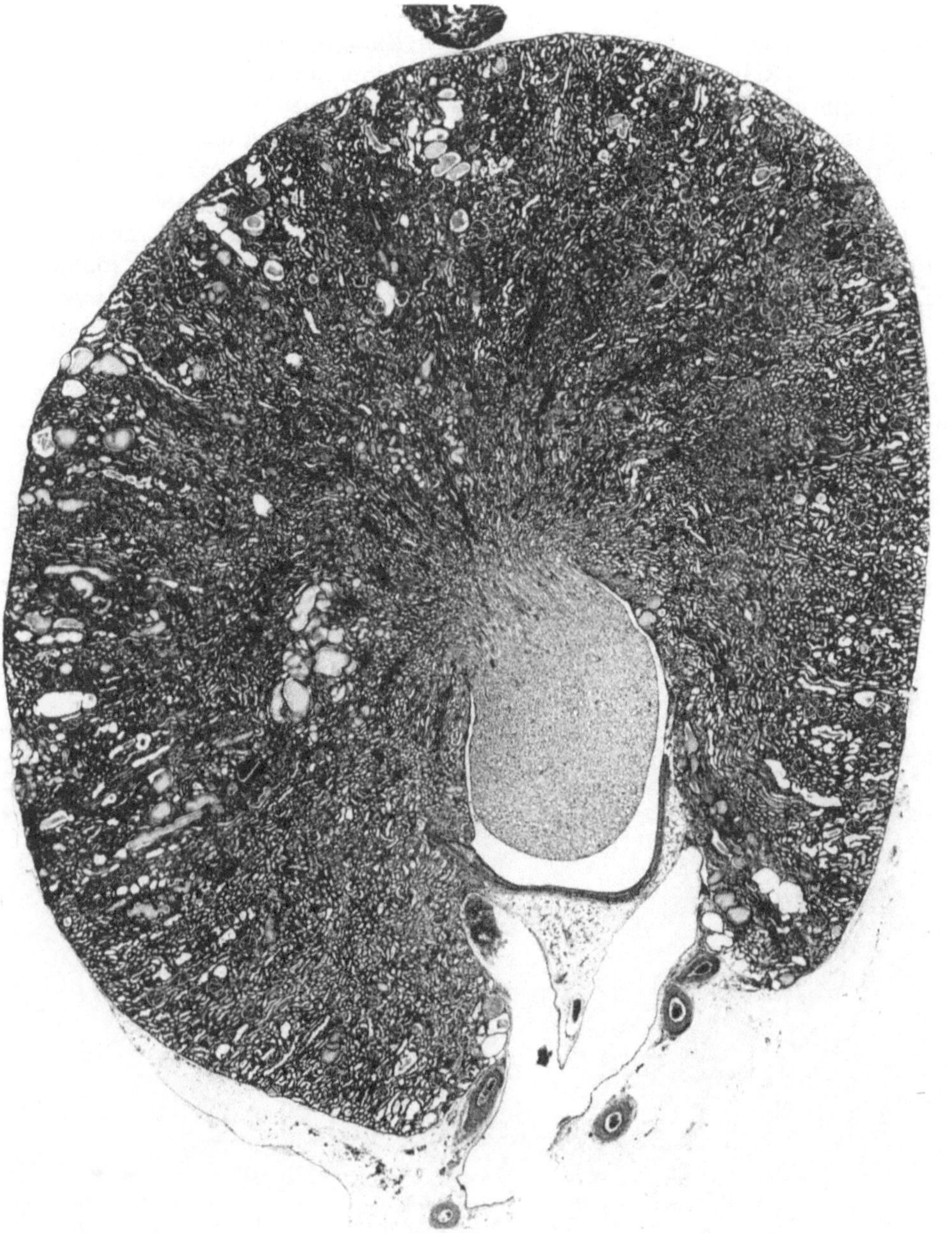

FIGURE 4.77. Kidney from an aging male (WAG × BN)F_1 rat illustrating moderate chronic progressive glomerulonephropathy. This was one of the more severely affected kidneys in this study. (HPS; magnification × 32.)

one of the more severely affected kidneys in this study. Only two rats died with more extensive lesions and both of these were female F_1 rats (ages 28 and 35 months) that had severe, bilateral renal disease. Although the incidence of glomerulonephropathy increased with age, the severity remained mild. Too few examples of moderate to severe renal disease were present to plot by life-table calculations; therefore, the age-associated incidence was not determined.

d. Inflammatory Lesions

Of 124 male WAG/Rij rats, 9 died with suppurative pylonephritis. All nine had severe, chronic, suppurative prostatitis (see Section II.N) that apparently resulted in an ascending urinary tract infection.

Cystitis, ureteritis, and pyelitis were common in BN/Bi rats. They appeared to be secondary to ureter or bladder tumors, urolithiasis, or both.

e. Miscellaneous Lesions

Additional nonneoplastic lesions were observed, but they were not specifically studied. Examples of such lesions include hemosiderin deposits in tubular epithelial cells, solitary cortical cysts, focal calcium deposition, and the appearance of PAS- positive and PAS-negative lipofuscin-like granules in the epithelium of renal tubules. Some of these lesions may be age associated, but additional studies are needed to establish this.

2. Neoplastic Lesions

a. Kidney Tumors

Neoplastic lesions of the kidney were found in three male rats. A 19-month-old male WAG/Rij rat died with multiple bilateral cortical adenomas and adenocarcinomas, a 24-month-old male WAG/Rij had a solitary cortical adenoma (Figures 4.78 and 4.79), and a 36-month-old male (WAG × BN)F_1 had a solitary cortical adenocarcinoma. The tumors tended to be circumscribed nodules that were well demarcated from the normal renal tissue. The epithelial cells formed nests or tubules, were round to polygonal in shape, and had coarsely or finely

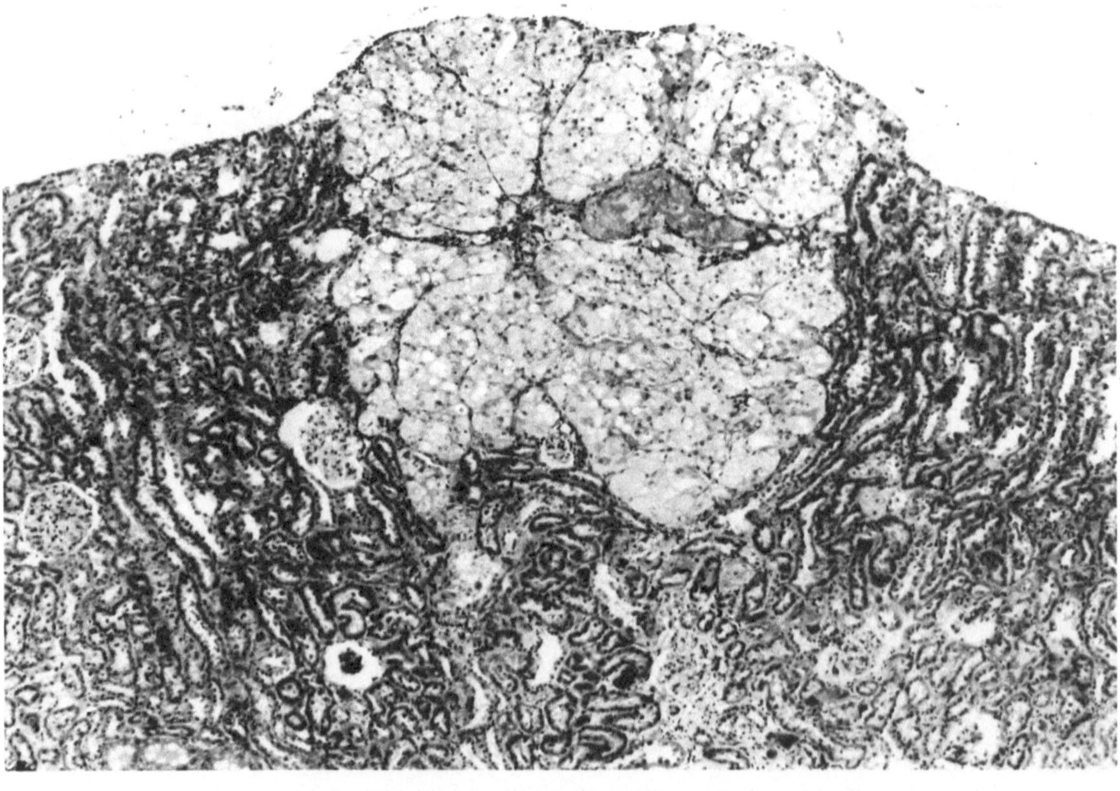

FIGURE 4.78. Adenoma in the renal cortex. (HPS; magnification × 50.)

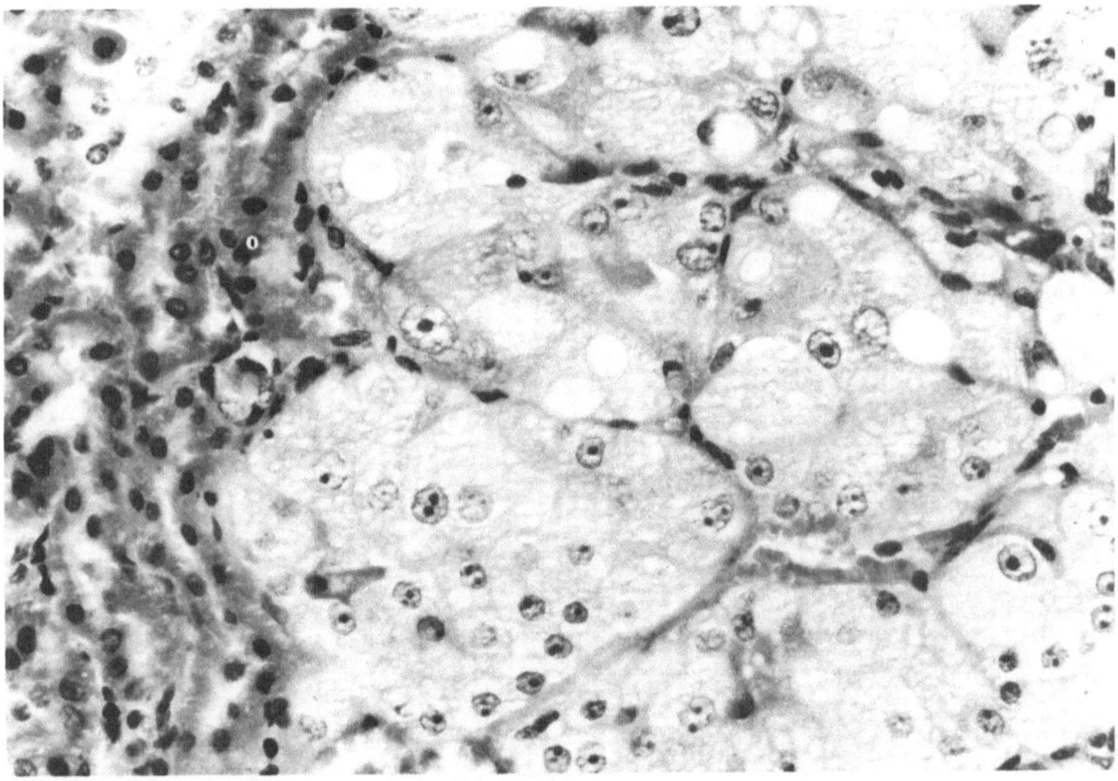

FIGURE 4.79. Higher magnification of the renal cortical adenoma shown in Figure 4.78. (HPS; magnification × 465.)

vacuolated pink cytoplasm with the nuclei round and uniform in size. The difference between the adenomas and adenocarcinoma was subjective. The tumors diagnosed as carcinomas had similar cells but focal hemorrhage or necrosis was observed, mitotic figures were present, and they were larger than the adenomas.

b. Ureter and Bladder Tumors

Neoplastic and "preneoplastic" lesions of the urinary bladder and ureter of BN/Bi rats included a wide range of benign and malignant lesions. The lesions were graded using the same classification as Boorman and Hollander[19] used in their original paper describing the urothelial tumors in BN/Bi rats. The lesions were graded as P1S for noninvasive cancer, P1 or P2 for invasion into the submucosa up to the muscular tunics, P3 where muscle was clearly invaded, and P4 where the entire thickness of the ureter or bladder wall was invaded by the tumor. In addition, hyperplasia was recognized and was either diffuse or nodular (focal or multifocal). The difference between nodular hyperplasia and P1S cancer was often difficult to distinguish. Hyperplasias usually had normal or nearly normal cells (Figure 4.80), while P1S cancer usually exhibited disorganization of the normal urothelial architecture, mitoses, tinctorial variability among the cells, and often some cellular or nuclear pleomorphism (Figure 4.81).

The lesions in the ureters were invariably in the distal half and were often confined to the distal one third of the ureter. The lesions ranged from diffuse or nodular hyperplasias and from P1S to P4 carcinomas. The carcinomas were usually squamous cell carcinomas that readily invaded the submucosa and muscular wall (Figures 4.82 and 4.83). Therefore, most of the ureter tumors were classified as P2 to P4 squamous cell carcinomas. In most rats the ureter carcinomas were unilateral, but several rats had bilateral tumors.

The bladder tumors were more difficult to classify as benign or malignant. Most were pap-

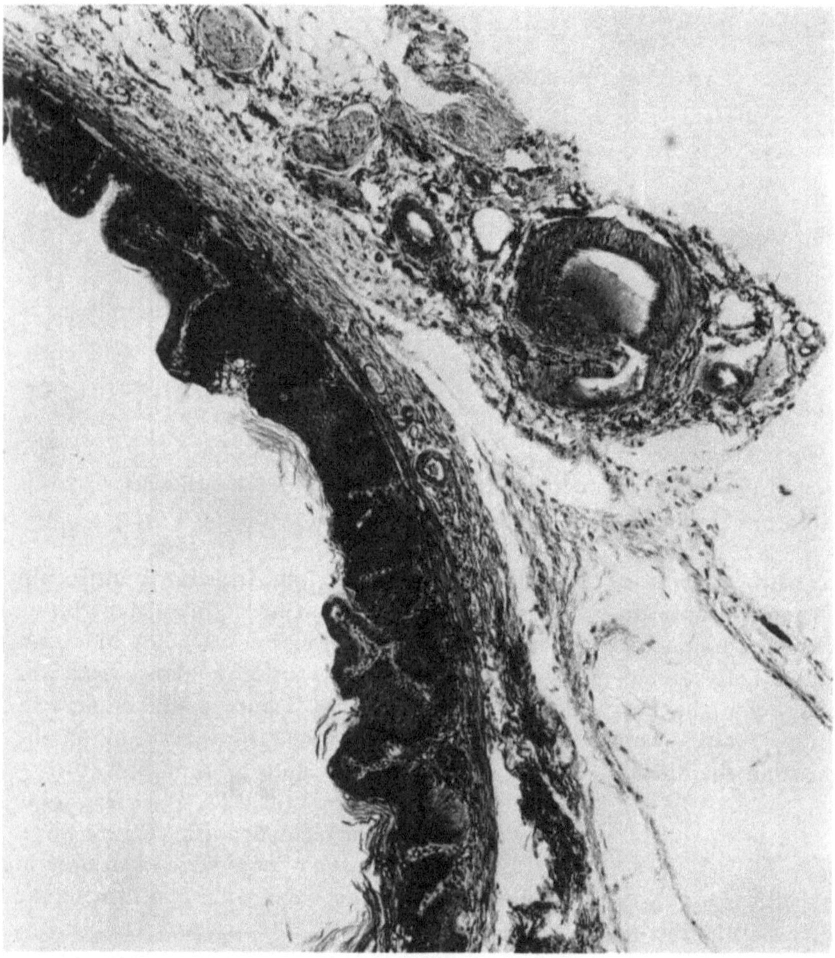

FIGURE 4.80. Ureter from a female BN/Bi rat showing hyperplasia and hyperkeratosis. Note the normal epithelium in the upper right-hand corner. (HPS; magnification × 80.)

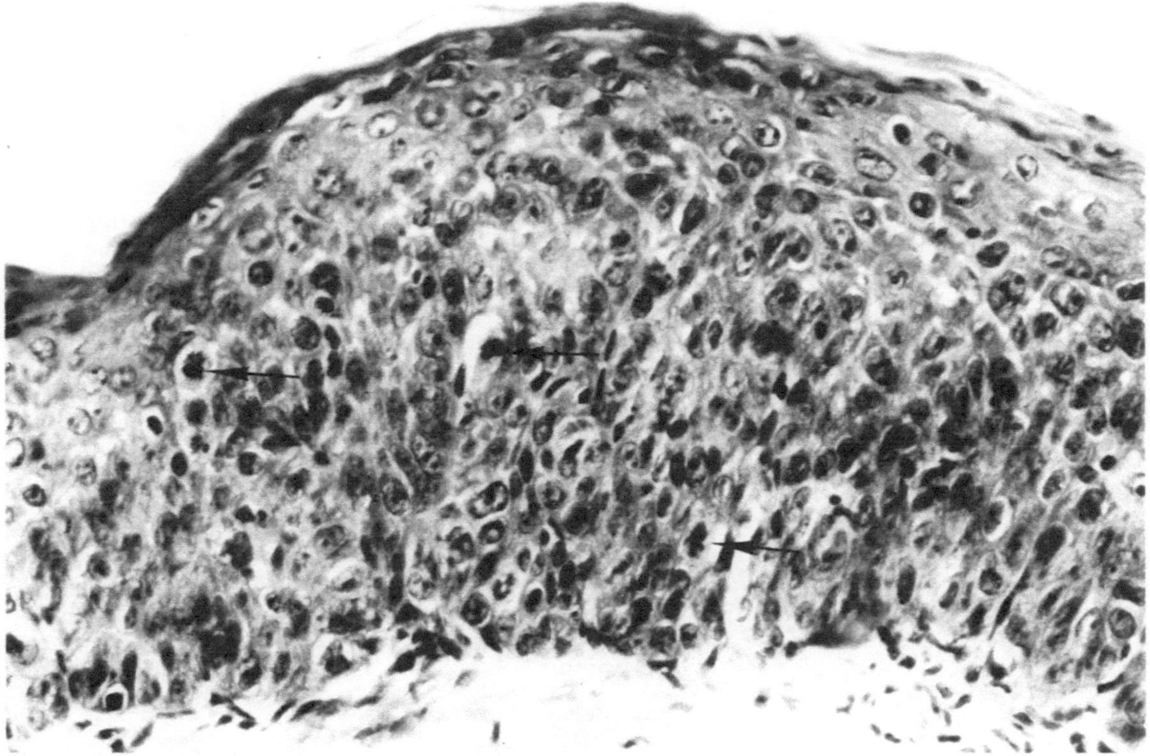

FIGURE 4.81. Ureter from a female BN/Bi rat with noninvasive carcinoma. Note the cellular pleomorphism and mitotic figures (↑). (HPS; magnification × 465.)

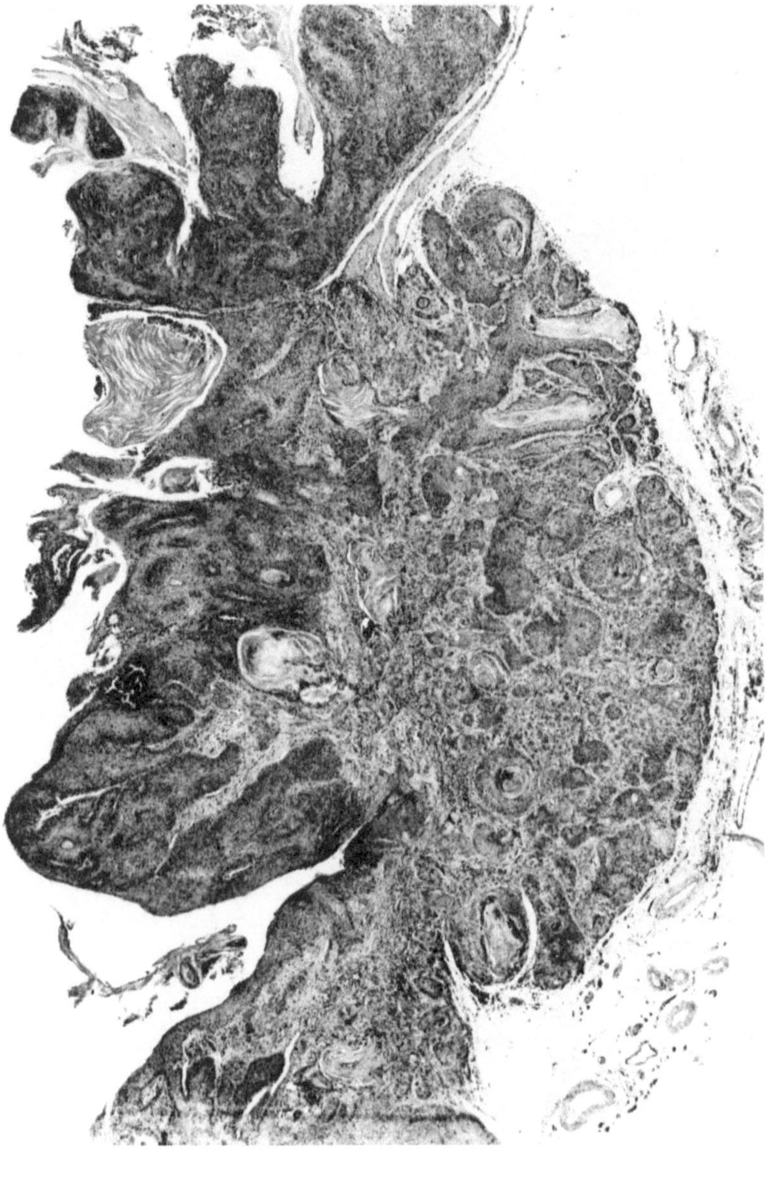

FIGURE 4.82. Squamous cell carcinoma in the ureter of a female BN/Bi rat. (HPS; magnification × 33.)

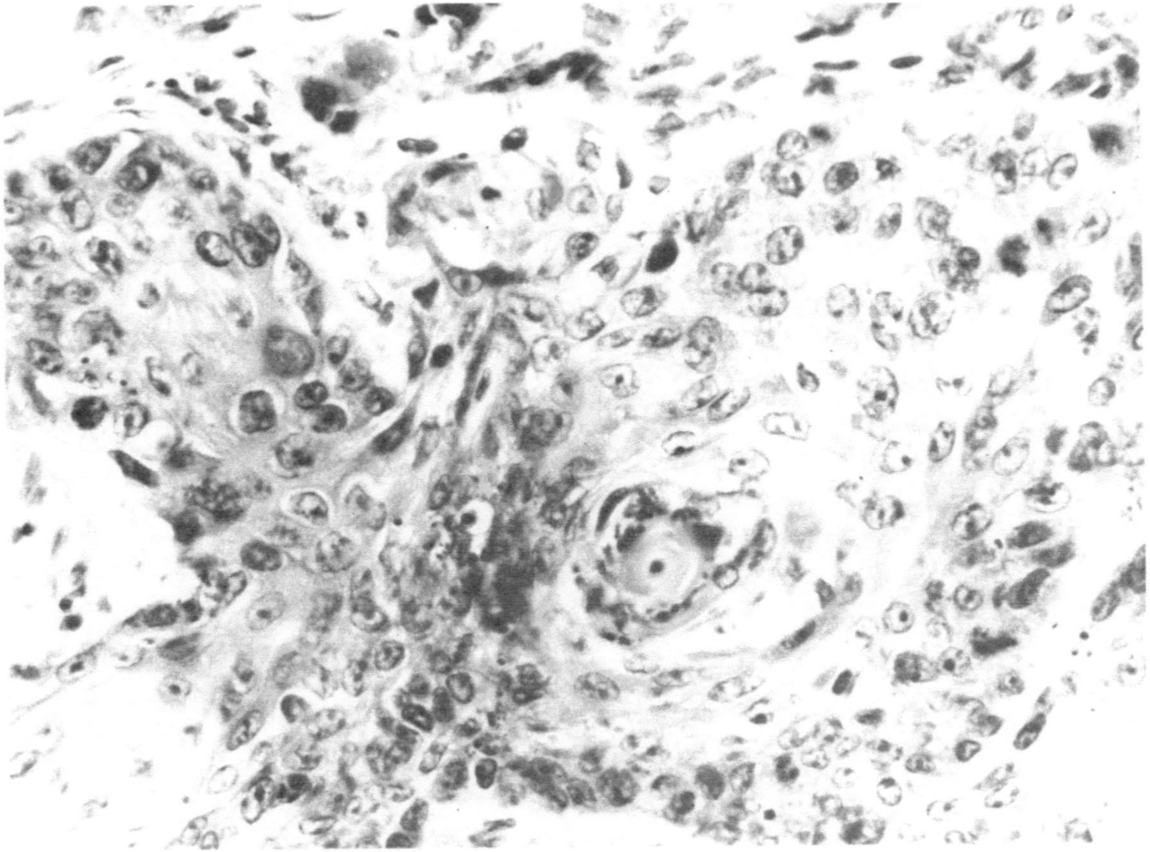

FIGURE 4.83. Higher magnification of Figure 4.82, showing an epithelial pearl and squamous epithelial cells. (HPS; magnification × 500.)

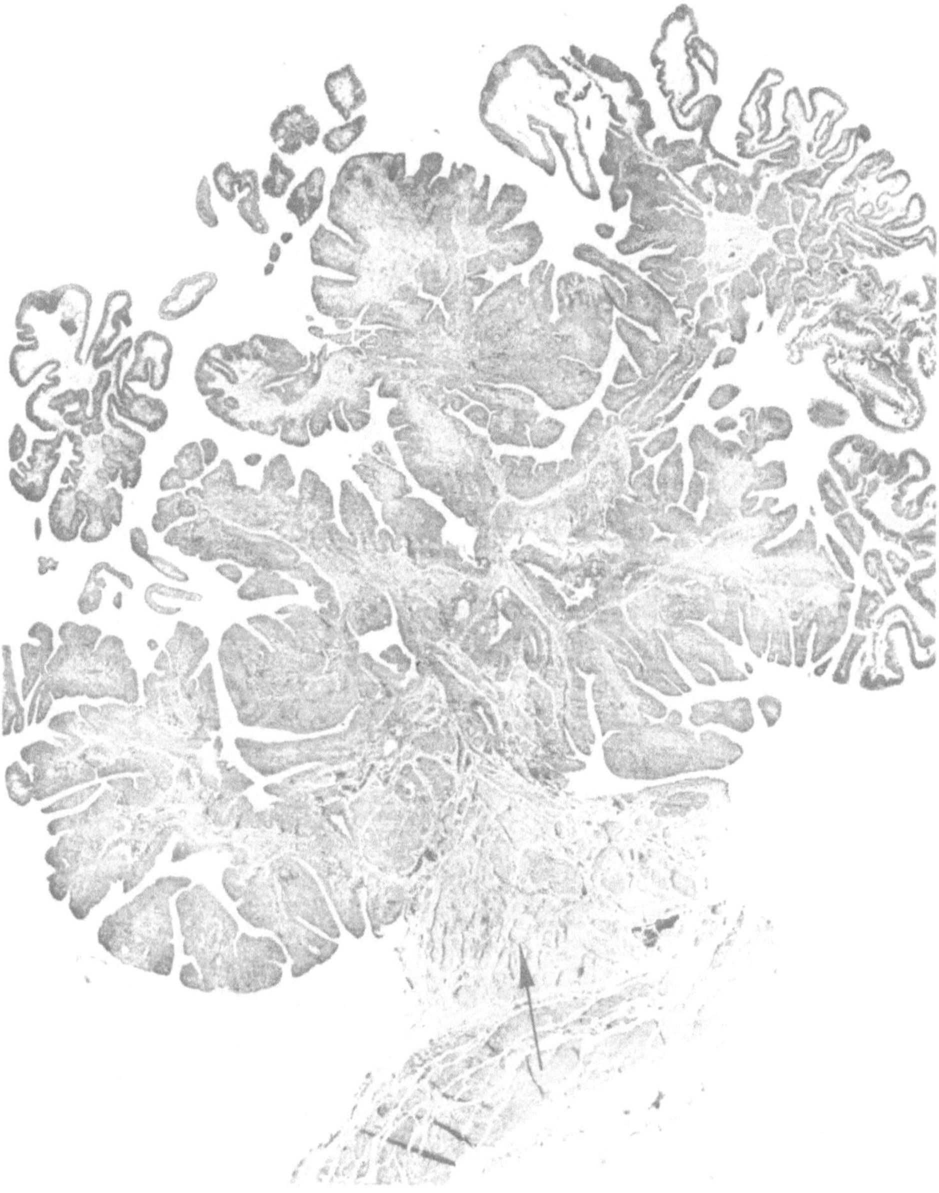

FIGURE 4.84. Papillary adenocarcinoma in the bladder of a male BN/Bi rat. Note the cellular infiltration into the stalk of the tumor (↑). (HPS; magnification × 15.)

illary and contained only a few layers of urothelium over the papillary fronds of connective tissue. Most had areas of "piling-up" of cells, focal cellular atypia and pleomorphism and some showed clear invasion of the papillary stalk (Figure 4.84) and submucosa. The second type of bladder carcinomas were tumors consisting of transitional epithelium (Figures 4.85 and 4.86). They readily invaded surrounding tissues and were usually graded as P2 to P4 carcinomas. At times, focal squamous metaplasia was observed; however, unlike the ureters, bladders showed few squamous cell carcinomas. Diffuse and nodular hyperplasias were common in rats with tumors, but were rarely found in rats without tumors. Finally, multiple lesions were common in an individual bladder.

Metastases were observed in some rats with

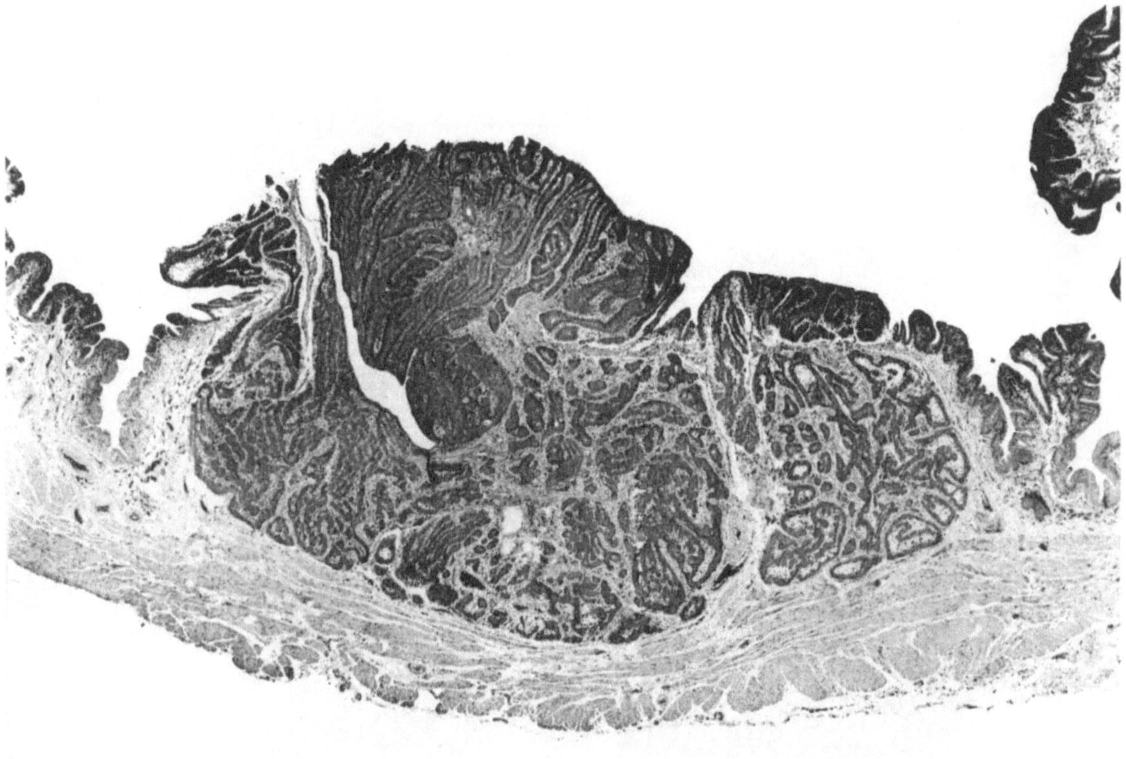

FIGURE 4.85. Transitional cell carcinoma in the bladder of a male BN/Bi rat. (HPS; magnification × 23.)

urothelial tumors of both the ureter and bladder (Table 4.20). The squamous cell tumors of the ureters in the males and females were all widely metastatic with metastases to the lymph nodes and lung. The two metastatic bladder cancers metastasized to local lymph nodes.

The incidence of urothelial tumors in BN/Bi rats is shown in table 4.20; 22% of the female BN/Bi rats died with neoplastic lesions of the ureter in contrast to 9% of the males. On the other hand, 35% of the males had cancer of the bladder in contrast to only 3% of the females.

The age-associated incidence is shown in Figure 4.87. There was a clear peaking of ureter tumors in males and females and bladder tumors in males. The rats dying in the oldest age periods were apparently at less risk of dying with these tumors than were their younger cohorts.

One papillary tumor from the bladder of a male BN/Bi rat and three squamous cell carcinomas from the ureter of three different female BN/Bi rats were transplanted subcutaneously into 6-week-old BN/Bi rats. The transplanted tumors grew to sizes of 2 to 4 cm within 2 months. All showed extensive and widespread local invasion and all metastasized in the recipient host to the regional lymph nodes and lungs. The three squamous cell tumors from the ureter retained their squamous features in the transplant and in the distant metastases of the transplant. The transplanted papillary bladder tumor grew as a transitional cell carcinoma without squamous metaplasia. Even in the metastases of this tumor in the recipient host, squamous differentiation was not observed.

In contrast to the BN/Bi strain, only one urothelial tumor was found in the other rats. A solitary, well-differentiated papilloma of the bladder was found in a 40-month-old male F_1 rat. Urothelial tumors were not recognized in any of the WAG/Rij or the female (WAG × BN)F_1 rats.

3. Discussion

Spontaneously occurring chronic progressive

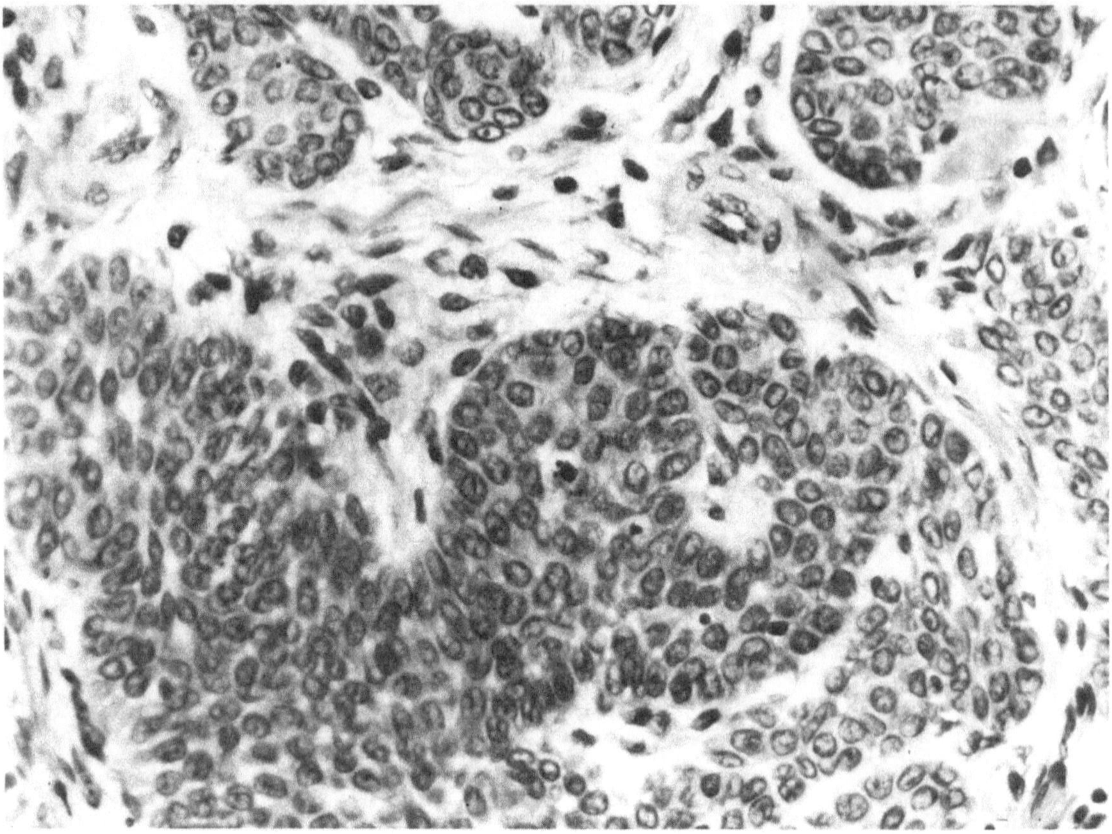

FIGURE 4.86. Higher magnification of the same tumor illustrated in Figure 4.85 to show the transitional cell appearance of the tumor cells. (HPS; magnification × 500.)

TABLE 4.20

Urothelial Tumors in the Ureter and Bladder From 236 Female and 74 Male BN/Bi Rats

Location of tumor	Sex	No. with tumor	%	Mean age (range) in months	No. with metastatic urothelial carcinoma	Age (in months) of rats with metastatic ureter or bladder carcinoma
Ureter carcinoma	Female	53	22	30 (17—39)	4	30 (26—33)
	Male	7	9	25 (16—35)	2	25, 26
Bladder carcinoma	Female	8	3	33 (27—40)	1	27
	Male	26	35	27 (16—43)	1	27

glomerulonephropathy has been reported in most aging rat strains. The light microscopic lesions have been described by several investigators.[2,13,45,52,74,156] The ultrastructural features of the disease have been documented,[45,74,101] scanning electron microscopy of the glomerulus evaluated,[6,7] and immunofluorescent studies reported.[16,45,46,77a] Recent review articles have alluded to the pathogenesis, gross and histological findings, ultrastructural features, and immunofluorescent studies.[4,41]

Many factors are capable of altering the onset and progression of the disease. As early as 1928, Newburgh and Curtis[122] were able to produce renal injury in rats by altering protein levels in the diet. Since then, others have changed the progression and severity of renal lesions in rats by changing various dietary

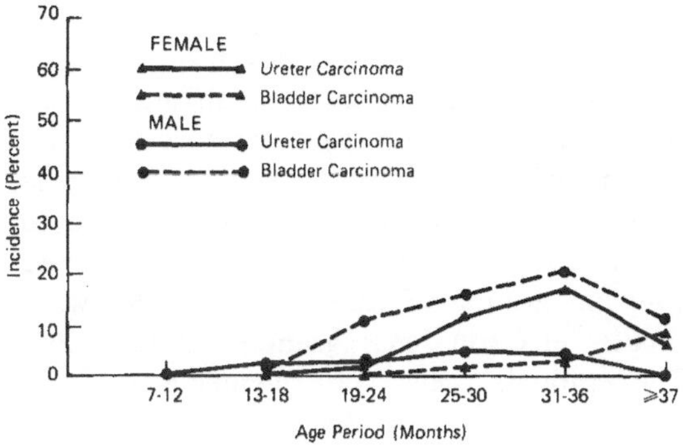

FIGURE 4.87. Percent incidence with age of bladder and ureter carcinomas in 236 female and 74 male BN/Bi rats.

factors.[10,105,123,150] The important role of nutrition in the development and progression of this disease has been adequately reviewed.[4,41]

Another factor that can influence the onset and progression of chronic glomerulonephropathy is the strain of rat. The BN/Bi, WAG/Rij, and (WAG × BN)F_1 rats in this study did not develop severe renal disease. A partially inbred stock of Sprague-Dawley rats was also produced and maintained at the REP Institutes. They were fed the same commercial diet as the aging rats reported in this study. Despite this, "typical" chronic progressive glomerulonephropathy has been observed.[83] Snell[156] noted that BUF and OM rats had less chronic renal disease than the MS20 and WN strains. Bolton et al.[16] also noted the importance of the genetic background of rats and the progression of chronic kidney lesions and proteinuria. Therefore, the relatively low background of chronic renal disease in the aging BN/Bi, WAG/Rij, and (WAG × BN)F_1 rats of this study is probably a strain-dependent phenomenon.

Congenital hydronephrosis in BN/Bi rats was studied by Cohen et al.,[40] who found a 31% incidence of the lesion. They also demonstrated the heritable nature of the condition in this strain. Newborn WAG/Rij and (WAG × BN)F_1 male rats have not been studied to determine the prevalence of congenital hydronephrosis, but it is likely that the + and + + hydronephrosis in these rats is congenital rather than an acquired age-associated lesion.

The aging BN/Bi rats described in this section had a 39% incidence of hydronephrosis in females and a 43% incidence in males. The higher incidence of severe hydronephrosis in aged BN/Bi rats is certainly secondary to the high incidence of ureter and bladder tumors and urolithiasis. In all groups, the males had more hydronephrosis than females of the same strain and the right kidney was more often affected than the left kidney in (WAG × BN)F_1 and male WAG/Rij rats.

Urothelial tumors in BN/Bi rats were previously reported by Boorman and Hollander[23] and have been discussed in two additional publications from the Institute for Experimental Gerontology.[24,28] The high incidence of spontaneously occurring urothelial cancers in this strain is in sharp contrast to the low incidence in the WAG/Rij and (WAG × BN)F_1 rats and to most other rat strains as reviewed by others.[1,77,163] The high incidence of spontaneous urothelial tumors in this strain, their ease of transplantability, and the nearly identical pathologic features of the tumors to those in humans make this strain a potential model to study urothelial cancers. It may also be useful to use this strain for testing bladder carcinogens and to compare results with known low-incidence strains.

In addition to urothelial cancers, urolithiasis was common in BN/Bi rats. They were the only rats to develop large stones that caused urinary obstruction. Therefore, a possible correlation could not be excluded between the occurrence of urolithiasis and the development of urothelial cancers in this strain. Boorman and Hollander[23] showed an apparent correlation between the two conditions in BN/Bi rats. Burek (see Chapter 6), however, was unable to show a similar correlation in 14- and 21-month-old female rats that were killed. In addition, Boorman[16a] took stones from BN/Bi rats that died with urothelial tumors and surgically placed them in the bladders of young BN/Bi rats. He did not find any increase in the incidence of bladder cancers in these rats compared to controls. Therefore, it is clear that both conditions, urolithiasis and urothelial cancers, occur in the BN/Bi rat strain. It remains unclear, however, if the stones predispose these rats to the development of urinary tract cancer. Additional studies are clearly needed to resolve this problem.

L. Hematopoietic System

1. Nonneoplastic Lesions

a. Thymus

The thymus was atrophic in all of the aging rats examined, but the degree of atrophy was strain, sex, and age dependent. Thymic involution in female BN/Bi rats was characterized by a loss of cortical lymphocytes and an apparent proliferation of epithelial structures, especially in rats older than 2 years (Figure 4.88). The majority of the gland was composed of a fibrovascular stroma, variable amounts of epithelial structures, and occasional small nests or foci of cortical lymphocytes. The epithelial elements formed cords and tubules that were surrounded by a basement membrane. A central lumen was present that contained a pink, homogeneous product. This product and the apical portion of many epithelial cells gave a strongly positive periodic-acid-Schiff reaction. The stroma between the epithelial structures was composed of blood vessels, supporting connective tissue, numerous plasma cells, some lymphocytes and macrophages, and occasionally mast cells.

Ultrastructural studies on the thymus of aged female BN/Bi rats revealed secretory activity in the thymus epithelium. Two types of granules were present in the apical portion of the cytoplasm. One was a large, membrane-bound mucous-like granule (Figure 4.89); the other was a smaller electron-dense granule (Figure 4.90). The epithelial cells were connected to each other by desmosomes. Individual cells had pinocytotic vesicles along the luminal margins, microtubules, prominent endoplasmic reticulum, and numerous mitochondria. Many lymphoid cells and macrophages were seen between the epithelial cells and within the lumen of cords and tubules.

Thymic atrophy in female WAG/Rij and (WAG × BN)F_1 rats was similar to that in the female BN/Bi, but the epithelial cells were less prominent or formed small cystic structures. Their thymuses also had a loss of cortical lym-

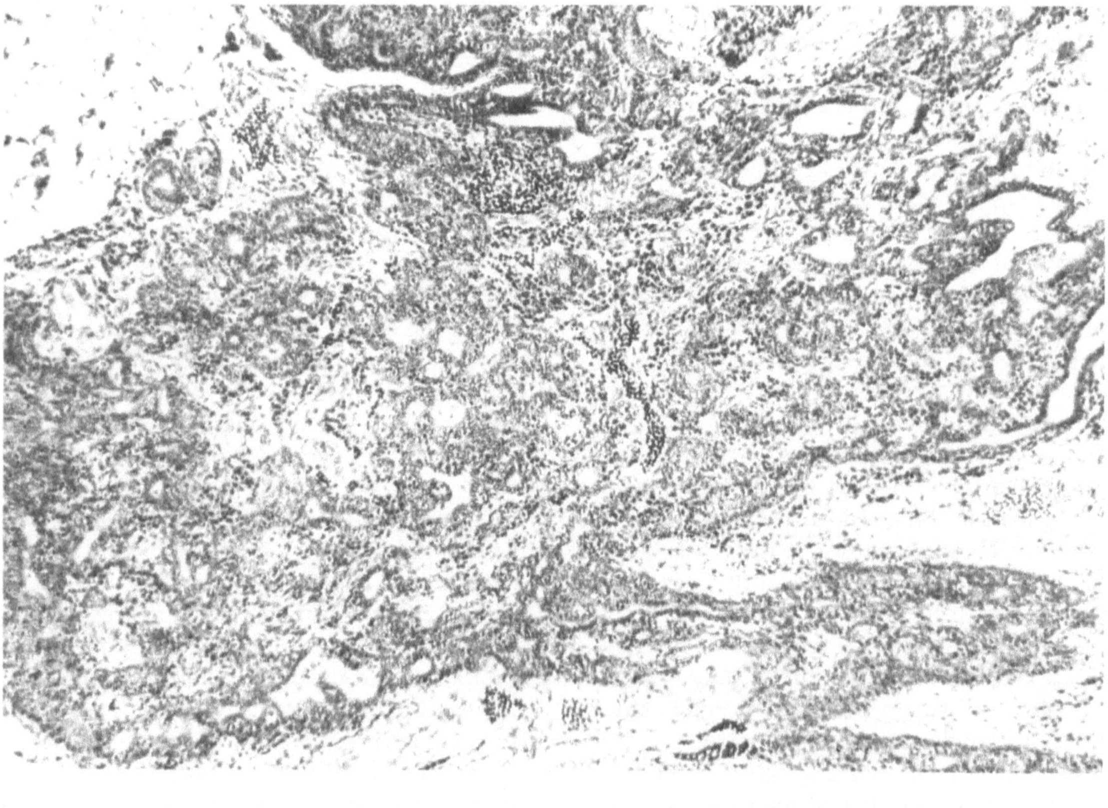

FIGURE 4.88. Atrophic thymus from a BN/Bi female with prominent epithelial cords and tubules. (HPS; magnification × 80.)

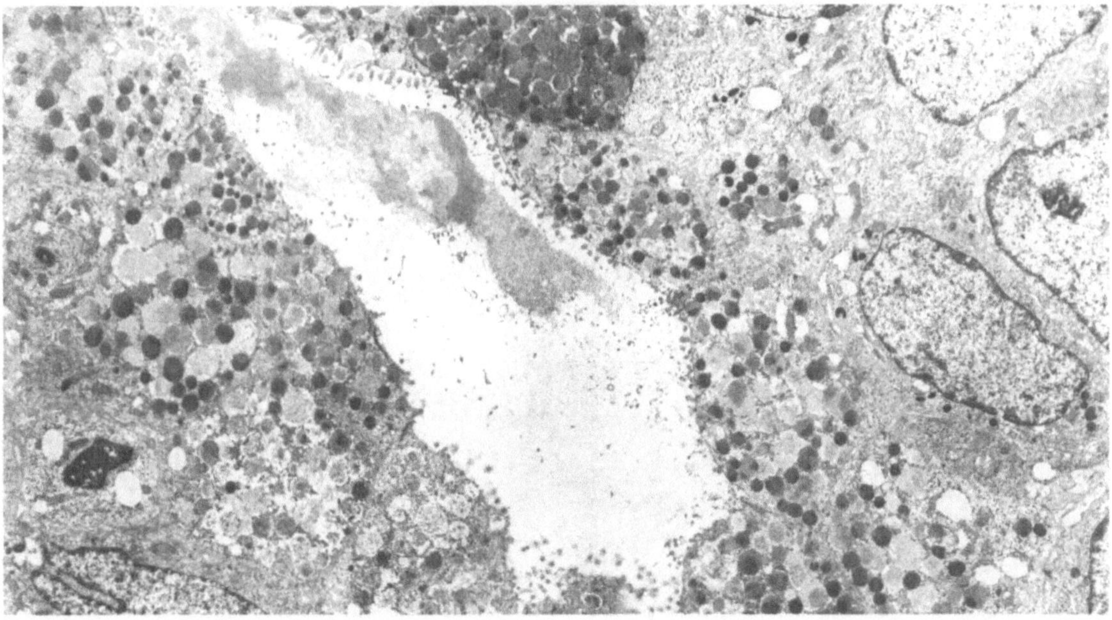

FIGURE 4.89. An epithelial tubule in the thymus from a BN/Bi female showing the numerous secretory granules in the cytoplasm of the epithelial cells. (Magnification × 5021.)

phocytes. Epithelial cords and tubules could be seen in most, but they were less prominent than in the BN/Bi females. Many of these females had one or more small cystic dilatations in the thymus (Figure 4.91). Their structures were lined by low-cuboidal nonciliated epithelium, contained a pink, homogeneous product, and appeared to be derived from the thymic epithelium. Such cystic structures were not a feature of the atrophic female BN/Bi thymuses despite the clearly proliferative appearance of the epithelium in that strain.

Male BN/Bi and (WAG × BN)F_1 rats showed a greater degree of atrophy than the females. Most of these males had thymuses composed of a few nests of cortical lymphocytes and only rarely were clearly identifiable epithelial cords or tubules present as small noncystic structures.

The most severe atrophy occurred in male WAG/Rij rats. The WAG/Rij males that died, regardless of their age, usually had severe, nearly complete atrophy of the thymus (Figure 4.92). The cortex was often absent and all that remained were small fibrovascular structures with a few scattered plasma cells and lymphocytes.

b. Lymph Nodes

Age-associated changes in lymph nodes were impossible to accurately assess based on the available routine material. Therefore, only a summary and brief discussion of impressions and trends will be presented.

In general, lymph nodes were inactive. There was an age-associated decrease in the number of germinal centers per lymph node. One of the most striking changes was that the medullary cords were usually densely packed with numerous plasma cells that often extended into the paracortical zones. The subcapsular and medullary sinuses had many hemosiderin containing macrophages as well as macrophages with phagocytized red blood cells. In many of the old rats, large histocytic cells with pink to red, coarsely granular cytoplasm were seen (Figure 4.93).

Several lymph nodes of (WAG × BN)F_1 rats had dilated, cystic sinusoids (Figure 4.94). The cysts varied from microscopic to 8 mm in size, usually contained a few cells and pink homogeneous material, and were lined by endothelial cells. They were found in one female 40 months old and in ten (15%) of the males. In males the mean age was 35 months with a range of 22 to

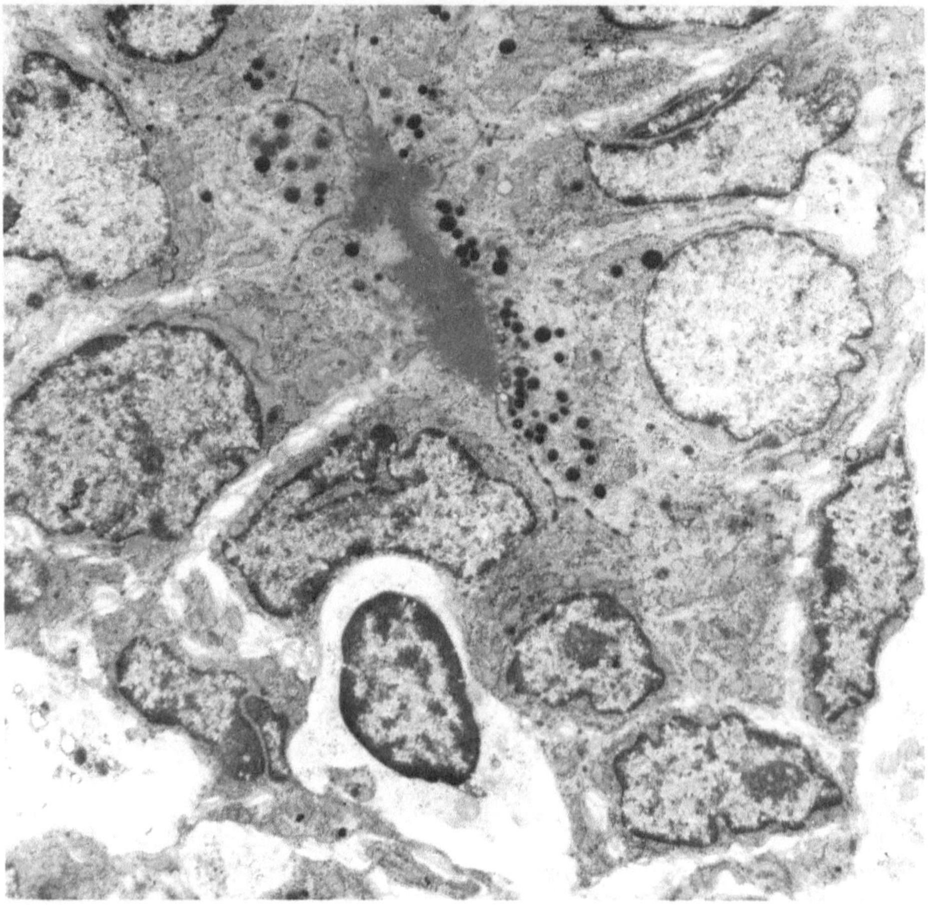

FIGURE 4.90. An epithelial tubule in the thymus from a BN/Bi female showing small dark granules confined to the apical portion of the epithelial cell cytoplasm. (Magnification × 6650.)

44 months. These tumors appear to be age associated, but their significance is unknown.

One 32-month-old male BN/Bi died with mesenteric lymph nodes that were grossly enlarged. Microscopically, the nodes were composed of multiple blood-filled cystic structures consistent with mesenteric disease.

c. Spleen

As was true with the lymph nodes, age-associated lesions in the spleen were neither readily observed nor adequately assessed. Increased hemosiderin accumulations, extramedullary hematopoiesis, or both were seen in the spleens of some aged rats. However, these changes were usually present only in rats with a large tumor located elsewhere in the body.

2. Neoplastic Lesions

Lymphoreticular tumors were observed in 6% of the female and 15% of the male BN/Bi rats, 7% of the female and 3% of the male WAG/Rij rats, and 3% of female and 7% of male F_1 rats. The types of tumors seen are shown in Table 4.21.

Myelomonocytic leukemias were common in BN/Bi rats. The individual cells were pleomorphic and blast-like and had round to slightly lobulated nuclei. Cytoplasm was usually observed and seemed to appear finely granular. As it was not possible to clearly differentiate these tumors as unequivocal myeloid or monocytic tumors, they were lumped into the category of myelomonocytic until additional histochemical and ultrastructural studies could be done. In general, these tumors infiltrated the red pulp of the spleen, sinusoids, and portal triads of the liver; the interstitial septae of the lungs, and the bone marrow. They did not form solid tumor masses, however.

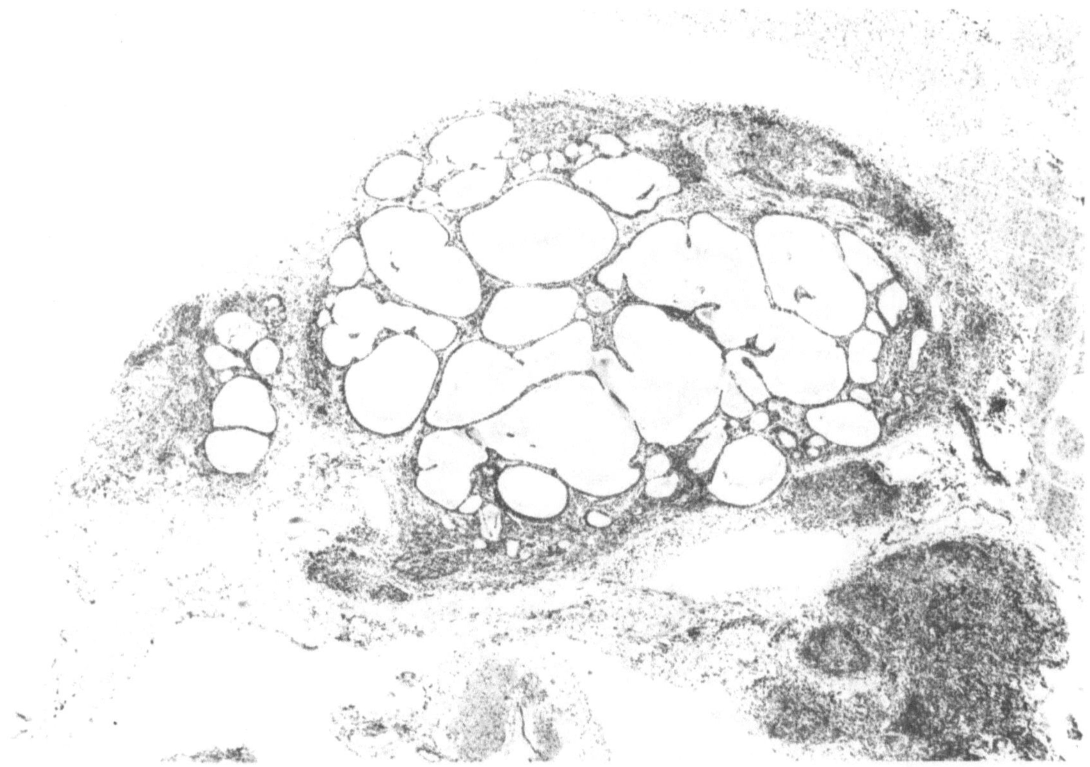

FIGURE 4.91. Atrophic thymus from a WAG/Rij female with cystic epithelial tubules filled with pink homogeneous material. (HPS; magnification × 40.)

Six tumors of the lymphoid cells were observed. They formed large masses in lymph nodes or other organs and were composed of lymphocytes or lymphoblasts.

Four thymomas were found. All were large masses in the anterior mediastinum and consisted of sheets of small lymphoid cells. No epithelial structures were recognized.

Seven histiocytic sarcomas were present. They were composed of sheets of large histiocytic-like cells. Their cytoplasm was abundant and pale, and phagocytized red blood cells or cellular debris was recognized. All were widespread throughout the body and formed large, firm masses.

A 29-month-old female WAG/Rij rat had a solitary plasmacytoma in a mesenteric lymph node. The entire node was replaced by well-differentiated plasma cells.

Finally, three tumors were considered unclassifiable because the cell of origin could not be determined.

Most of the lymphoreticular neoplasms occurred in rats with an average age of 28 months or older. Two exceptions to this statement were seen, however. First, the myelomonocytic leukemias in BN/Bi males were all present in rats younger than 30 months. The mean age was 21 months despite the fact that the 50% survival age was 28 months for the necropsied rats (see Table 3.3). Second, the four lymphoreticular tumors in male WAG/Rij rats occurred in rats 24 months old or younger.

3. Discussion

The organs of the lymphoreticular system were difficult to assess with regard to the age-associated nonneoplastic lesions. To accurately evaluate most of the observed changes would require morphometric evaluation at several different age groups. Despite this limitation, several patterns of changes were recognized. The most consistent were the thymic epithelial cell proliferations in BN/Bi females, cystic epithelial changes in thymuses of both the WAG/Rij and (WAG × BN)F_1 females, more complete

FIGURE 4.92. Severe thymic atrophy in a WAG/Rij male. (HPS; magnification × 19.)

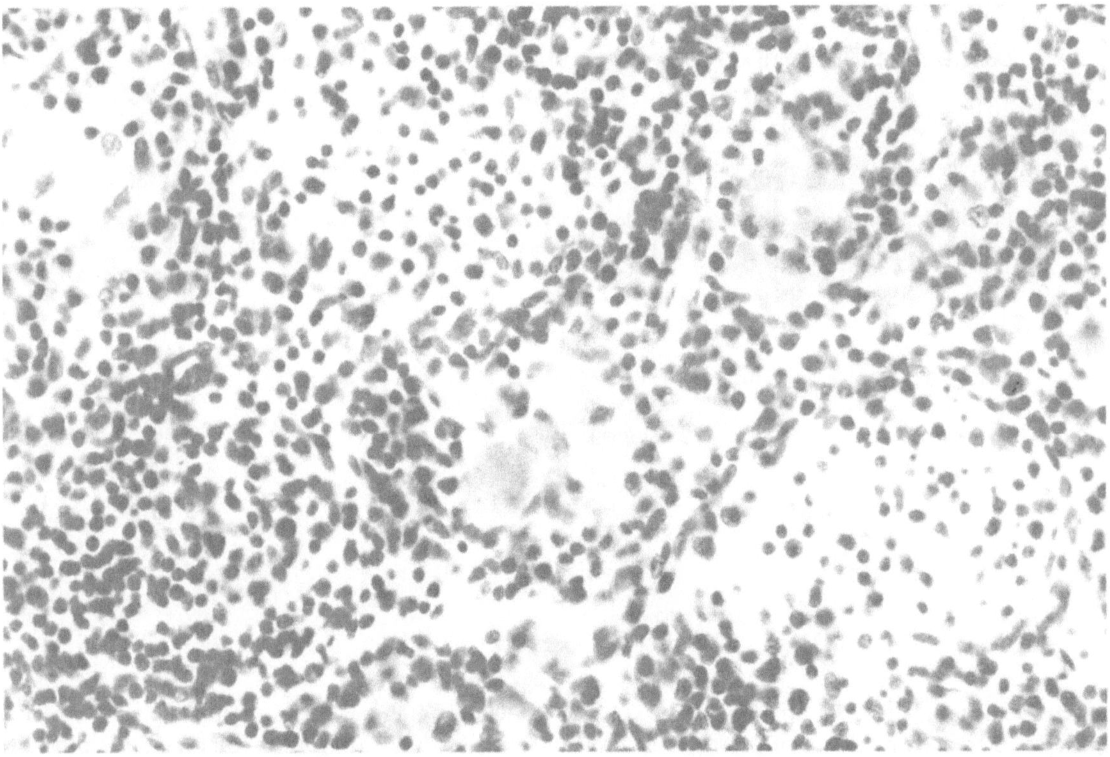

FIGURE 4.93. Large histiocytic cells filled with pink to red, coarsely granular cytoplasm in a lymph node from a BN/Bi female. (HPS; magnification × 465.)

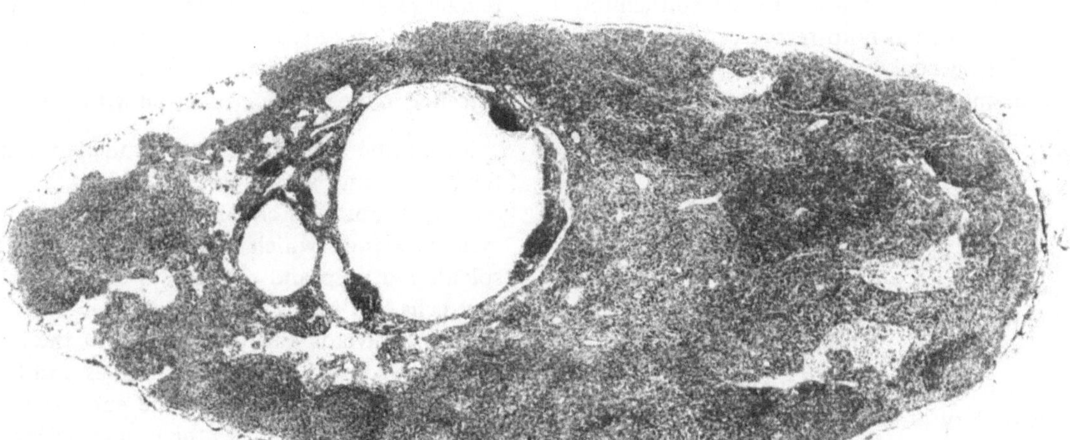

FIGURE 4.94. Lymph node from a WAG/Rij male showing a small cystic sinusoid. (HPS; magnification × 30.)

TABLE 4.21

Occurrence of Lymphoreticular Tumors in Aging BN/Bi, WAG/Rij, and (WAG × BN)F_1 Rats

Strain	Sex	No. examined	Diagnosis	No. of each diagnosis	%	Mean age (range) in months
BN/Bi	Female	236	Myelomonocytic leukemia	12	5	28 (19—34)
			Lymphosarcoma	2	1	32, 33
			Unclassifiable	1	<1	36
	Male	74	Myelomonocytic leukemia	8	11	21 (17—26)
			Lymphosarcoma	2	3	32, 39
			Unclassifiable	1	1	29
WAG/Rij	Female	101	Thymoma	4	4	36 (32—44)
			Histiocytic sarcoma	2	2	36, 37
			Plasmacytoma	1	1	29
	Male	124	Histiocytic sarcoma	2	2	20, 24
			Lymphosarcoma	1	1	21
			Unclassifiable	1	1	16
F_1	Female	68	Myelomonocytic leukemia	2	3	17, 31
	Male	67	Histiocytic sarcoma	3	4	35 (28—42)
			Lymphosarcoma	1	1	29
			Thymoma	1	1	33

and earlier onset of thymic atrophy in WAG/Rij males, increased numbers of plasma cells in the lymph nodes of all groups of rats, and decreased size and number of germinal centers in lymph nodes and spleen.

Proliferation of thymic epithelium in rats has been described by Ross and Korenchevsky,[139] Plagge,[126] and Cherry et al.[37] These studies have also shown that hormones such as estrogen can result in a proliferation of thymic epithelium, while others such as testosterone can cause a decrease. Cherry et al.[37] also found epithelial cysts in the thymuses of nearly all rats more than 1 year of age. The BN/Bi female rats reported in this thesis had the most impressive thymic epithelial proliferations, but few cystic structures. The WAG/Rij and (WAG × BN)F_1 females, on the other hand, clearly had less epithelial proliferation, but epithelial cysts were frequent.

Thymic epithelial cells in aging BN/Bi females have been studied in greater detail.[30,115] They have reported additional ultrastructural studies and autoradiography studies following pulse labeling with tritiated leucine. They have shown that the epithelial cells secrete a product. In addition, lymphocytes and macrophages appear to migrate freely into and out of the epithelial tubules and acinar structures.

WAG/Rij males had the most severe thymic atrophy and the onset seemed to occur earlier than in the other groups of rats. This observation led to additional studies to determine if such morphological observations could be correlated with functional studies. Such studies were done by Kruisbeek et al.[103] They were able to show that T-lymphocytes from 18-month-old WAG/Rij male rats had significantly lower stimulation to Con A and PHA than WAG/Rij female rats 18 months of age. Therefore, these preliminary results suggested that the morphological observations did, in fact, correlate with a decline in the immune function with age.

Neoplastic lesions occurred in the lymphoreticular system, as shown in Table 4.21. The most common was classified as a myelomonocytic leukemia, which seemed to begin in the splenic red pulp and resulted in circulating tumor cells and infiltration of the liver sinusoids and portal triads and interstitium of the lung. Only sporadically were lymph nodes and bone marrow involved. The cell type and organ involvement were similar to the mononuclear cell leukemia reported by Maloney et al.[117a]

Recent reviews of spontaneous lymphoreticular tumors in rats include those by Squire and Goodman,[163] Swaen and van Heerde,[169] and Gössner et al.[72]

M. Female Reproductive System

1. Nonneoplastic Lesions

a. Ovaries

Ovaries from all females over 18 months old were atrophic, but the degree of atrophy was variable. To discuss the findings thoroughly would require a much more detailed discussion than can be presented here; therefore, only a few salient features of ovarian atrophy and some strain differences will be presented.

Atrophy of the ovaries in BN/Bi, WAG/Rij, and (WAG × BN)F_1 rats was characterized by a decreased number of normal or mature follicles, an increased prominence or actual increased number of interstitial cells, and a general appearance of inactivity (Figure 4.95). In most, only a few follicles were present and they were usually located around the periphery. Ova were seldom observed in follicles, but they were found in a few rats and even in some older than 40 months. Similarly, only a few atretic follicles were present. Copora lutea were seen infrequently and when present numbered between one and three. They were seen occasionally in all age groups. The majority of the atrophic ovaries consisted of various amounts of haphazardly arranged interstitial cells and numerous phagocytic cells with their cytoplasm filled with yellow or golden-brown lipofuscin-like pigment (Figure 4.96) and a fibrovascular stroma.

In addition to the typical pattern of atrophy that was observed in all groups of rats, there were a few findings that were clearly strain dependent. For example, approximately 40% of the female WAG/Rij rats had areas in their ovaries that had an apparent proliferation of the granulosa-thecal cell component of interstitial cells (Figure 4.97) with an increase in Sertoli-like tubules and cords (Figure 4.98). Such areas were only rarely found in (WAG × BN)F_1 or BN/Bi rats. Similarly, nearly all WAG/Rij rats with active-appearing ovaries had distinct glandular structures lined by Sertoli-like cells

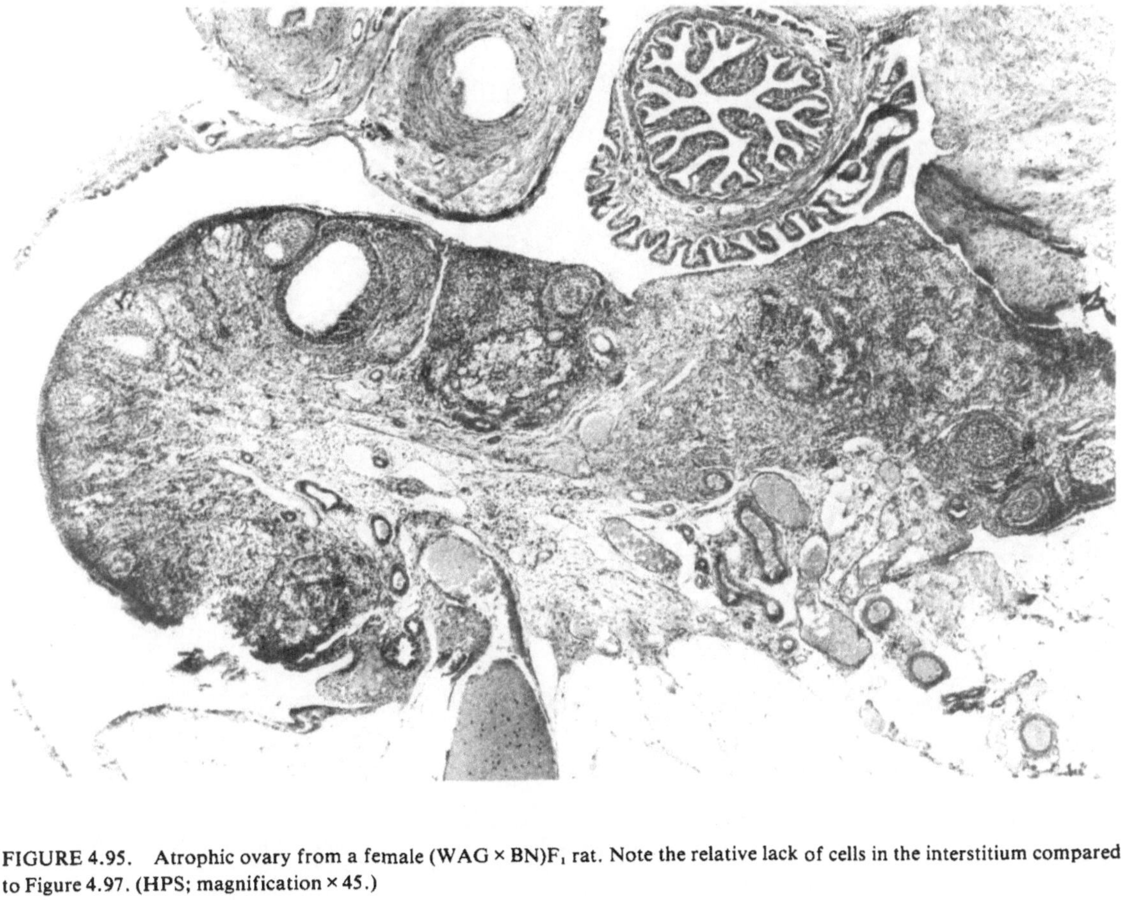

FIGURE 4.95. Atrophic ovary from a female (WAG × BN)F_1 rat. Note the relative lack of cells in the interstitium compared to Figure 4.97. (HPS; magnification × 45.)

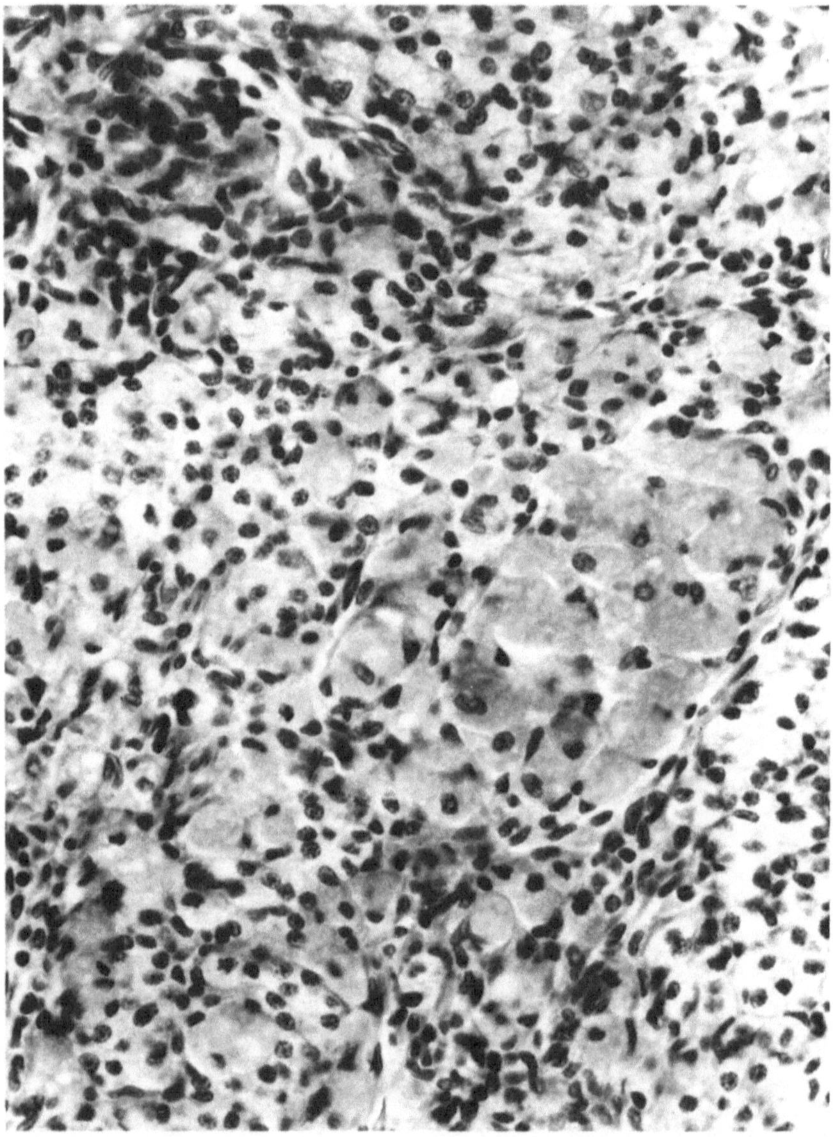

FIGURE 4.96. Pigmented cells in the interstitium of the ovary. (HPS; magnification × 500.)

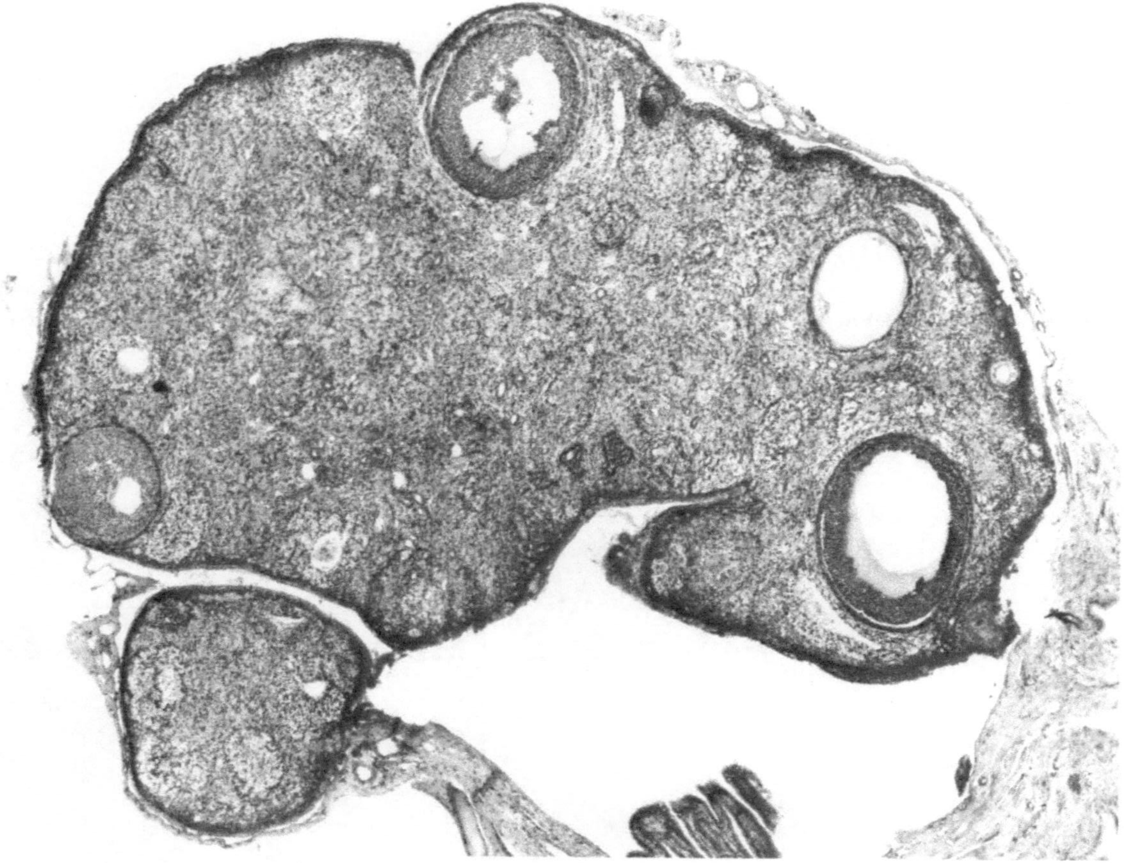

FIGURE 4.97. Atrophic ovary from a female WAG/Rij rat. Note the prominence of the cells in the interstitium. Nests of Sertoli-like cells and clumps of pigmented cells are present. (HPS; magnification × 45.)

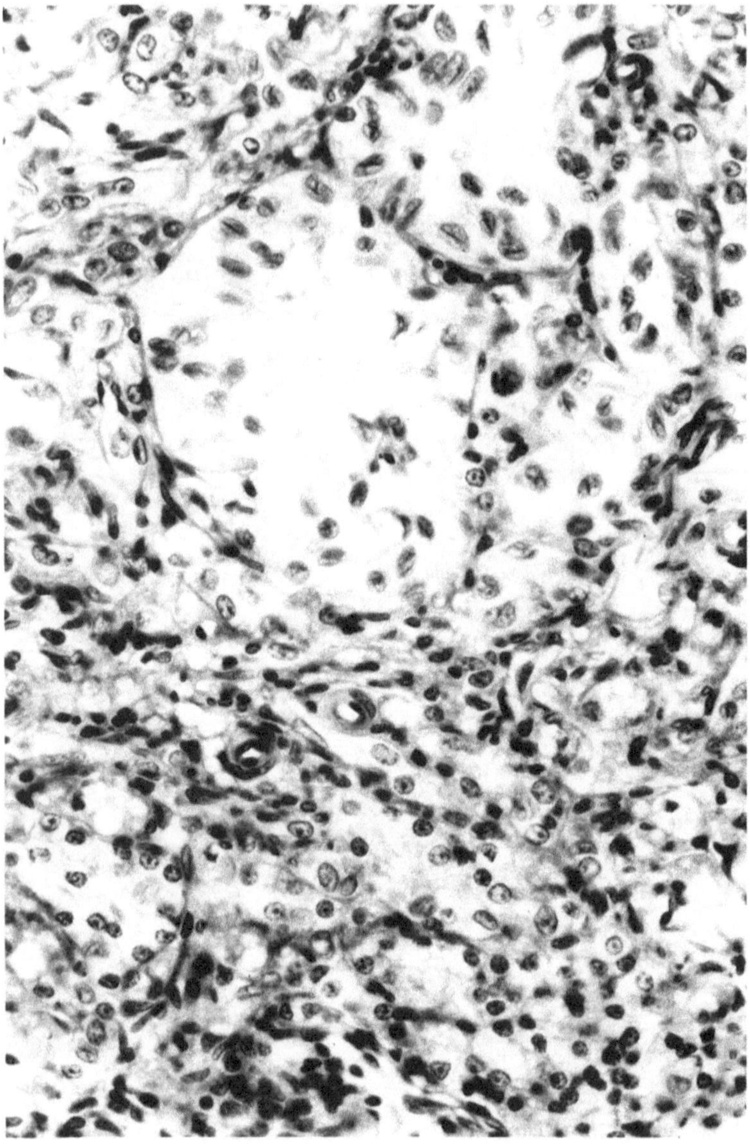

FIGURE 4.98. Nests of Sertoli-like cells (right) and interstitial cells (left) in the ovary of a WAG/Rij female. (HPS; magnification ×465.)

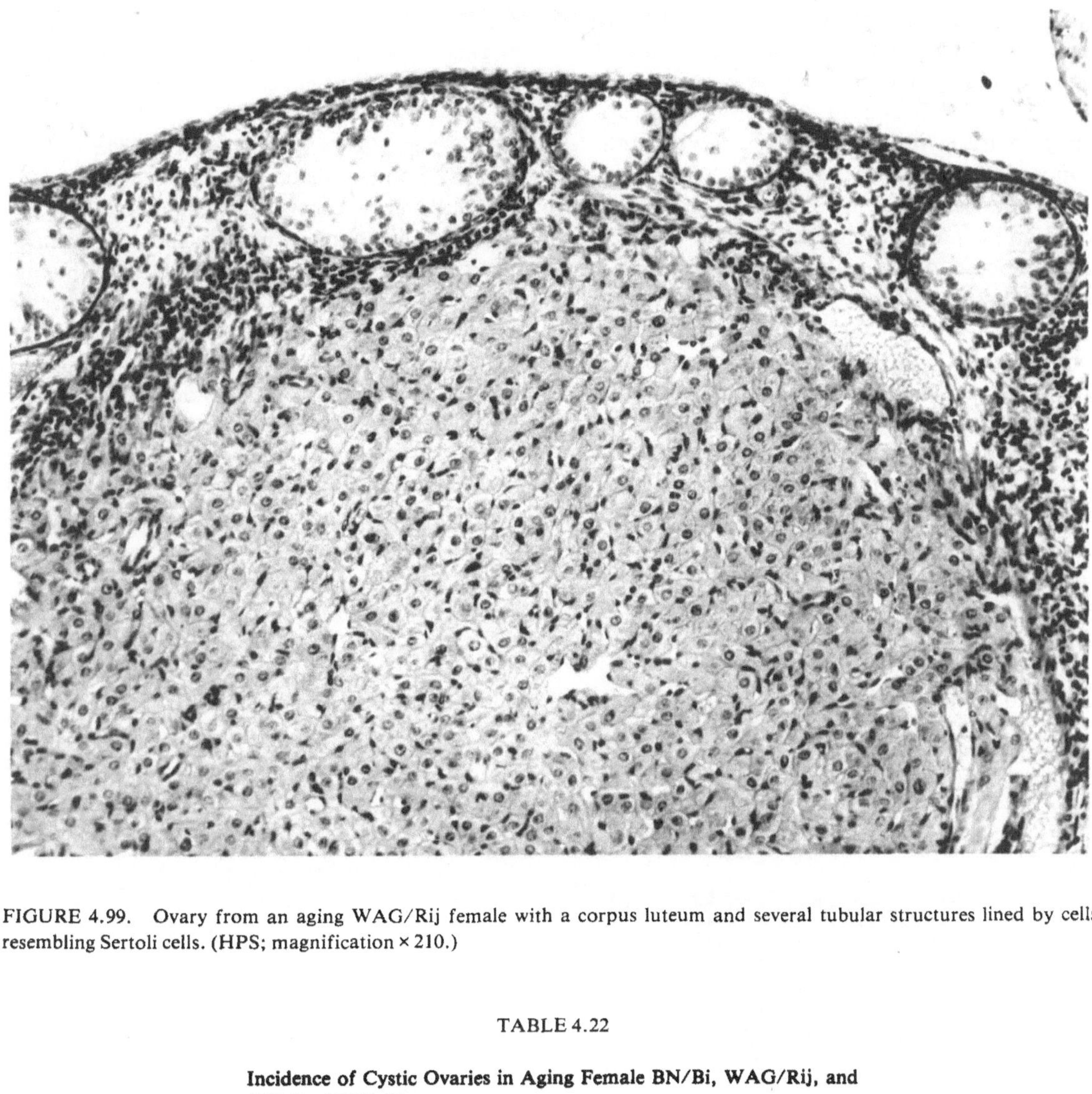

FIGURE 4.99. Ovary from an aging WAG/Rij female with a corpus luteum and several tubular structures lined by cells resembling Sertoli cells. (HPS; magnification × 210.)

TABLE 4.22

Incidence of Cystic Ovaries in Aging Female BN/Bi, WAG/Rij, and (WAG × BN)F_1 Rats

Strain	No. examined	No. with ovarian cysts	%	Mean age (range) in months
BN/Bi	236	7	3	32 (25—38)
WAG/Rij	101	55	54	34 (26—44)
F_1	68	13	19	30 (25—38)

(Figure 4.99). Such structures were rarely seen in (WAG × BN)F_1 or BN/Bi rats. At times, the Sertoli-like cells were very prominent in the areas of granulosa-thecal cell proliferation and often appeared to constitute the major cell type in these areas.

Ovarian cysts were another strain-dependent finding in atrophic ovaries of aging female rats. As shown in Table 4.22, 54% of the female WAG/Rij rats, 19% of the female F_1 rats, and 3% of the female BN/Bi rats died with ovarian cysts. They were grossly visible and ranged in size from about 5 mm to 2 cm. However, most were about 1 cm in diameter. They were usually unilocular, contained clear to amber-colored fluid, and appeared to involve the ovary. Most were unilateral but some were bilateral. Microscopically, it was difficult to localize the site of origin of all of the cysts. Some were clearly in the region of the ovarian stalk, but some appeared to arise within the ovary. All were similar in appearance irrespective of location, however. They were lined by simple squamous or rarely simple cuboidal epithelium and the wall was composed of smooth muscle, connective tissue, or both (Figure 4.100). None of these cysts contained follicular or luteal cells lining them. Therefore, most or all were morphologically consistent with paraovarian cysts. Rarely, true follicular cysts were seen, but they were small and lined by follicular epithelial cells of variable thickness. Such follicular cysts were not included in the tabulation of Table 4.22 or Figure 4.101.

Regardless of their origin, ovarian cysts were common in WAG/Rij females and relatively uncommon in BN/Bi females (Table 4.22). The age-associated incidence (Figure 4.101) shows a peak incidence of 70% in female WAG/Rij rats over 37 months of age. The incidence in female F_1 rats showed a plateau between 15 to 20% after 24 months of age.

b. Oviduct, Uterus, Cervix, and Vagina

Many nonneoplastic lesions were seen in the oviduct, uterus, cervix, and vagina of these aging rats. The changes, usually sporadic or nonspecific, were found in less than 10% of the rats and were not tabulated to determine age-associated incidence and relative prevalence in the different rat strains. Such lesions included pyometra, metritis, cystic endometrial hyperplasia, and mucometra. Mild inflammations were also seen and were characterized by a few inflammatory cells such as neutrophils and lymphocytes in the lumen, mucosa, or submucosa of the uterus, cervix, or vagina.

2. Neoplastic Lesions

a. Ovary

Seven primary tumors of the ovary were found in this series, as shown in Table 4.23.

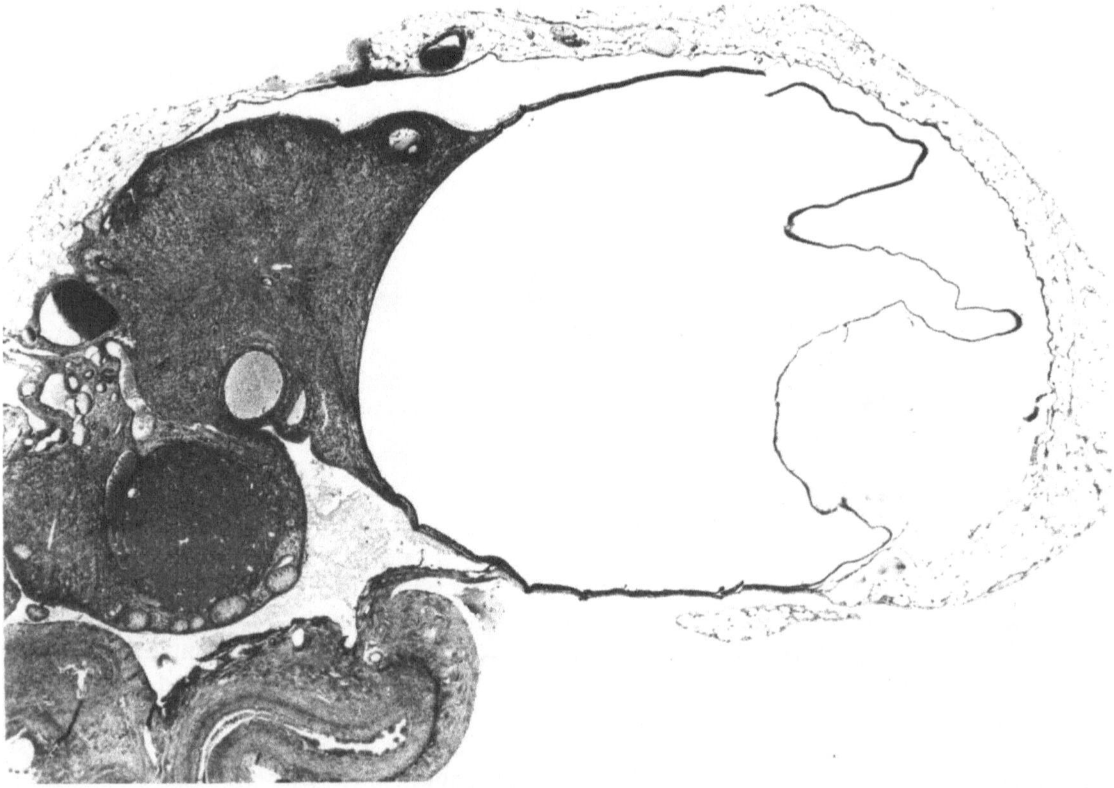

FIGURE 4.100. Cyst in the ovary of a female WAG/Rij rat. (HPS; magnification ×48.)

Grossly they were unilateral, firm, white nodules about 1 to 1.5 cm in diameter. All were diagnosed as granulosa-theca cell tumors. They were composed of a mixture of granulosa cells and thecal cells, but the thecal element appeared to be the most prominent. The cells were usually spindle shaped and were arranged in whorls or sheets separated by bands of collagen. The individual cells were variable in size and shape. Some had vacuolated, pink cytoplasm and indistinct cell boundaries, while other cells were fibroblast-like. Mitoses were often numerous, but no metastases were found.

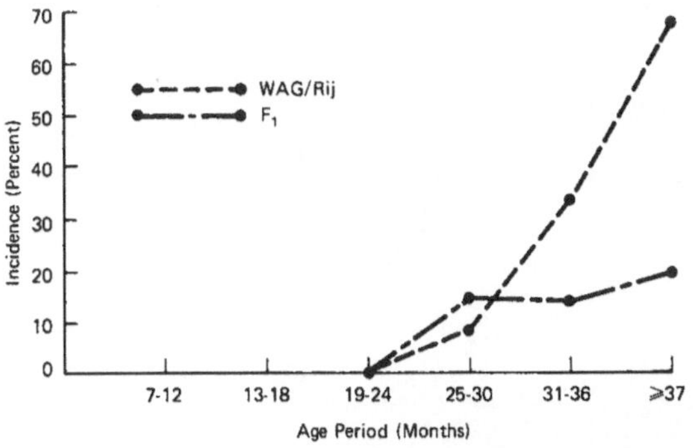

FIGURE 4.101. Percent incidence with age of ovarian cysts in 101 female WAG/Rij and 68 female (WAG × BN)F_1 rats.

b. Oviduct

Only one tumor was identified as clearly originating in the oviduct. It was a leiomyoma in a 30-month-old female BN/Bi rat. The tumor was in the wall of the oviduct, was composed of well-differentiated smooth muscle cells, and was clearly demarcated from the normal musculature.

A second tumor appeared to arise in the region of the oviduct of a 34-month-old WAG/Rij rat. The tumor was composed of undifferentiated spindle-shaped mesenchymal cells, had numerous mitoses, and had infiltrated the ovary, oviduct, and portions of the uterine horn, so that the precise origin could not be determined.

c. Uterine Horns

The most common tumors in the uterine horns were endometrial stromal polyps which were found in 17% of the WAG/Rij, 4% of the BN/Bi, and 3% of the F_1 rats (Table 4.24). Grossly, the polyps were seen as localized swellings or thickenings of the unopened uterine horn. When the uterus was opened, the polyps

TABLE 4.23

Incidence of Granulosa-thecal Cell Tumors in the Ovary of Aging Female BN/Bi, WAG/Rij, and (WAG × BN)F_1 Rats

Strain	No. examined	No. with granulosa-thecal tumors	%	Mean age (range) in months
BN/Bi	236	3	1	37 (35—39)
WAG/Rij	101	3	3	38 (31—46)
F_1	68	1	1	30

TABLE 4.24

Incidence of Endometrial Polyps in the Uterus of Aging Female BN/Bi, WAG/Rij, and (WAG × BN)F_1 Rats

Strain	No. examined	No. with polyps	%	Mean age (range) in months
BN/Bi	236	9	4	33 (25—37)
WAG/Rij	101	17	17	34 (24—39)
F_1	68	2	3	25, 31

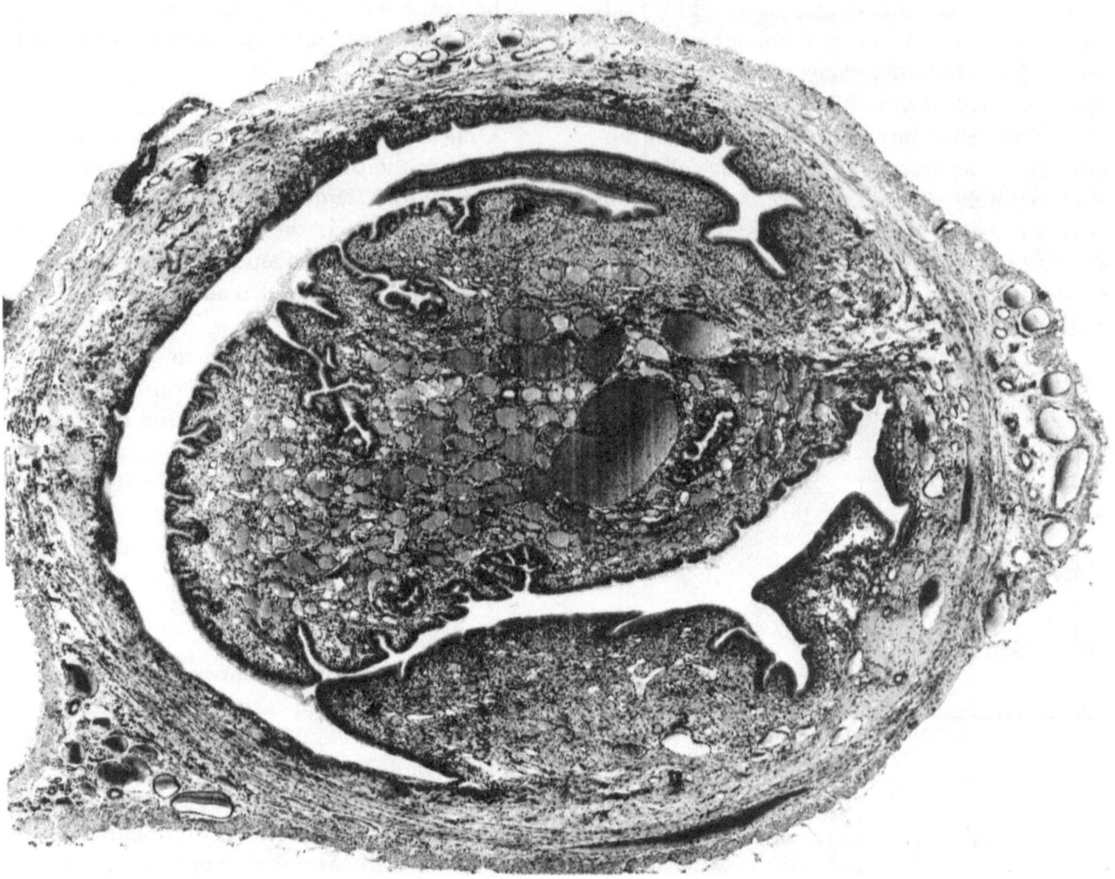

FIGURE 4.102. Uterus with an endometrial polyp. (HPS; magnification × 38.)

were firmly attached to the uterus by a stalk. Their color varied; some were white or tan and others were dark red and hemorrhagic. The size ranged from a few millimeters to more than 1.5 cm in diameter. Most were solitary, but multiple polyps in one rat were not uncommon.

Microscopically they consisted of polypoid masses lined by normal or nearly normal endometrial epithelial cells. Their stroma varied in appearance, however. The typical polyp had a fibrovascular stroma with abundant collagen and blood vessels (Figure 4.102). Some were composed of smooth muscle cells in addition to the fibrous and vascular elements. Mitoses were observed in the stroma of some lesions; focal necrosis and inflammatory cells were also seen. The larger polyps often had focal loss or ulceration of the epithelial cell lining.

The female WAG/Rij rats were the only group to develop sufficient numbers of endometrial polyps to determine the age-associated incidence. This is shown in Figure 4.103 with the peak being 24% in rats older than 37 months.

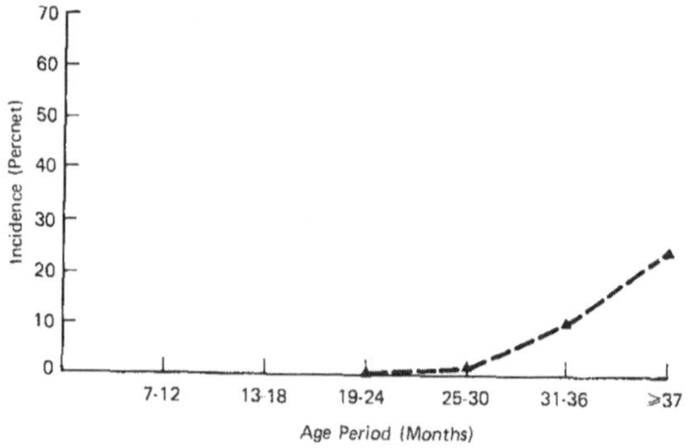

FIGURE 4.103. Percent incidence with age of endometrial polyps in 101 female WAG/Rij rats.

The occurrence of tumors (other than endometrial polyps) at different ages, the different diagnoses, and the number of tumors with metastases are shown in Table 4.25. Of the ten uterus tumors recognized, only two were in rats younger than 30 months. The histological features of the leiomyomas, leiomyosarcomas, and round cell sarcomas were similar to those observed in the cervix and vagina and are briefly described in Section II.M.2.e.

In addition to the smooth muscle tumors, three other sarcomas were found. One was in a 36-month-old BN/Bi rat and appeared histologically as an undifferentiated sarcoma composed of small spindle cells, numerous mitoses, and extensive local invasion of adjacent tissues. The second was a malignant hemangioendothelioma in a 40-month-old F_1 rat, which contained numerous large and small channels or spaces that were filled with blood and were lined by pleomorphic endothelial cells. It was locally invasive, but metastases were not observed. The third occurred in a 37-month-old female WAG/Rij and was diagnosed as a fibroleiomyosarcoma because it was composed of a mixture of fibrous and smooth muscle elements.

Two WAG/Rij females (Table 4.25) developed well-differentiated adenocarcinomas from the endometrium of the uterus. Both were composed of gland-like nests and clusters of epithelial cells and had locally invaded surrounding tissue including the myometrium. Despite the local invasion, distant metastases were not found.

d. Cervix and Vagina

The incidence of spontaneous tumors of the cervix, vagina, or both was 19% in BN/Bi rats, 3% in F_1, and 0% in the WAG/Rij rats (Table 4.26), with half of the tumors in rats older than 30 months.

TABLE 4.25

Incidence and Type of Tumor (Other than Endometrial Polyps) in the Uterus of Aging Female BN/Bi, WAG/Rij, and (WAG × BN)F_1 Rats

Strain	No. examined	Type of tumor	No. of each tumor	%	Mean age (range) in months	No. with metastatic tumor	Age (in months) of rats with metastatic uterine tumor
BN/Bi	236	Undifferentiated sarcoma	1	<1	36	0	—
WAG/Rij	101	Adenocarcinoma	2	2	35, 36	0	—
		Fibroleiomyosarcoma	1	1	37	0	—
		Leiomyosarcoma	2	2	19, 36	1	19
		Round cell sarcoma	1	1	38	0	—
F_1	68	Leiomyoma	1	1	31	0	—
		Malignant hemangioendothelioma	1	1	40	0	—
		Round cell sarcoma	1	1	25	0	—

TABLE 4.26

Incidence and Type of Tumors in the Cervix, Vagina, or Both from Aging Female BN/Bi, WAG/Rij, and (WAG × BN)F_1 Rats

Strain	No. examined	No. of rats with 1 or more tumors	%	Type of tumor	No. of each tumor	Mean age (range) in months	No. with metastatic tumor	Age (in months) of rats with metastatic cervix or vaginal tumor
BN/Bi	236	44	19	Squamous cell carcinoma	6	30 (24—36)	1	26
				Leiomyoma	4	34 (27—40)	0	—
				Leiomyosarcoma	9	31 (17—37)	1	26
				Round cell sarcoma	20	26 (18—38)	1	36
				Diffuse stromal sarcoma	5	27 (24—30)	0	—
				Anaplastic sarcoma	2	31, 34	2	31, 34
WAG/Rij	101	0	0	—	—	—	—	—
F_1	68	2	3	Round cell sarcoma	2	25, 33	0	—

Forty mesenchymal cell tumors were recognized in this study and the salient histological features are summarized below. Leiomyomas were composed of elongated spindle cells with pink cytoplasm, distinct cell boundaries, and long slender nuclei with well-rounded ends. Leiomyosarcomas were similar to leiomyomas, but had numerous mitoses, exhibited cellular pleomorphism, and had locally invaded adjacent tissues (Figure 4.104). Round cell sarcomas were composed of round to polygonal cells with indistinct cell boundaries and had an epithelioid appearance with pale pink or vacuolated cytoplasm (Figures 4.105 and 4.106). The exact cell of origin for these tumors is unknown, but they seemed to be a variant of leiomyosarcoma. Anaplastic sarcomas were made up of pleomorphic cells with numerous, often bizarre mitoses and many multinucleated giant cells. Diffuse stromal sarcomas arose in the submucosa, were usually diffuse rather than localized, and were composed of small round or slightly spindle-shaped cells that were loosely arranged. Giant cells, mitoses and invasion or extension into the outer muscular tunics were also features of this tumor.

In addition to mesenchymal cancers, six BN/Bi rats had squamous cell carcinomas of the cervix and vagina. These lesions were composed of sheets and cords of squamous cells, keratin pearls were frequent and mitoses common, and local invasion was seen in all cases. Some rats also had focal areas of atypical hyperplasia of the cervical and vaginal epithelium; these were usually associated with an underlying sarcoma.

Two cervicovaginal sarcomas from two different aging BN/Bi females were transplanted into 6-week-old female BN/Bi recipients. Both of the primary tumors were typical round-cell sarcomas. The transplanted tumors grew rapidly reaching sizes of several centimeters in diameter within 4 to 6 months. Histologically, the

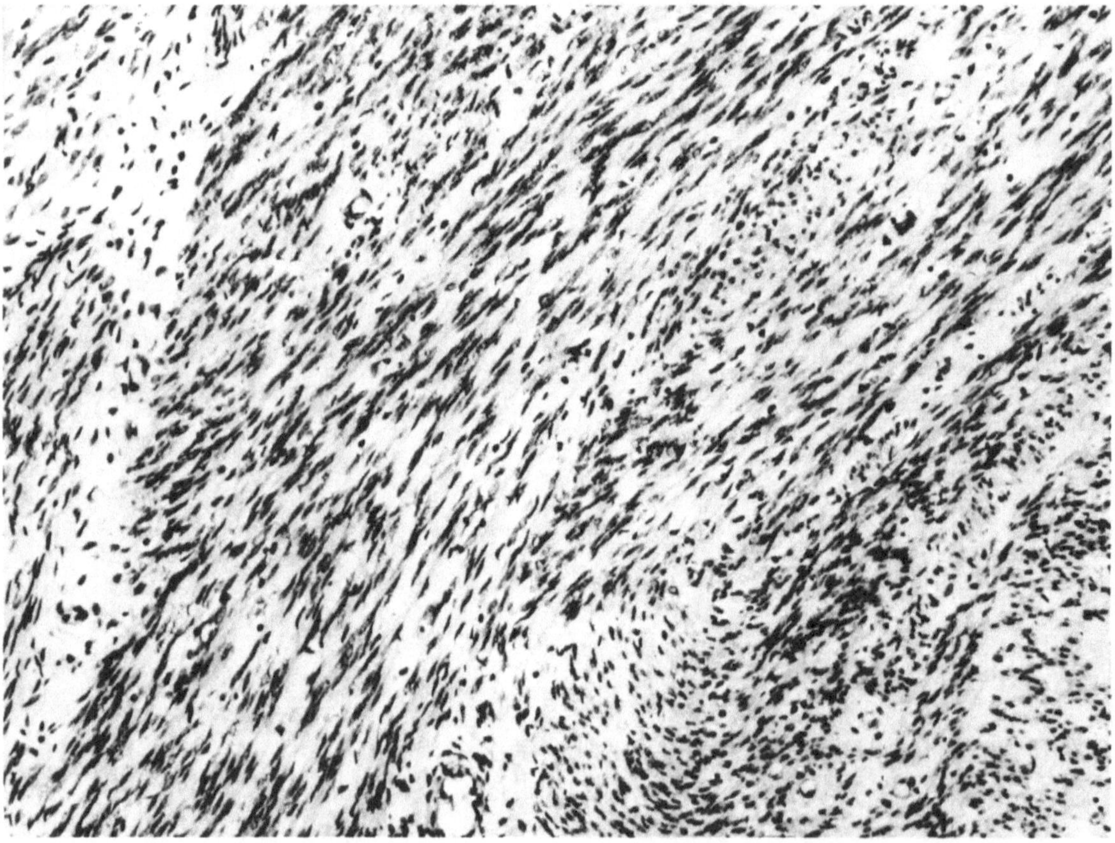

FIGURE 4.104. Leiomyosarcoma in the cervix of a female BN/Bi rat. (HPS; magnification × 210.)

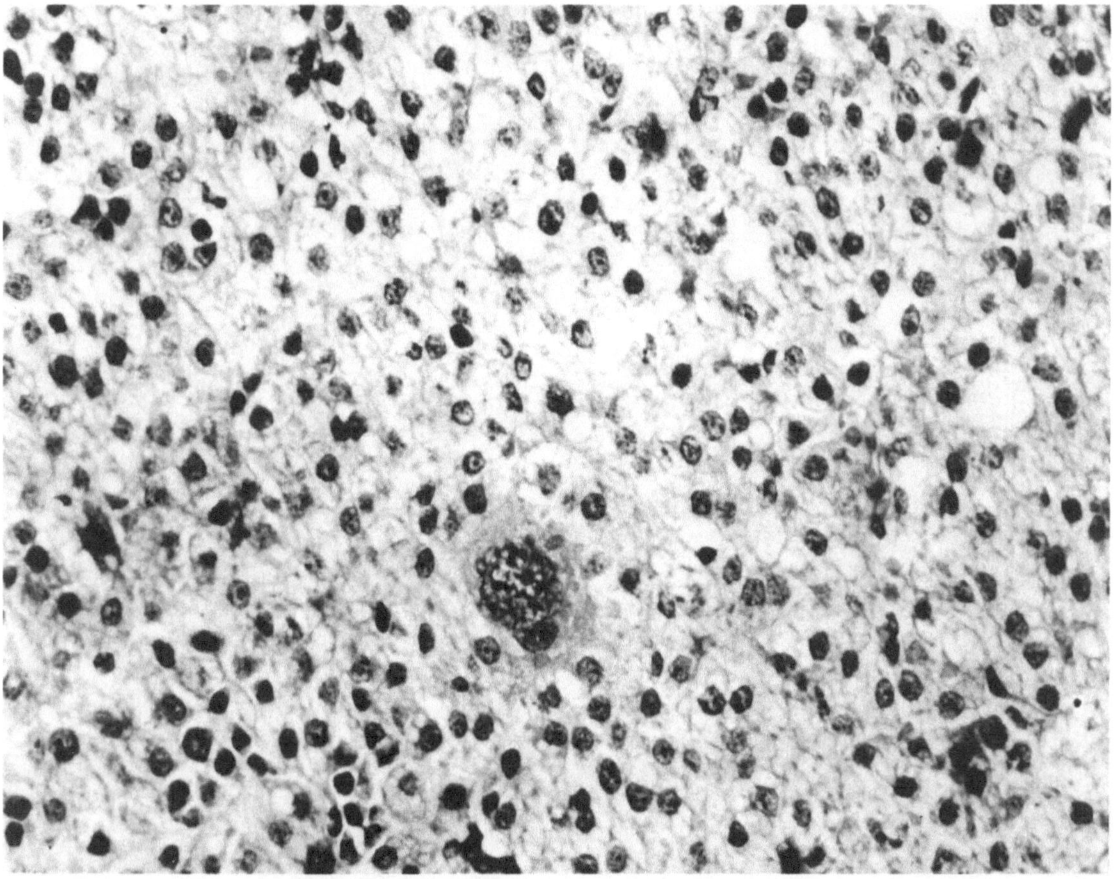

FIGURE 4.105. Round cell sarcoma in the vagina of a female BN/Bi rat. Note the vacuolated appearance to the tumor cells. (HPS; magnification × 500.)

transplanted tumors were locally invasive and were composed of sheets of elongated spindle cells with long slender nuclei having blunt ends. The morphological pattern of the transplants resembled leiomyosarcomas rather than the round-cell pattern of the primary tumors (Figure 4.107)

3. Discussion

A detailed review of the age-associated nonneoplastic changes in the female reproductive system is beyond the scope of this discussion. A great deal of individual variation and strain variations occurs in aging female rats. Thung[171] reviewed the literature and presented a detailed description of the microscopic structure of the ovaries of aging mice and also discussed observations in aging rats. Others have tried to correlate histologic observations with functional changes in aging rat ovaries as reviewed by Crumeyrolle-Arias et al.,[48] and still others[4,147] have summarized morphological and functional changes in the reproductive system of aging female rats. Finally, others have tried to correlate changes in the ovaries, uterus, mammary glands, and pituitary glands with abnormal estrus cycles in rats[116] and have evaluated the reproductive capacity of aging female rats.[86]

The present study showed changes in aging female BN/Bi, WAG/Rij, and (WAG × BN)F_1 rats that were similar to the observations of the investigators noted in the previous paragraph. Some rats had atrophic ovaries with no obvious follicular or luteal elements. Most had some evidence of ovarian activity based on the presence of one or more corpora lutea and follicles with ova. The interstitial elements varied among the three strains. WAG/Rij females

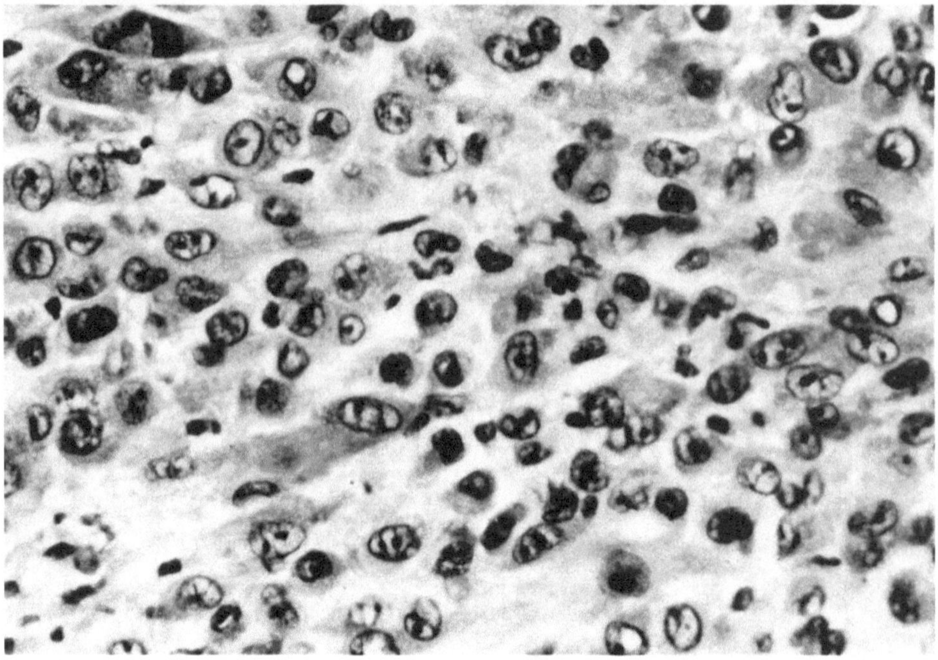

FIGURE 4.106. Round cell sarcoma in the vagina of a female BN/Bi rat that is less vacuolated and more pleomorphic than that illustrated in Figure 4.105. (HPS; magnification × 500.)

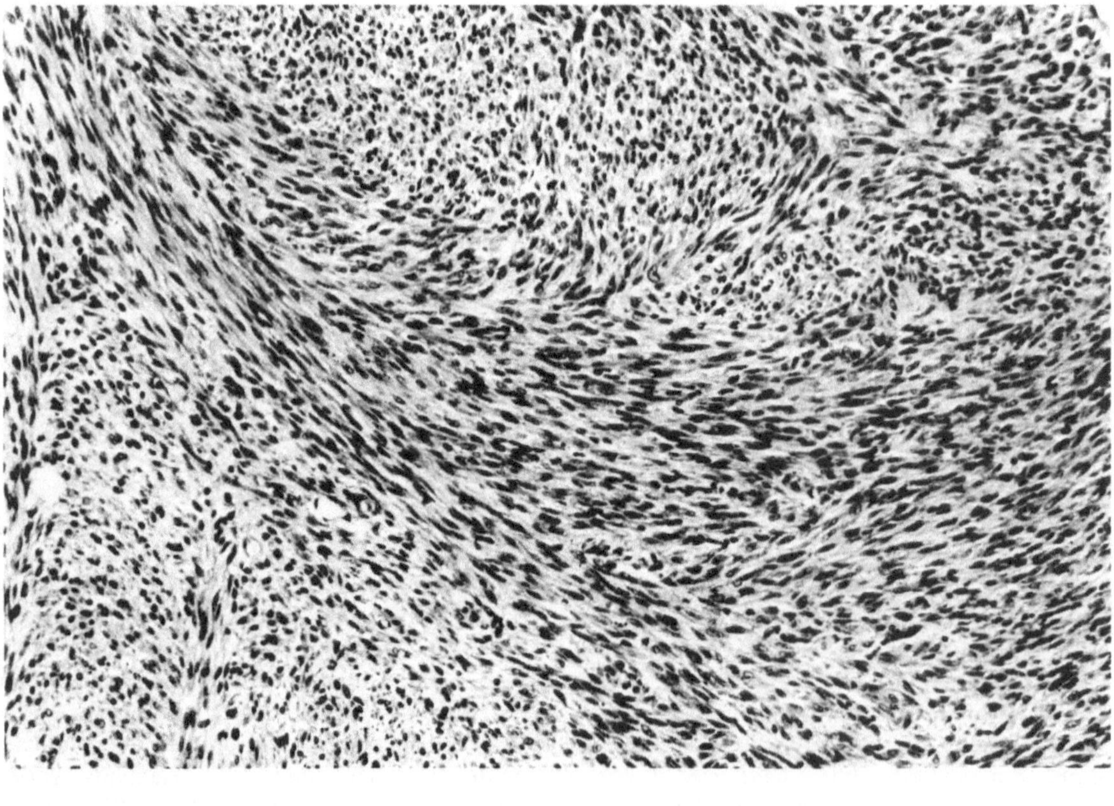

FIGURE 4.107. Pattern suggestive for a leiomyosarcoma in a transplanted round cell sarcoma from a BN/Bi rat. The original (primary) tumor was histologically similar to that illustrated in Figure 4.105. (HPS; magnification × 195.)

clearly had a proliferation of granulosa-thecal cell elements, while this was not common in BN/Bi and (WAG × BN)F_1 rats. Similarly, WAG/Rij had nests of Sertoli-like cells which seemed to proliferate along with the granulosa-thecal cells in the ovarian interstitial tissue. Such proliferation was not common in BN/Bi or (WAG × BN)F_1 rats, but it has been seen in other rat strains.[48,171]

No attempt was made to correlate the morphological appearance of the ovaries, uterus, mammary glands, and pituitary glands of these rats as was done in the study by Meites and Huang.[116] Such a correlation is clearly needed however.

Burek et al.[27] first reported a high incidence of tumors in the cervix and vagina of BN/Bi rats and described their morphological patterns. They also believed that round cell sarcomas were probably variants of smooth muscle tumors. As part of this present study, two round cell sarcomas from BN/Bi rats were transplanted in young rats and grew rapidly. The transplanted tumors resembled smooth muscle tumors rather than the pattern of round cell sarcoma as seen in the primary. Although more transplantation studies are needed, these results give additional support to the smooth muscle origin of the round cell sarcomas.

Baba and Von Haam,[8] Carter and Ird,[36] Squire and Goodman,[163] (in press) and Altman and Goodman[1] have recently reviewed the literature on tumors of the female rat reproductive system. With the exception of BN/Bi rats, spontaneous tumors of the cervix and vagina are not common. Tumors of the uterus are common, however, with endometrial polyps and smooth muscle tumors most frequently observed.

Tumors of the uterus, cervix, and vagina occurred in the rats of this series. Endometrial polyps were common in WAG/Rij females (19%), but were less frequent in BN/Bi (4%) or (WAG × BN)F_1 (3%) females. Of the WAG/Rij females, 6% had tumors of the uterus, but 0% in the cervix and vagina. In contrast, <1% of BN/Bi females had uterine tumors, but 19% had tumors in the cervix, vagina, or both. The F_1 hybrid females had a few tumors in both the uterus (3%) and cervix or vagina (3%).

N. Male Reproductive System

1. Nonneoplastic Lesions

a. Testicle and Epididymis

The most common age-associated change in the testicles of aging male rats was unilateral or bilateral atrophy. The peak incidence reached 100% in rats over 18 months. The severity was variable, especially in rats younger than 24 months, with a range from slight, multifocal to total atrophy of the tubules. Tubular atrophy was characterized in the early stages by degenerative changes of the spermatogonia, often with the formation of multinucleated spermatids (Figure 4.108). Advanced atrophy consisted of partial to complete loss of germinal cells, leaving only Sertoli cells (Figure 4.109). In some tubules, even the Sertoli cells were lost. Atrophic tubules were smaller than normal with greater space between individual tubules, which often appeared edematous and had prominent interstitial cells. At times, hyperplasia of interstitial cells was observed. Unexpectedly, strain differences were found. Both interstitial edema and interstitial cells were more prominent in WAG/Rij and (WAG × BN)F_1 males than in the BN/Bi males.

Focal or multifocal intratubular calcium deposits were common in atrophic testicles (Figure 4.110). They appeared to start first in a localized tubule that was filled with masses of degenerating and necrotic sperm. Calcium formed in the center of the necrotic debris and as the calcification progressed, the sperm and tubular epithelial cells disappeared, leaving only a calcified mass.

Periarteritis nodosa of testicular arteries was observed in some atrophic testicles. The lesion has been described (See figure 4.63) and the overall incidence in the aging rats has been presented elsewhere in this chapter (Section II.I). When present in testicles, periarteritis was focal or multifocal and occurred only in testicles that were atrophic.

A few sporadic findings such as sperm granulomata and spermatoceles were also found, but were not specifically studied.

b. Accessory Sex Glands

Acute inflammation, chronic inflammation, or both were very common in the prostate, coagulating glands, ampullary glands, and seminal vesicles of male WAG/Rij and (WAG ×

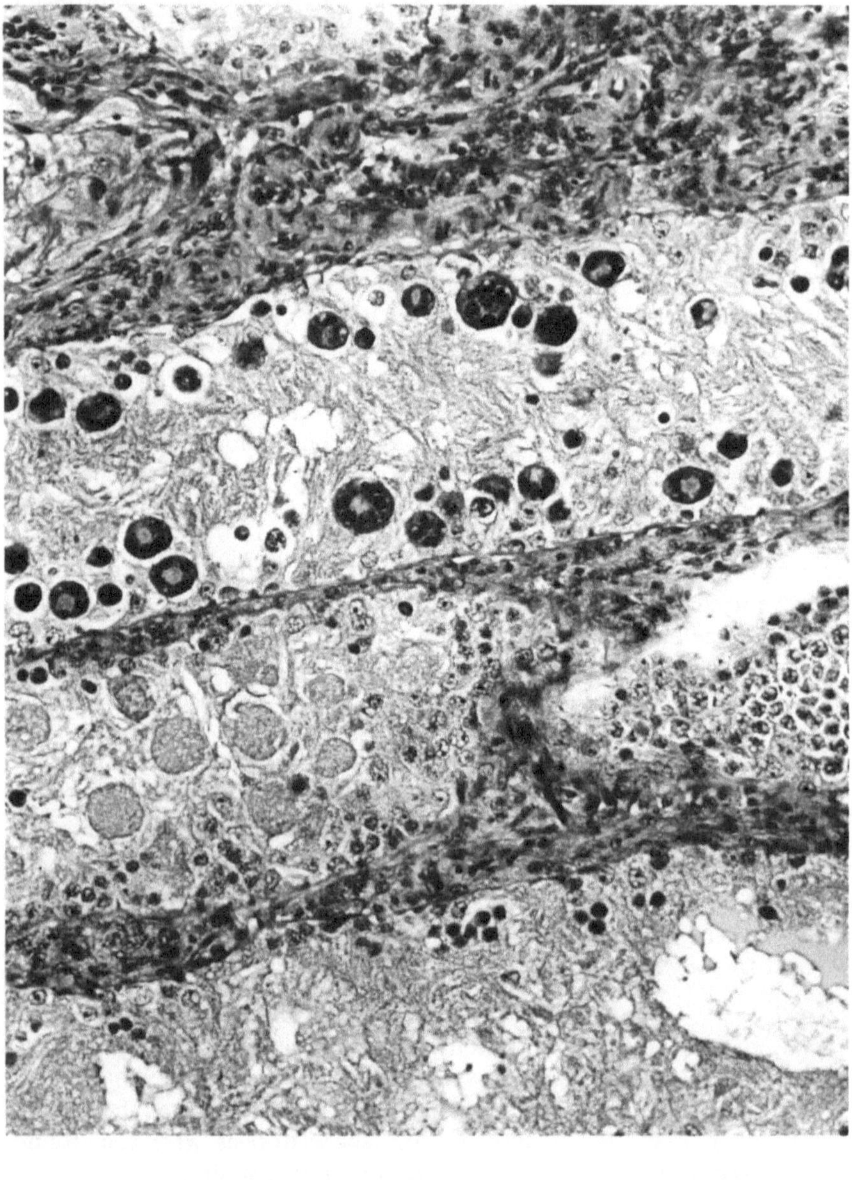

FIGURE 4.108. Giant cells (multinucleated spermatids) in the tubule of an atrophic testicle. (HPS; magnification × 210.)

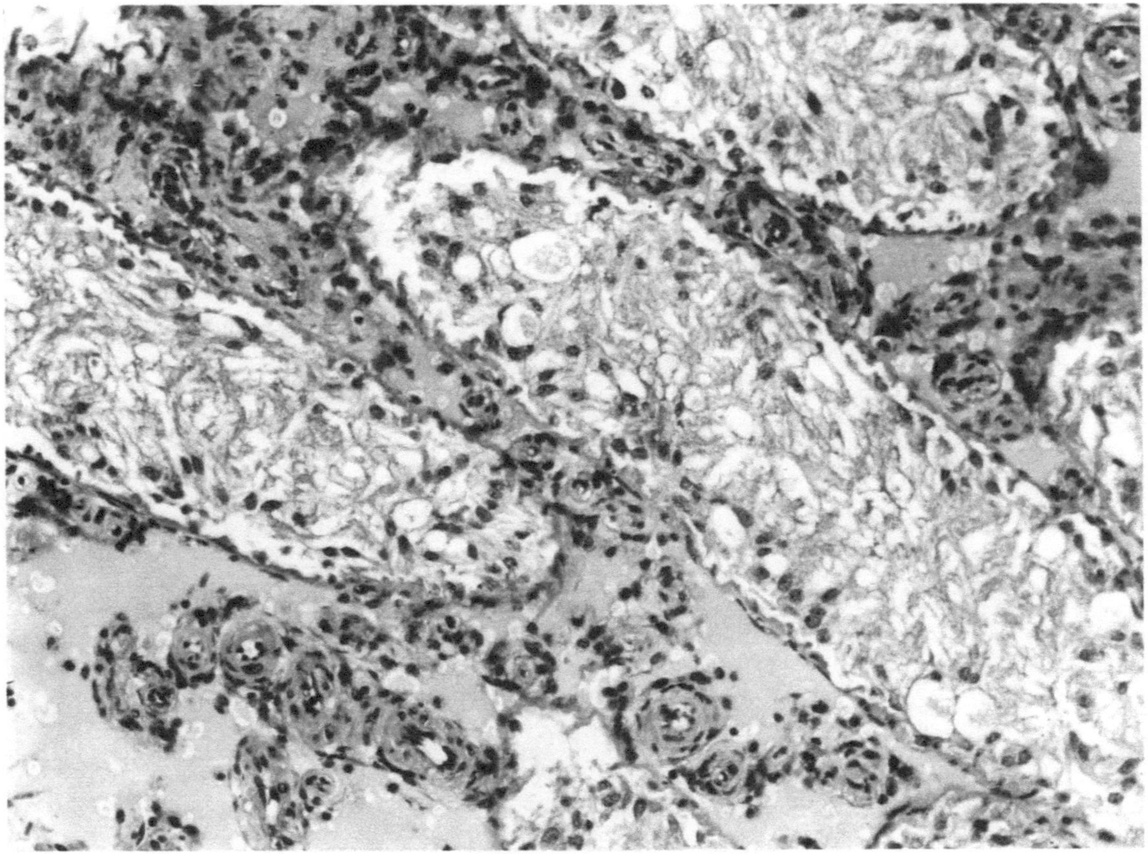

FIGURE 4.109. Testicle with loss of germinal cells and numerous Sertoli cells within the tubules. Note the interstitial edema, especially in the lower half of the photograph. (HPS; magnification × 210.)

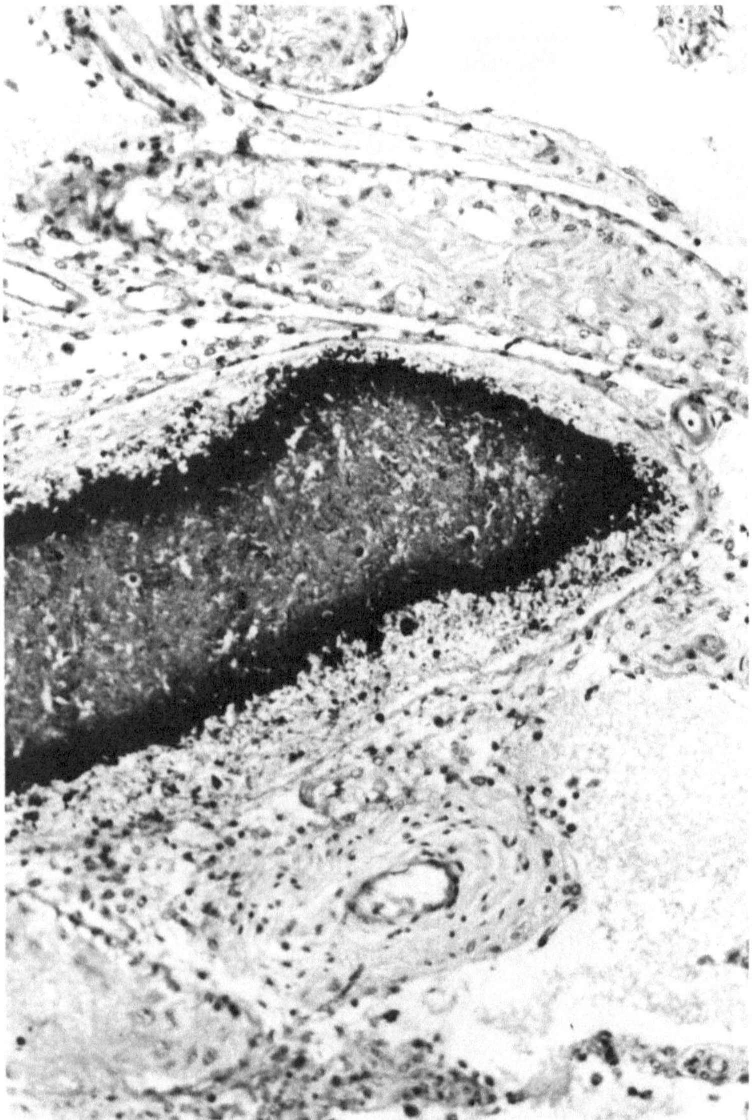

FIGURE 4.110. Testicle with focal mineralization in the lumen of a tubule. (HPS; magnification × 195.)

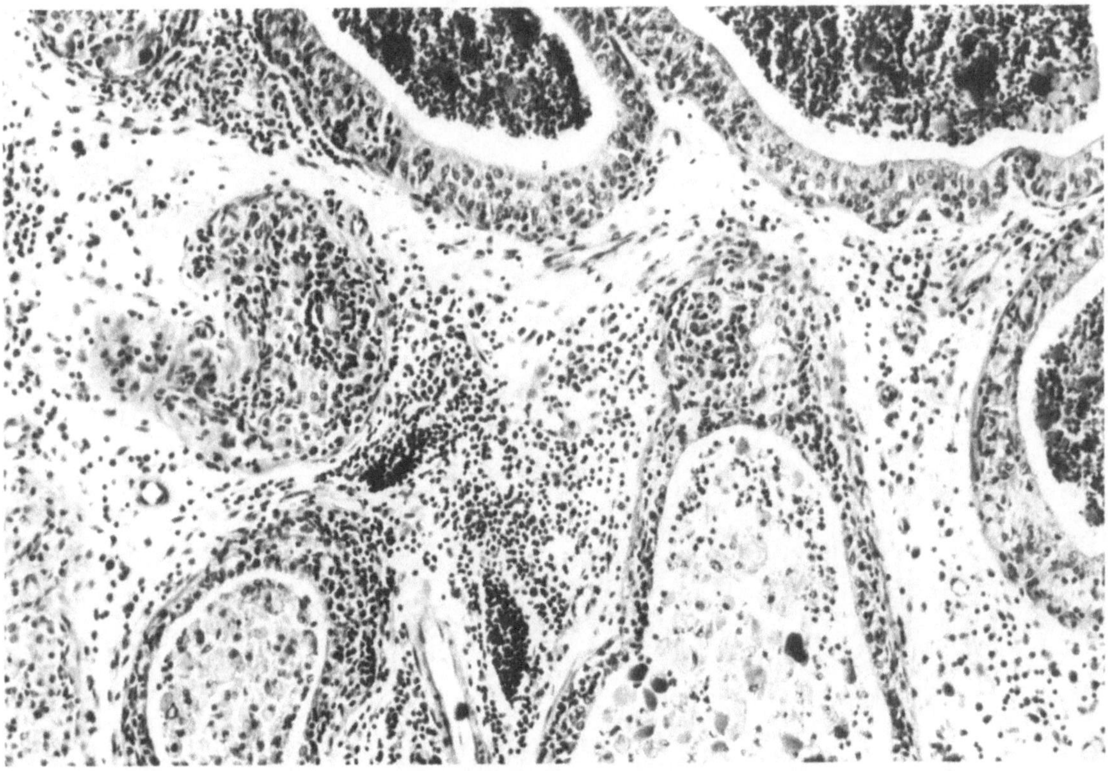

FIGURE 4.111. Prostate with mononuclear cell inflammation in the interstitium and inflammatory cells and necrotic debris in the lumina of alveoli. (HPS; magnification × 195.)

BN)F_1 rats. A wide range of lesions was recognized, and it would require a much more detailed study to sort out all of the observed changes than can be given here. Briefly, the most common reaction was a diffuse, multifocal, or focal lymphocytic and plasmacytic infiltration of the involved gland or glands (Figure 4.111). Suppurative inflammation was present in more severe cases. The lumina of the involved glands were either normal or filled with necrotic debris and intraluminal microabscesses were frequently present (Figure 4.112). Severely inflamed glands (usually the prostate, coagulating glands, or both) often had large abscesses (Figure 4.113). Additionally, in glands with severe inflammation and tissue destruction, granulomatous inflammation with Langhans'-type giant cells was present. Accompanying the chronic inflammation, focal hyperplasia or metaplasia of one or more accessory glands occasionally occurred.

Table 4.27 shows the incidence of abscess formation in the prostate glands of the male rats. Clearly, the male WAG/Rij and (WAG × BN)F_1 rats had more prostatic abscesses than male BN/Bi rats. In rats without abscesses, there was considerably more chronic inflammation in WAG/Rij and (WAG × BN)F_1 males than in the BN/Bi.

Noninflammatory, age-associated lesions were also found in the accessory sex glands, but these were even more difficult to evaluate. Most of the aged male rats had accessory genitalia that were smaller than normal, i.e., atrophic. The seminal vesicles of some males had coagulated proteinaceous material in their lumina. Some rats had several different changes that could be seen in the prostate, seminal vesicles, or coagulating glands. Some areas appeared normal, while in other regions the epithelium was lower than normal and still other areas had apparently hyperplastic changes. The content of the prostatic alveoli was also variable. Some alveoli were empty, while others were filled with pink homogeneous material. Some alveoli contained amphophilic to basophilic structures,

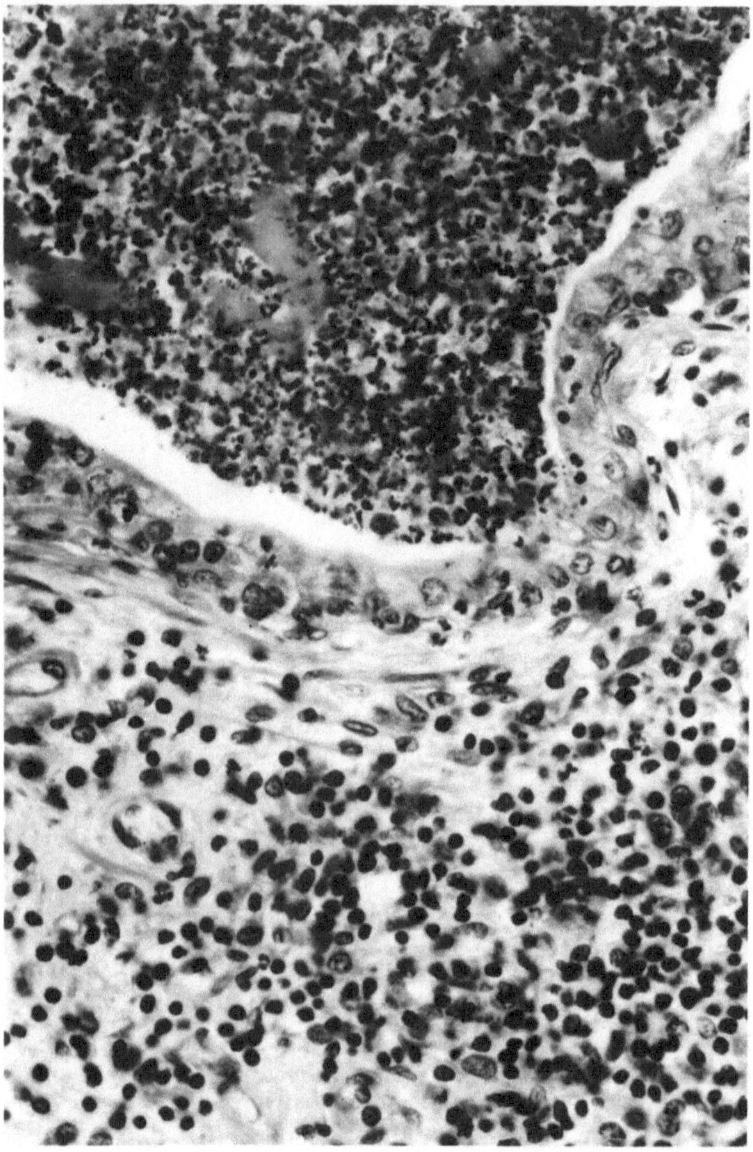

FIGURE 4.112. Higher magnification of same prostate illustrated in Figure 4.111. The interstitial cells are predominantly lymphocytes and plasma cells, while the alveolus is filled with necrotic debris. (HPS; magnification × 465.)

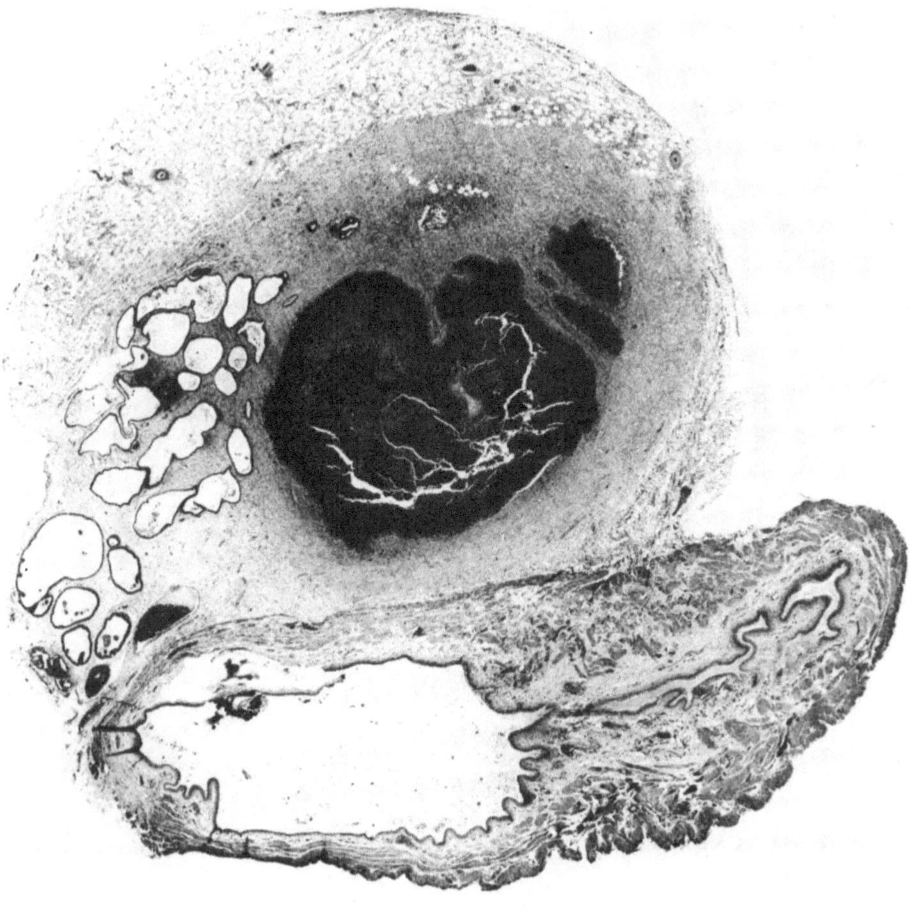

FIGURE 4.113. Prostate gland that is nearly replaced by a large abscess. The organ in the lower portion of the photograph is the urinary bladder. (HPS; magnification × 10.)

TABLE 4.27

Incidence of Severe Suppurative Prostatitis with Abscess Formation in Aging Male BN/Bi, WAG/Rij, and (WAG × BN)F_1 Rats

Strain	No. examined	No. with prostatitis	%	Mean age (range) in months
BN/Bi	74	1	1	34
WAG/Rij	124	45	36	23 (9—31)
F_1	67	14	21	34 (24—44)

similar to corpora amylacea. The structures were not laminated, however, and were variable in size and shape (Figure 4.114).

2. Neoplastic Lesions

Tumors of the male reproductive system were infrequently observed (Table 4.28). Eight interstitial (Leydig) cell tumors were found. All were histologically similar, consisting of well-differentiated epithelial cells that formed sheets and were closely packed together. Individual cells had indistinct cell boundaries, pink cytoplasm that was often finely vacuolated, and round to oval nuclei that contained a nucleolus. The tu-

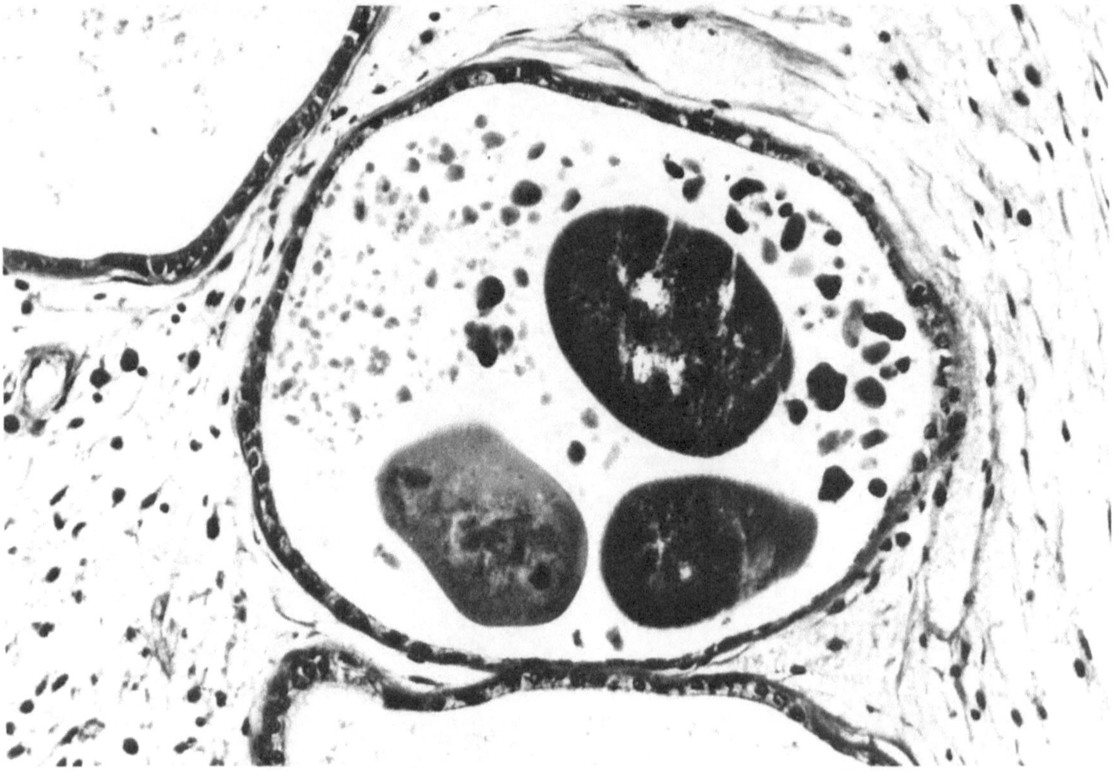

FIGURE 4.114. Corpora amylacea-like bodies in a prostatic alveolus. (HPS; magnification × 300.)

TABLE 4.28

Interstitial Cell Tumors in the Testicles from Aging Male BN/Bi, WAG/Rij, and (WAG × BN)F_1 Rats

Strain	No. examined	No. with interstitial cell tumors	%	Mean age (range) in months
BN/Bi	74	0	—	—
WAG/Rij	124	4	3	28 (26—29)
F_1	67	4	6	36 (29—39)

mor cells appeared to infiltrate between and separate the testicular tubules. At times, compressed and degenerating tubules surrounded by tumor cells were present.

A lipoma composed of well-differentiated fat cells occurred in a testicle of a 22-month-old BN/Bi rat. Finally, a small hemangioendothelioma consisting of blood-filled, endothelial lined spaces was present in the testicle of a 33-month-old BN/Bi rat. It had an irregular border and infiltrated between the tubules, but it was well differentiated and confined to the testicle.

Five mesotheliomas, arising from the mesothelium of the tunica vaginalis and genital omentum, were found in four male WAG/Rij and one male (WAG × BN)F_1 (Table 4.29). The mesotheliomas were located on the serosal surface of the testis, epididymis, and genital omentum and were unilateral. They were composed of multiple, papillary fronds arising from the serosal surface (Figure 4.115). The papillary

TABLE 4.29

Mesotheliomas of the Scrotal Mesothelium in Aging Male BN/Bi, WAG/Rij, and (WAG × BN)F_1 Rats

Strain	No. examined	No. with mesothelioma	%	Mean age (range) in months
BN/Bi	74	0	—	—
WAG/Rij	124	4	3	26 (19—29)
F_1	67	1	1	40

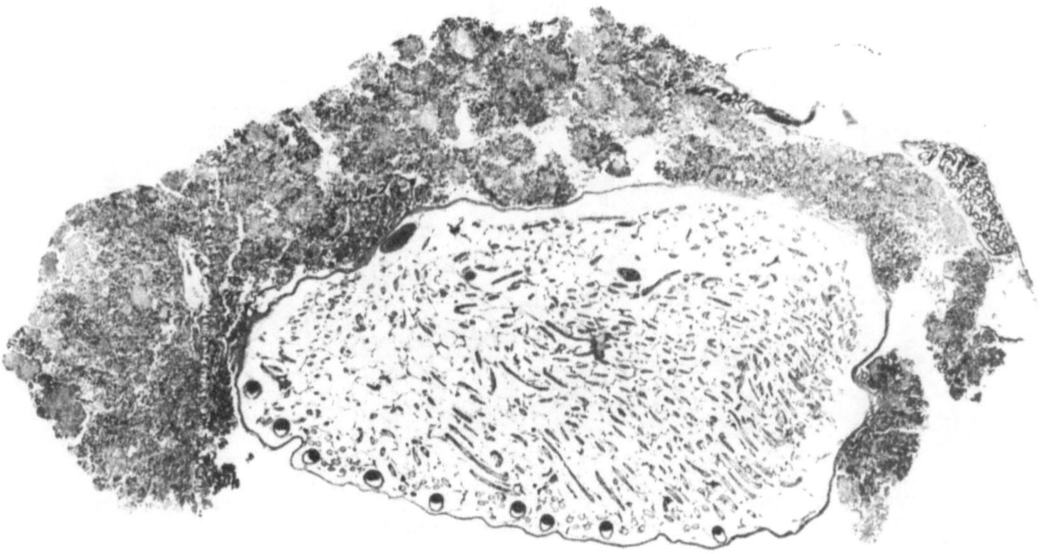

FIGURE 4.115. Severely atrophic testicle surrounded by a mesothelioma. (HPS; magnification × 6.)

structures had a supporting connective tissue stalk with blood vessels and were lined by mesothelial cells. The lining usually consisted of one row of cells, but at times piling-up of cells occurred. Mitoses were infrequent (Figure 4.116). Distant metastases, direct extension beyond the scrotal sac, or invasion into surrounding structures was not observed.

Primary tumors of the accessory glands were not found, but a malignant schwannoma was seen in one 38-month-old male F_1 rat. The cells were elongated and had indistinct cell boundaries and pink cytoplasm. The nuclei were oval to spindle shaped. Palisading and swirling could be found in several regions of the tumor. It had infiltrated most of the organs in the caudal portion of the abdominal cavity, including the seminal vesicles, coagulating glands, prostate, and urinary bladder. It was impossible to determine the site of origin, and although highly invasive, distant metastases were not found.

3. Discussion

The organs of the male reproductive system of the aging BN/Bi, WAG/Rij, and (WAG × BN)F_1 rats all had some degree of atrophy. The severity and histological appearance of the various organs showed a great deal of individual variability, however. As a result, many nonneo-

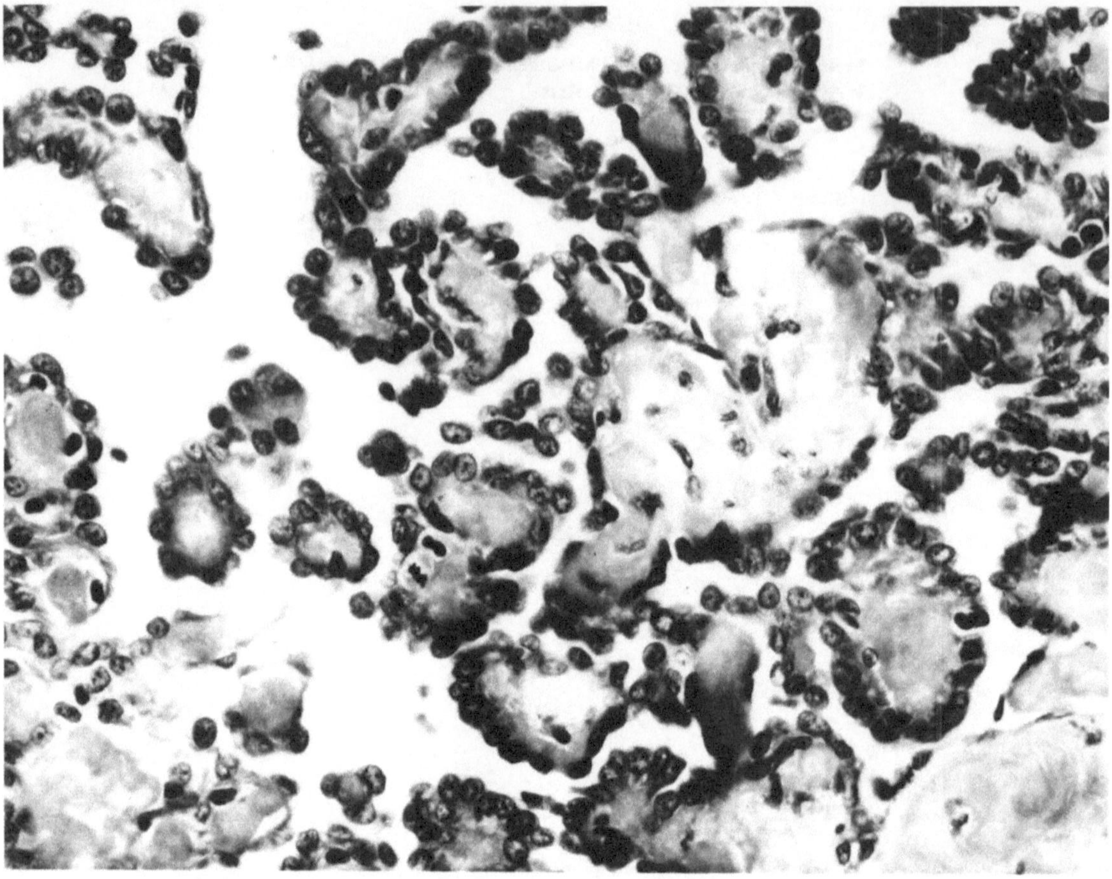

FIGURE 4.116. Higher magnification of the mesothelioma illustrated in Figure 4.115, showing a fibrovascular stroma that is lined by cuboidal cells. Note the mitotic figure (↑). (HPS; magnification × 500.)

plastic age-associated changes were seen in these rats, including the following testicular changes: tubular giant cells, intertubular edema, increased prominence and proliferation of interstitial cells, intratubular mineralization, periarteritis nodosa, and atrophy. Similarly, the accessory genitalia had several types of age-associated lesions including atrophy, focal hyperplasia, chronic inflammation, suppurative inflammation, and corpora amylacea-like bodies.

Similar lesions in aging rats have been well documented by several investigators such as Coleman et al.[42] for Fischer rats, Kociba et al.[99] for Sprague-Dawley (Spartan substrain) rats, and Kroes et al.[102] for Wister rats. Also, such lesions have been discussed in review articles.[4,147]

The severe inflammatory lesions of the accessory genitalia in the male WAG/Rij and (WAG × BN)F_1 rats needs further study. Special stains including Giemsa, PAS, and silver stains were done on some of these lesions, but microorganisms were not demonstrated. An infectious etiology cannot be eliminated, however, because bacteriologic culturing has not been done. Sixty males died with abscesses in their prostate glands. Others had severe suppurative inflammation of the prostate, seminal vesicles, or coagulation glands. Still others had less severe inflammatory changes. Despite this, only nine males had an apparent ascending urinary tract infection that was associated with the reproductive system inflammation.

The early mononuclear inflammation, especially in the prostate, may have another explanation, namely an autoimmune reaction. The entire series of lesions seen in WAG/Rij and

(WAG × BN)F_1 rats and the near absence of these changes in BN/Bi males indicate it is probably a strain-dependent phenomenom. The reaction seemed to begin first with an infiltration of lymphocytes and plasma cells. Later, the process seemed to become more extensive and involved the prostatic alveoli. Necrotic debris in the alveoli and the increased number of neutrophils and macrophages occurred later in more severely affected glands. As the process continued, rupture of alveoli, which released their contents into the surrounding tissue, may have resulted in a severe suppurative inflammatory response with numerous macrophages and giant cells. The early mononuclear cell response was similar to the experimentally induced allergic prostatitis in the dog prostate,[135] but additional studies are clearly indicated to confirm or eliminate this possibility.

Neoplastic lesions of the male reproductive system of rats have been reviewed.[1,59,118,163] From these reviews it is clear that the type of neoplastic lesion that is seen differs among the different rat strains. In general, the most commonly observed testicular neoplasms appear to be interstitial (Leydig) cell tumors, but some strains, such as Fischer rats,[149] may have an incidence over 80% and others, such as the BN/Bi strain reported in this study, a very low incidence.

Mesotheliomas of the tunica vaginalis were observed in five rats of this study. These tumors were similar to the mesotheliomas in male rats reported by Mawdesley-Thomas and Hague,[113] Snell,[156] Mostofi and Bresler,[118] and Gould.[73]

O. Brain

1. Nonneoplastic Lesions

The more common age-associated nonneoplastic lesions are discussed in this section. The incidence of each lesion was not determined, however, because more detailed systematic studies, often with special staining, would be required before most of these lesions could be accurately assessed. Despite this shortcoming, it was clear that all of the following lesions (except melanosis) had an increased incidence with age and all were very common in the oldest rats.

a. Vacuolation (Vacuolar Encephalopathy)

One of the most common age-associated lesions encountered in the brains was an idiopathic vacuolation. The incidence was high, with most rats older than 24 months having at least mild changes. The lesion consisted of clear spaces (vacuoles) in the white matter of the thalamic area, pons, midbrain, and cerebellum. In severe cases, the cerebrum was also involved. Some spaces appeared to contain a pale-staining homogeneous material, but most were empty (Figures 4.117 and 4.118). The spaces did not stain with PAS, Giemsa, Luxol Fast Blue, or silver stains, and a glial reaction was not associated with the vacuoles.

b. Lipofuscin

Intracytoplasmic lipofuscin accumulations were common in the neurons of brains. Lipofuscin was a pale, golden brown to yellow color with the HPS staining and was usually finely granular (Figure 4.119). It was strongly PAS and Ziehl-Nielson acid-fast positive and was found in neurons of the cerebrum, cerebellum, and midbrain. The quantity varied from rat to rat and from area to area in the same brain, but there was a clear trend for the incidence to increase with age. In rats older than 24 months, the incidence may reach 100%, but since PAS staining was not done routinely on all brains, the actual incidence could not be determined with certainty.

c. Basophilic PAS-positive Bodies

Irregular to spherical basophilic bodies were found in approximately 20% of the brains. They were usually found as bilateral deposits in the thalamic region, lacked an associated glial reaction, appeared to be extracellular, and were amorphous or laminated (Figure 4.120). Some were found in the grey matter of the cerebellum or rarely in the cerebrum. Two brains with typical bodies were used for special stains. The basophilic bodies were strongly PAS positive, were not metachromatic with Giemsa staining, and apparently contained little if any calcium, as suggested by negative or weakly positive reactions with Alizarin red and von Kossa stains. Special staining of additional cases would certainly be needed, however, to be certain that calcium is absent or minimal in all such deposits.

d. Melanosis

Melanin pigment was common in the brains

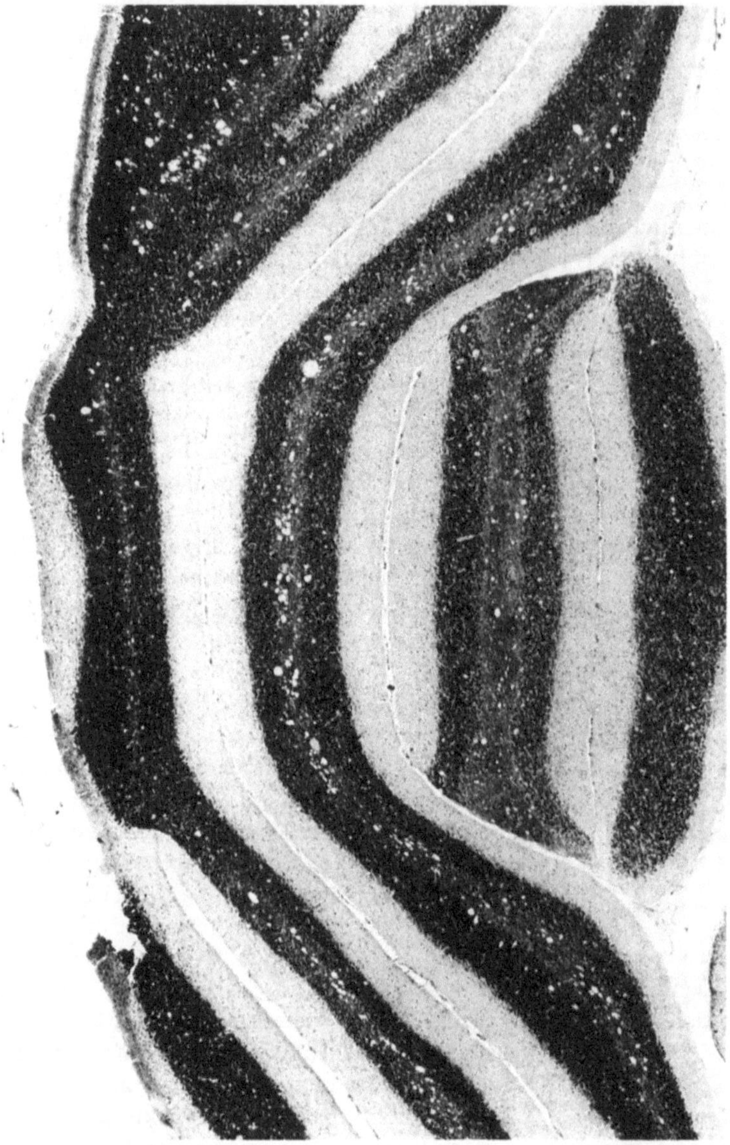

FIGURE 4.117. Vacuoles in the white matter of the cerebellum. (HPS; magnification × 39.)

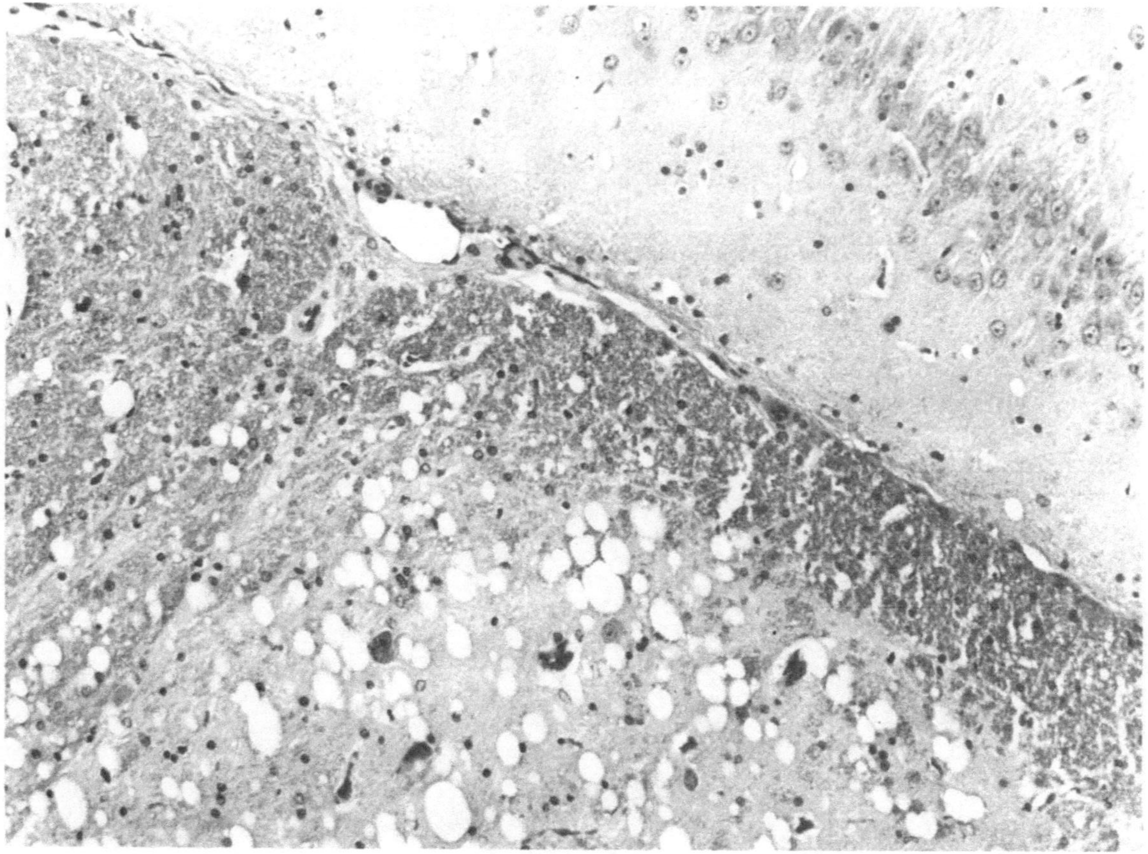

FIGURE 4.118. Vacuoles in the thalamic region of the brain. (HPS; magnification × 210.)

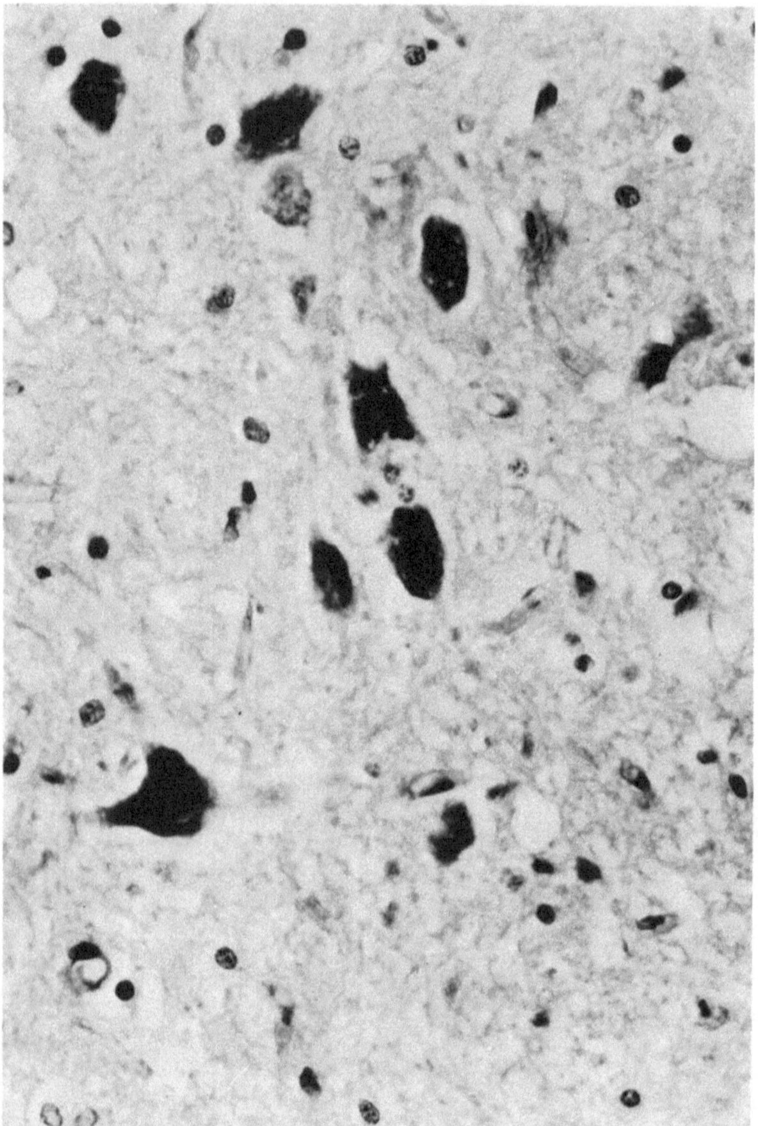

FIGURE 4.119. Lipofuscin pigment in the cytoplasm of neurons. (HPS; magnification × 500.)

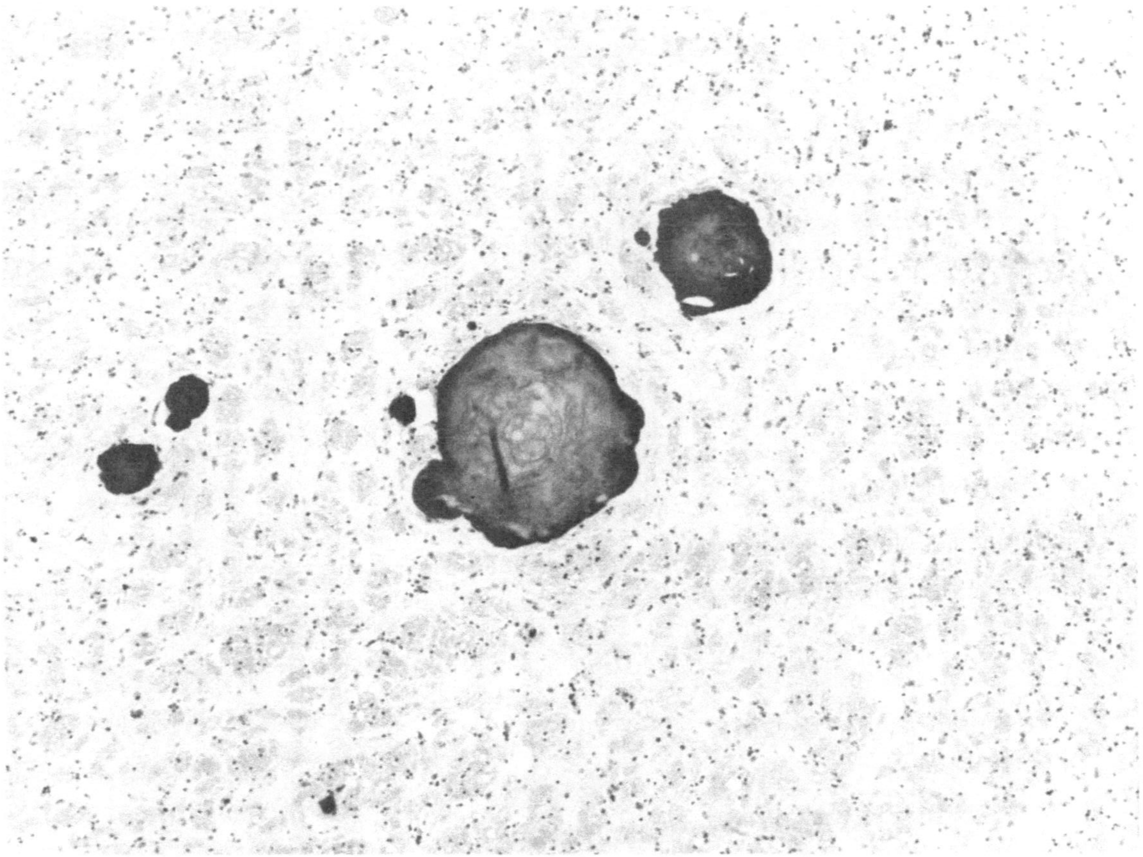

FIGURE 4.120. Basophilic bodies in the brain. Similar bodies in two different rats were strongly PAS positive and only weakly Alizarin red or von Kossa positive. (HPS; magnification × 80.)

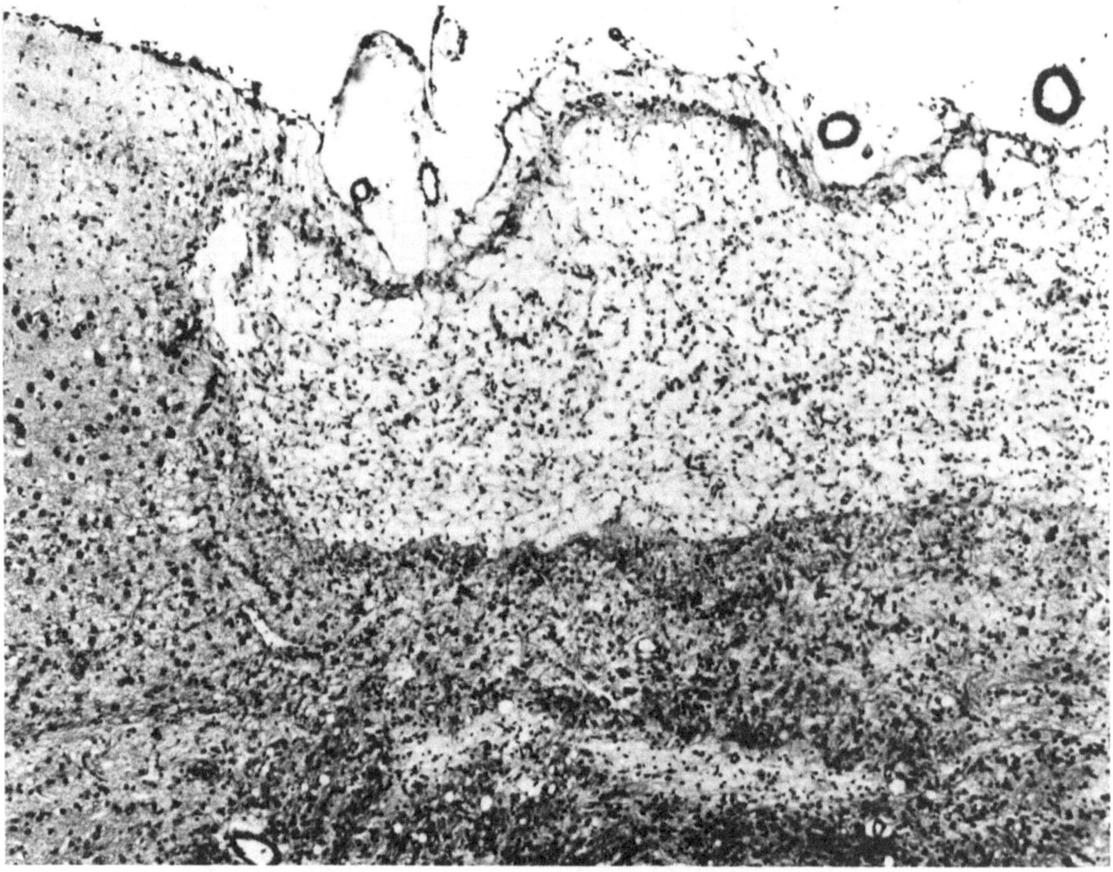

FIGURE 4.121. Area of encephalomalacia probably resulting because of an infarction in the outer layers of the cerebrum. (HPS; magnification × 80.)

from BN/Bi and (WAG × BN)F_1 rats. The most frequently affected sites were the meninges, olfactory lobes of the brain, and the pituitary gland. Melanosis was not an age-associated lesion because it was routinely found in young and old rats. It was not determined, however, if the amount of melanin increased with age.

e. Infarcts

Five female BN/Bi rats (ages 30, 30, 35, 38, and 39 months) had focal areas in the cerebrum consistent with old infarcts. The lesions consisted of large and small cystic spaces surrounded by delicate connective tissue trabeculae (Figure 4.121). Lipid- and hemosiderin-containing macrophages were numerous in and along the margins of the spaces (Figure 4.122). The margins of the lesion contained many glial cells (mostly astrocytes) and blood vessels and were sharply demarcated from the adjacent normal grey matter. In addition to old infarcts, many rats that died with leukemia had focal acute hemorrhages or hemorrhagic infarcts in the brain.

f. Hydrocephalus

Hydrocephalus was an age-associated lesion, but it invariably appeared secondary to a tumor of the brain or pituitary gland. It was common and had an increased incidence with age similar to that of the brain and pituitary tumors.

g. Multifocal Spongiform Encephalopathy

Two female WAG/Rij rats (ages 25 and 34 months) died with a severe, multifocal spongiform degeneration of the grey and white matter. The lesions were randomly distributed throughout the cerebrum, cerebellum, and mid-

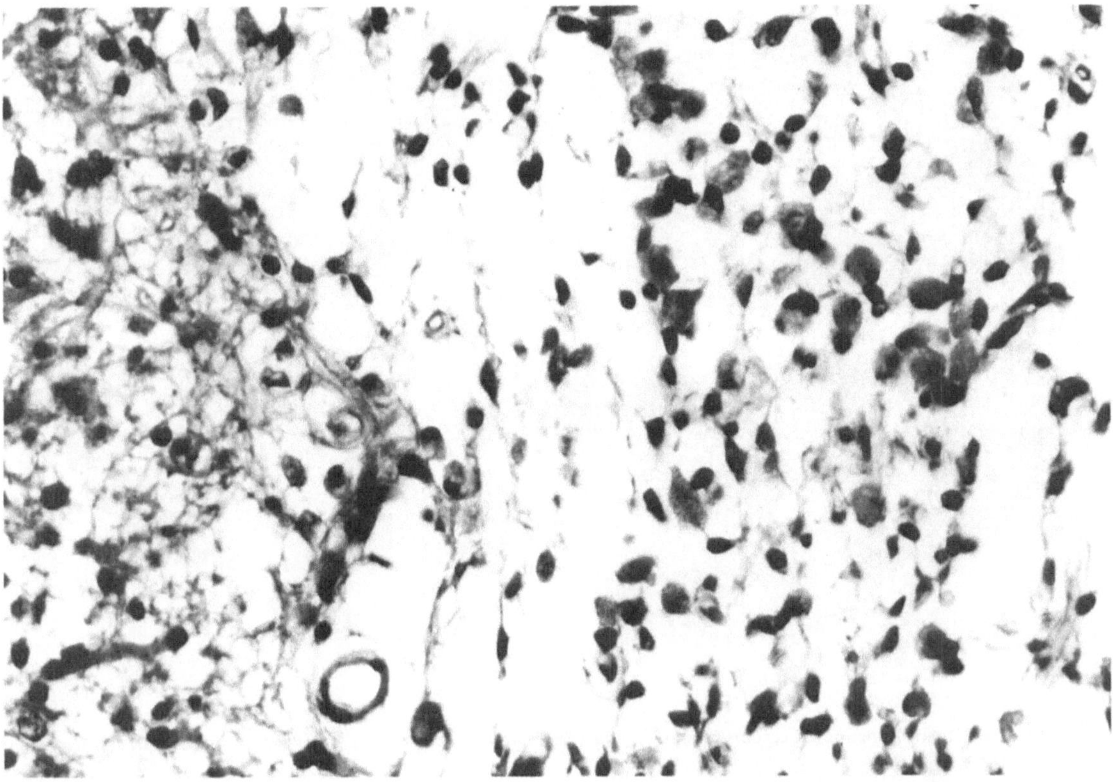

FIGURE 4.122. Higher magnification of the infarcted area illustrated in Figure 4.121, showing numerous hemosiderin-laden macrophages and the absence of normal cerebral tissue in the region. (HPS; magnification × 465.)

brain. However, they were most severe in the frontal and outer zones of the cerebral cortex. The lesions consisted of vacuolated foci with necrotic debris and a prominent glial reaction with numerous swollen astrocytes and variable numbers of macrophages (Figures 4.123 and 4.124). In the white matter, demyelinization was evident with HPS and PAS-Luxol Fast Blue stained sections. PAS-positive material was present in macrophages within the vacuolated areas and in macrophages around blood vessels that were some distance from the lesions. Unfortunately, spinal cords were not available from either of these rats, so it is not known if the spinal cord was also involved.

2. Neoplastic Lesions

a. Granular Cell Tumors

Granular cell tumors (granular cell myoblastoma) were the most common of the primary brain tumors. A few were grossly observed as solitary friable tumors, with a light pink to pale yellow or grey color and were sharply demarcated from the surrounding tissue.

The histologic features consisted of sheets or nests of closely packed round, polygonal, or elongated cells with pink, finely to coarsely granular cytoplasm (Figures 4.125 and 4.126). The nuclei were lightly vesicular and often contained a small nucleolus. The intracytoplasmic granules were PAS positive and diastase resistant. Mitotic figures were not common, but were observed in some tumors. Most were sharply demarcated from the surrounding tissue, but a few had perivascular infiltration, irregular edges, or both. The smallest tumors were confined to the meninges, while the larger tumors extended deep into the brain tissue, but all had contact with the leptomeninges at some point.

Electron-microscopic studies were performed on two granular cell tumors to determine their ultrastructural characteristics. Both tumors had similar features consisting of cells filled with many large and small membrane-bound gran-

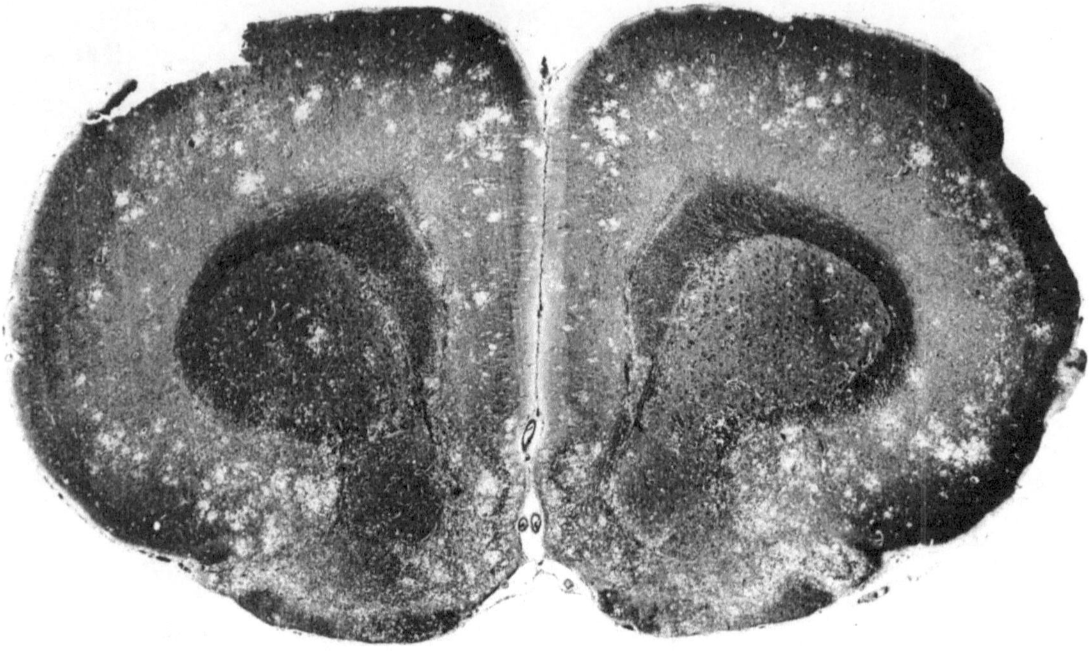

FIGURE 4.123. Rostral portion of cerebrum showing severe, bilateral leuko- and polioencephalopathy. (PAS-Luxol Fast Blue; magnification × 15.)

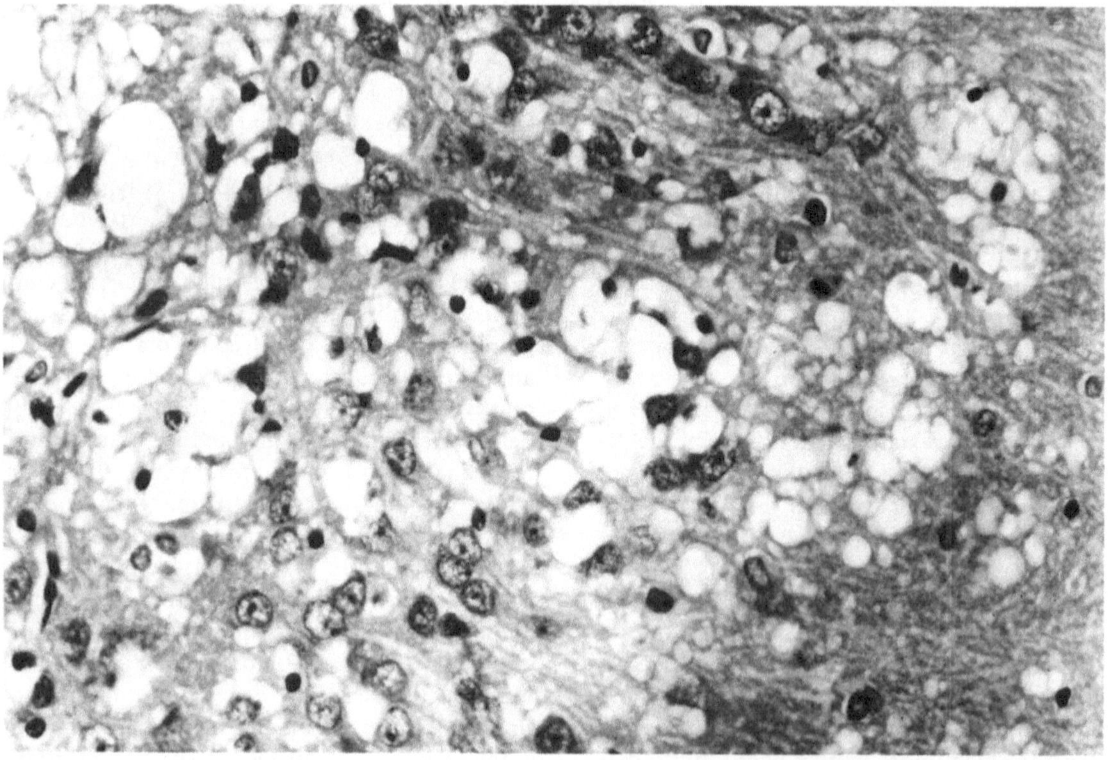

FIGURE 4.124. Higher magnification of the vacuoles from the same brain illustrated in Figure 4.123. (PAS-Luxol Fast Blue; magnification × 123.)

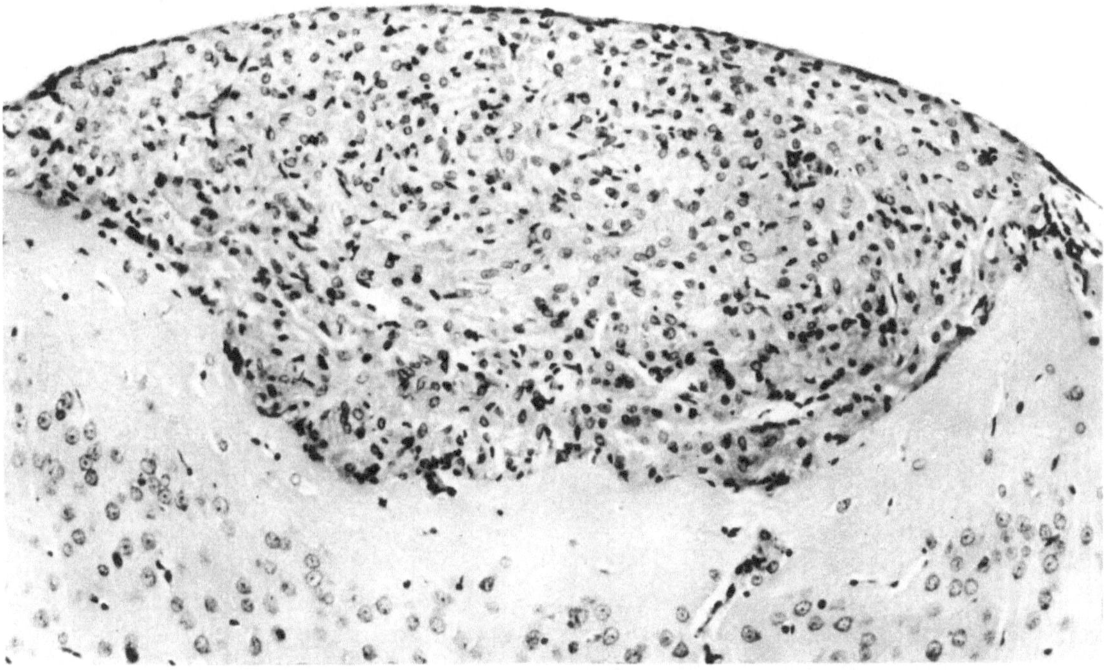

FIGURE 4.125. Granular cell tumor in the brain. (HPS; magnification × 195.)

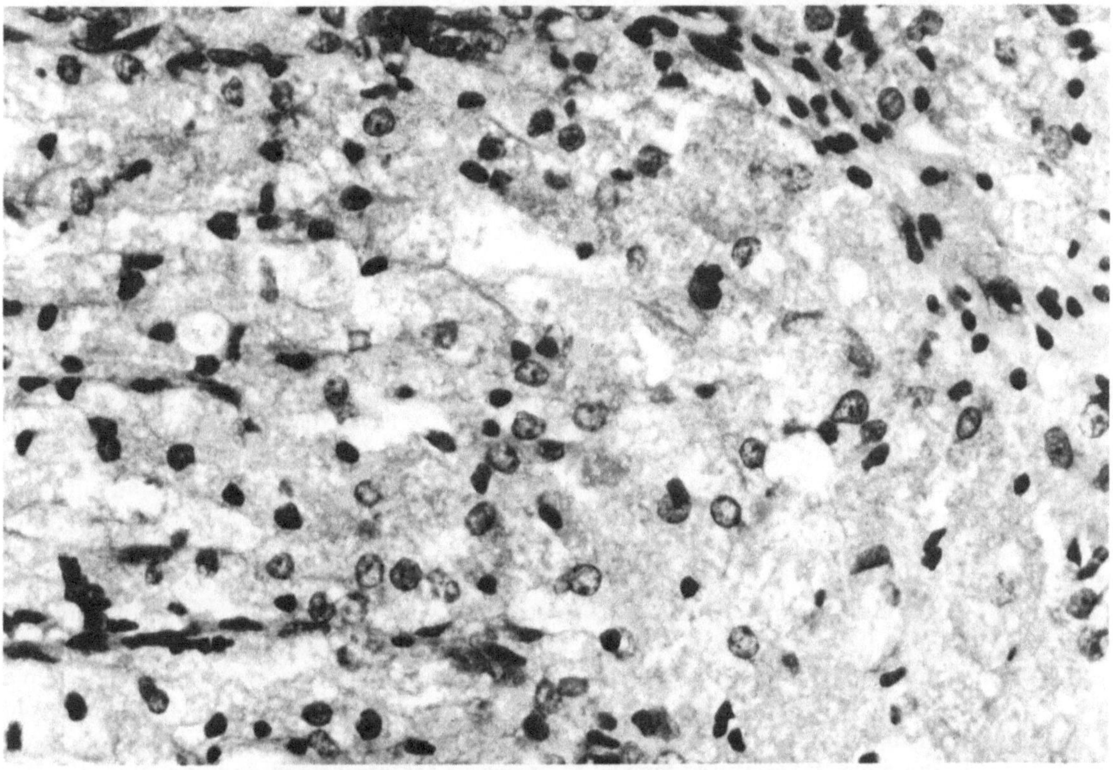

FIGURE 4.126. Granular cell tumor in the brain showing finely granular cytoplasm. (HPS; magnification × 465.)

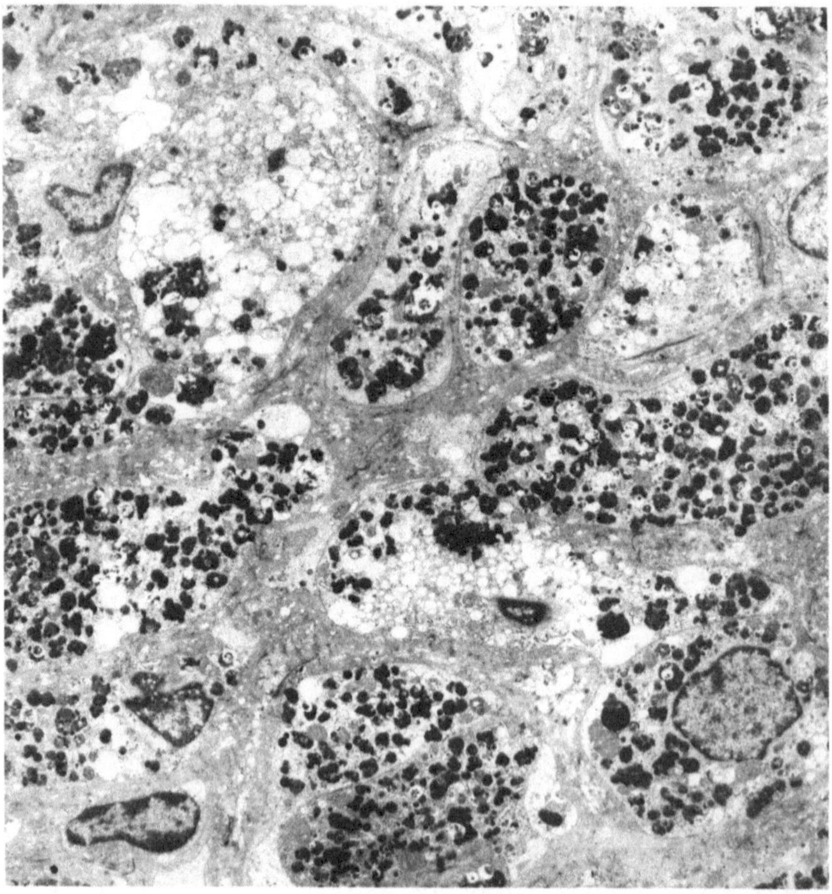

FIGURE 4.127. Electron photomicrograph of cells from a granular cell tumor in the brain. (Magnification × 3380.)

ules and vacuoles (Figures 4.127 and 4.128). The cells contained few mitochondria and had scant endoplasmic reticulum and occasionally a small golgi apparatus. Some granules had remnants of other cell organelles consistent with autophagocytosis. Some granules contained electron-dense, amorphous, lipid-like material and others contained finely to coarsely granular structures. Intracytoplasmic fibrils were found in some of the tumor cells. Collagen-producing fibroblasts and extracellular collagen were present between tumor cells.

Twenty granular cell tumors were found in the rat brains (Table 4.30), and it was clearly the most common brain tumor observed (excluding pituitary tumors). The incidence ranged from 2 to 7% in groups with tumors, but none was observed in the male WAG/Rij or female F_1 rats.

b. Glial Tumors

Nine tumors of glial cell origin were recognized in the brains as shown in Table 4.31. Astrocytomas occurred in two male (ages 22 and 28 months) and one female (age 41 months) WAG/Rij rats. All were large tumors, approximately 1 × 1 cm, that were diffuse, nonencapsulated, and highly infiltrative. They occupied a large area of the brain including the area around the lateral ventricles, thalamus, midbrain, and portions of the cerebellum. The cells tended to be moderately well-differentiated astrocytes with some cellular pleomorphism and numerous mitoses.

Glioblastoma multiforme was diagnosed in a 29-month-old female F_1 rat and a 34-month-old female WAG/Rij rat. The tumors were confined to the cerebellum and consisted of pleomorphic, anaplastic glial cells with numerous

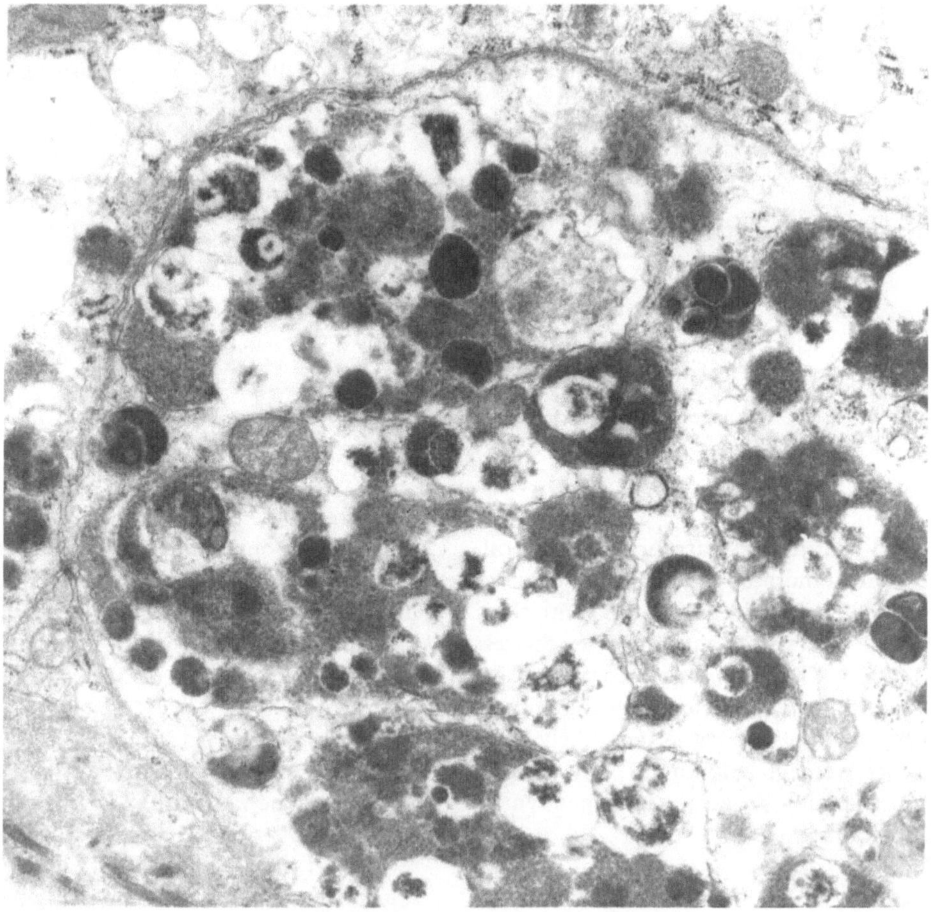

FIGURE 4.128. Higher power electronmicrograph of the cytoplasm of a cell from the granular cell tumor shown in Figure 4.127, illustrating the numerous large and small cytoplasmic granules. (Magnification × 18,900.)

TABLE 4.30

Incidence of Granular Cell Tumors in Aging BN/Bi, WAG/Rij, and (WAG × BN)F_1 Rats

Strain	Sex	No. examined	No. with granular cell tumors	%	Mean age (range) in months
BN/Bi	Female	236	11	5	32 (24—37)
	Male	74	2	3	30, 43
WAG/Rij	Female	101	2	2	33, 36
	Male	124	0	—	—
F_1	Female	68	0	—	—
	Male	67	5	7	34 (30—37)

multinucleated tumor cells and many mitotic figures.

A 22-month-old female BN/Bi and a 28-month-old female F_1 rat had an oligodendroglioma in the cerebrum. The tumor cells were uniform with small, dark, round nuclei surrounded by clear spaces.

Two female BN/Bi rats (ages 30 and 34

TABLE 4.31

Occurrence of Tumors, Other than Granular Cell Tumors, in the Brains of Aging BN/Bi, WAG/Rij, and (WAG × BN)F_1 Rats

Strain	Sex	No. examined	Type of tumor	No. of each tumor	Age (in months)
BN/Bi	Female	236	Ependymoma	2	30, 34
			Oligodendroglioma	1	22
	Male	74	None	—	—
WAG/Rij	Female	101	Astrocytoma	1	41
			Glioblastoma multiforme	1	34
	Male	124	Astrocytoma	2	22, 28
			Meningioma	1	24
F_1	Female	68	Oligodendroglioma	1	28
			Glioblastoma multiforme	1	29
	Male	67	None	—	—

months) had glial tumors surrounding a lateral ventricle. The tumors appeared to arise from the ependymal cells of the lateral ventricles. Pseudorosettes were frequent, but true rosettes were not found. The tumors were composed of astrocytic cells and oligodendrogliocytes in addition to ependymal-like cells. It is likely that both tumors were ependymomas, but mixed glial tumors arising from the subependymal plate could not be excluded.

c. Meningioma

A 24-month-old male WAG/Rij rat had multiple meningiomas (Table 4.31). All tumors were similar, consisting of cells arranged in lobulated masses in the meninges over the cerebrum, and contained numerous psammoma bodies. Cells were polyhedral and had indistinct cell boundaries, pale pink cytoplasm, and round to oval nuclei. Mitoses and local invasion were not observed.

d. Pineal Gland

Sections through the pineal gland were occasionally found in the routine preparations of the brain, and a tumor of the pineal gland was present in one of these random sections. It occurred in a 34-month-old female WAG/Rij rat. The pineal gland was only slightly enlarged (Figure 4.129), but the normal architecture was destroyed. It consisted of both large and small cells. The large cells were pleomorphic and had pink cytoplasm, indistinct cell boundaries, and large, round to elongated nuclei that occasionally contained a small nucleolus. Mitoses were common in these cells (Figure 4.130). The small cells had a round, darkly stained nucleus and little or no recognizable cytoplasm. Serial sections were made through this lesion, but local invasion into blood vessels, meninges, or brain parenchyma could not be demonstrated.

3. Discussion

A general review of all age-associated changes of the rat brain is not within the scope of this discussion. However, several excellent references are available that discuss many of the current topics about the aging nervous system.[15,65,114,136]

Nonneoplastic lesions were common in the brains of aging BN/Bi, WAG/Rij, and (WAG × BN)F_1 rats. The two most common were extracellular vacuolations and lipofuscin accumulations in neurons. The vacuolations appeared to be artifacts of tissue processing because they consisted only of holes without any glial or inflammatory response. The lesions were not recognized in younger rats (<6 months) of these strains, but were consistently present in most rats over 24 months old.[83] These changes were not found in young mice (<6 months old), but they increased in frequency with age, becoming very common in old mice (>24 months) of several different strains. Such lesions have been described by Garner et al.,[66] who frequently observed extracellular vacuolations in both rats and mice, and by Coleman et al.,[42] who saw them in Fischer rats. In addition, de Estable-Puig and de Estable-Puig[49] studied vacuolar degeneration in neurons of aging Sprague-Dawley rats by light and electron microscopy and reviewed the current

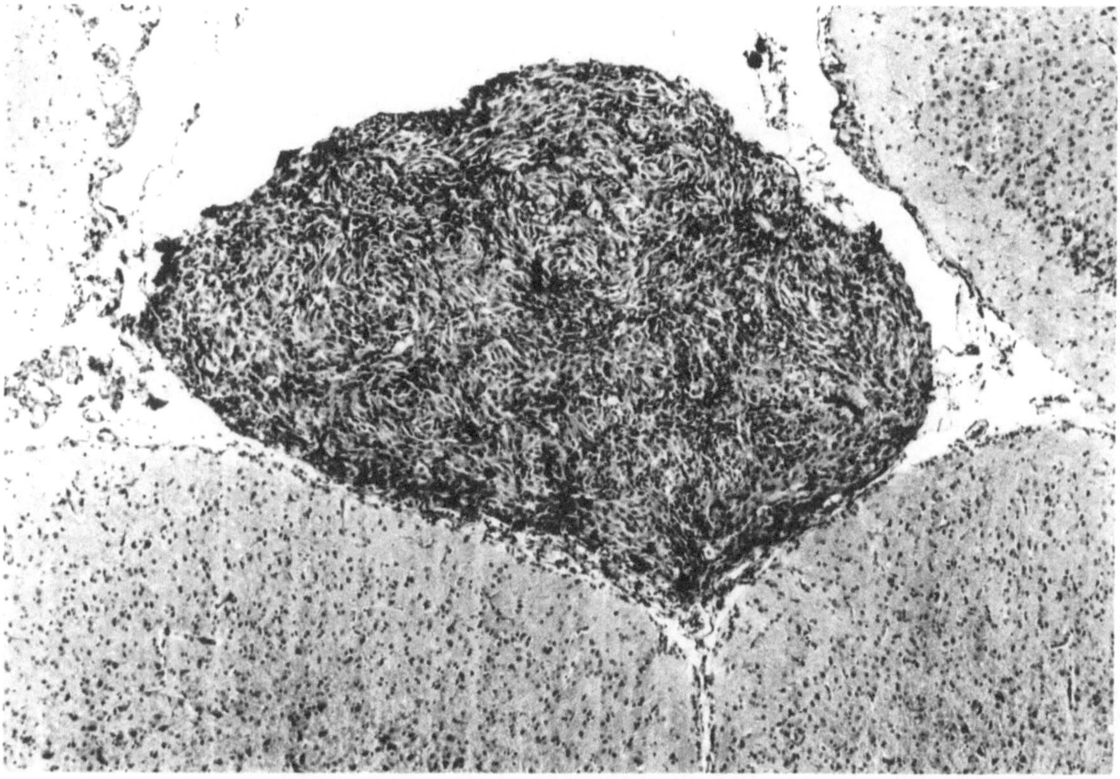

FIGURE 4.129. Pineal gland tumor. (HPS; magnification × 75.)

literature on vacuolations that develop in aging brains. The etiology and pathogenesis of these changes remain unknown, however, and the possibility that they are an artifact must still be considered.

Lipofuscin in the brains was PAS positive, acid-fast positive, and age associated. A discussion of the significance of lipofuscin in neurons of aging animals is not within the scope of this discussion. However, several published reviews on the subject are available.[4,127]

Tumors arising in the brain of rats have been recently reviewed by Mennel and Zülch,[117] Jänisch and Schreiber,[95] and others.[1,163] They have discussed the relative incidence of spontaneous brain tumors in different rat strains and described the types of tumors that may be seen after experimental induction.

In this study, 30 primary brain tumors (excluding the pituitary tumors) and 1 pineal gland tumor were present in 670 rat brains examined. Twenty neoplasms were granular cell (myoblastoma) tumors and they were the most common primary brain neoplasm. They apparently begin in the meninges and destroy adjacent brain tissue during enlargement. They are usually sharply demarcated from the surrounding brain tissue, but perivascular invasion into surrounding structures may occur. Electron-microscopic studies on two of these tumors showed them to be ultrastructurally similar to granular cell tumors described in humans.[157] Although they were not recognized in the aging female (WAG × BN)F_1 and male WAG/Rij rats of this study, they have been observed in female (WAG × BN)F_1 rats that were part of other long-term studies.[83]

These unusual neoplasms are not confined to rats at the Institute for Experimental Gerontology TNO. They have also been observed in Fischer 344,[82,149] Sprague-Dawley, (Hollander et al., 1976a), Osborne-Mendel,[82,163] Wistar,[102] BUF,[82] and Sprague-Dawley (Spartan substrain) rats[31a] and in a single rat.[173] Therefore, granular cell tumors have been found in most of the commonly used rat strains. They are a

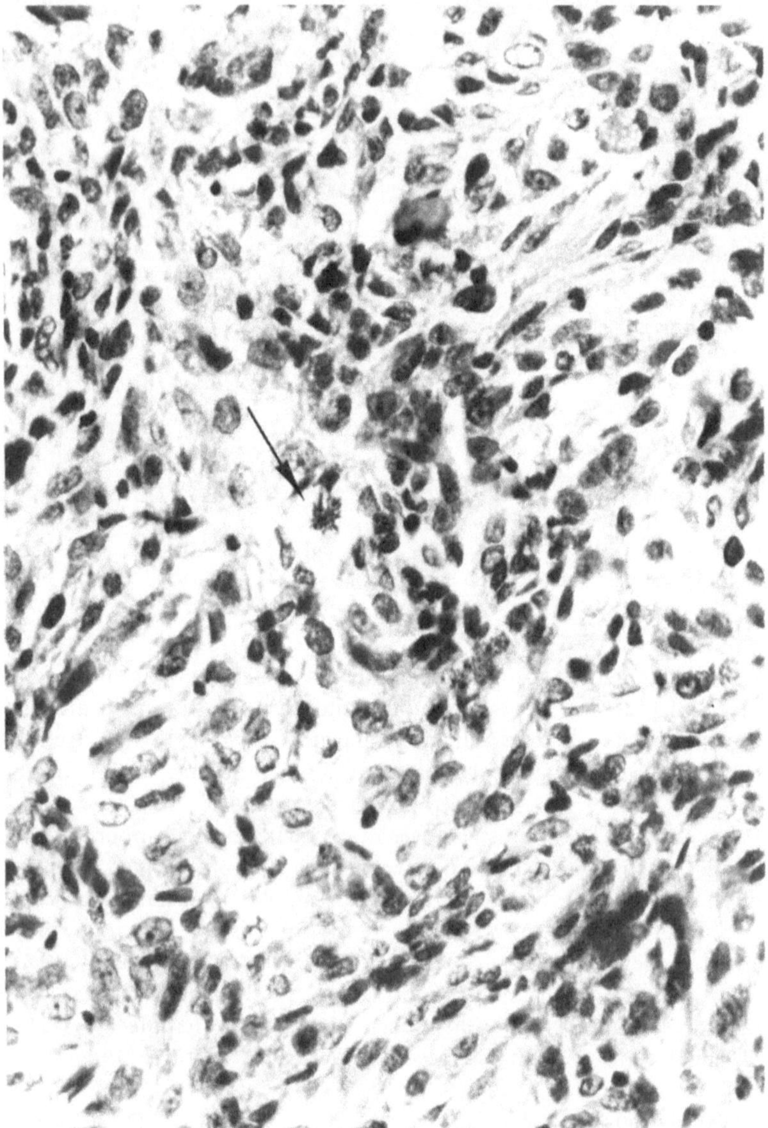

FIGURE 4.130. Higher magnification of the pineal gland tumor illustrated in Figure 4.129, showing cellular pleomorphism and a mitotic figure (↑). (HPS; magnification × 465.)

TABLE 4.32

Incidence of Clinically Severe Paresis or Paralysis in Aging BN/Bi, WAG/Rij, and (WAG × BN)F_1 Rats

Strain	Sex	No. examined	No. with paralysis	%	Mean age (range) in months
BN/Bi	Female	236	2	1	32, 36
	Male	74	1	1	30
WAG/Rij	Female	101	0	0	—
	Male	124	2	2	27, 31
F_1	Female	68	1	1	30
	Male	67	35	52	32 (25—44)

unique tumor that should be investigated in greater detail to determine the cell of origin. They may also serve as a potential model to compare with granular cell (myoblastoma) tumors in man.

P. Spinal Cord and Nerve Roots

1. Nonneoplastic Lesions

Paralysis (or severe paresis) of the posterior one half of the rat's body was clinically recognized in all three strains of aging rats. Severe clinical disease was often accompanied by loss of control of the tail, urinary incontinence, and atrophy of the skeletal muscles in the lumbar region and hind limbs. Table 4.32 shows the number of rats which had severe paresis or paralysis before they died spontaneously or were killed moribund. The BN/Bi and WAG/Rij strains had an incidence of less than 2%. Similarly, female (WAG × BN)F_1 rats had an incidence of only 1%. Male (WAG × BN)F_1 rats, however, had a higher incidence, with 52% developing this disease prior to death.

The age-associated incidence for male (WAG × BN)F_1 rats is shown in Figure 4.131. The age of onset of severe clinical disease began after 24 months with a striking increase in rats over 30 months old. The peak risk period was in the male (WAG × BN)F_1 rats older than 37 months where 80% died with this syndrome.

Microscopic lesions in the spinal cord consisted of demyelination, distended axon sheaths, swollen or absent axons, variable numbers of lipid-filled macrophages, and numerous swollen astrocytes (Figures 4.132 and 4.133). The lesions were limited to the white matter and were most severe in the lateral and ventral funiculi.

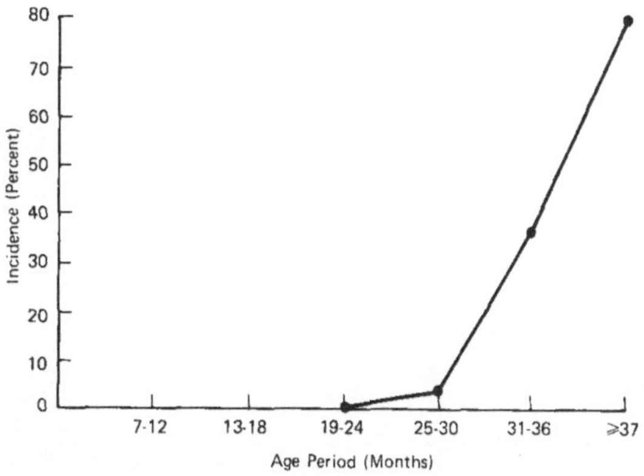

FIGURE 4.131. Percent incidence with age of paralysis in 67 male (WAG × BN)F_1 rats.

In the nerve roots, similar lesions were found (Figure 4.134). They consisted of mild to severe demyelination, distended myelin sheaths, swollen or absent axons, and lipid-filled macrophages (Figure 4.135). In addition, rhomboid clefts resembling cholesterol clefts, hemosiderin, and occasional hemorrhages were observed (Figure 4.136).

Muscles from the lumbar region or hind limbs were atrophic. Lesions consisted of fibers that were atrophic, rowing or clumping of muscular nuclei, swollen muscle fibers, loss of cross striations, and hyalin changes. Occasional fibers were fragmented.

Many of the vertebrae had focal or multifocal aseptic necrosis of bone. The lesions varied in size and location among individual vertebrae. Some rats had involvement of nearly all vertebrae, while only one or two were affected in other rats.

The dura and the epidural space were often

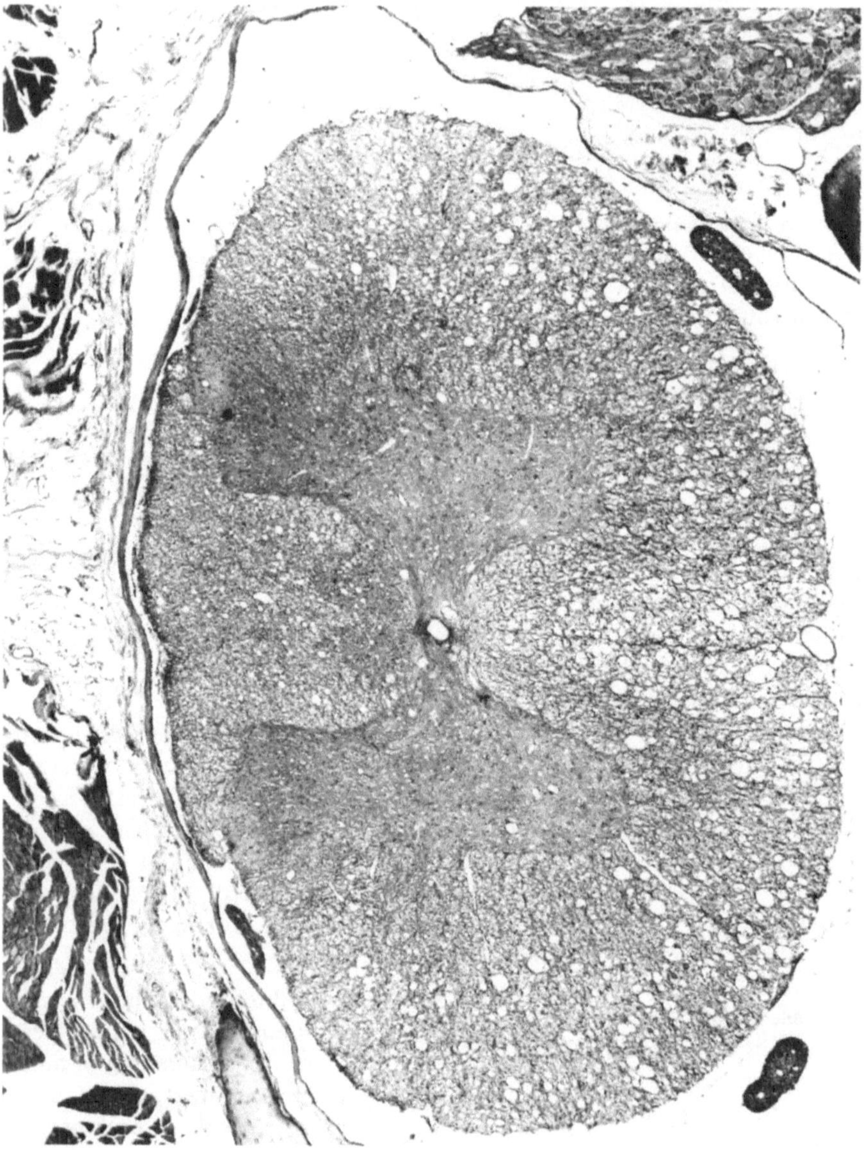

FIGURE 4.132. Spinal cord with distended myelin sheaths and loss of nerve axons in the ventral and lateral columns. (HPS; magnification × 43.) (From Burek, J. D., Van der Kogel, A. J., and Hollander, C. F., *Vet. Pathol.*, 13, 321, 1976. With permission of S. Karger, AG, Basel.)

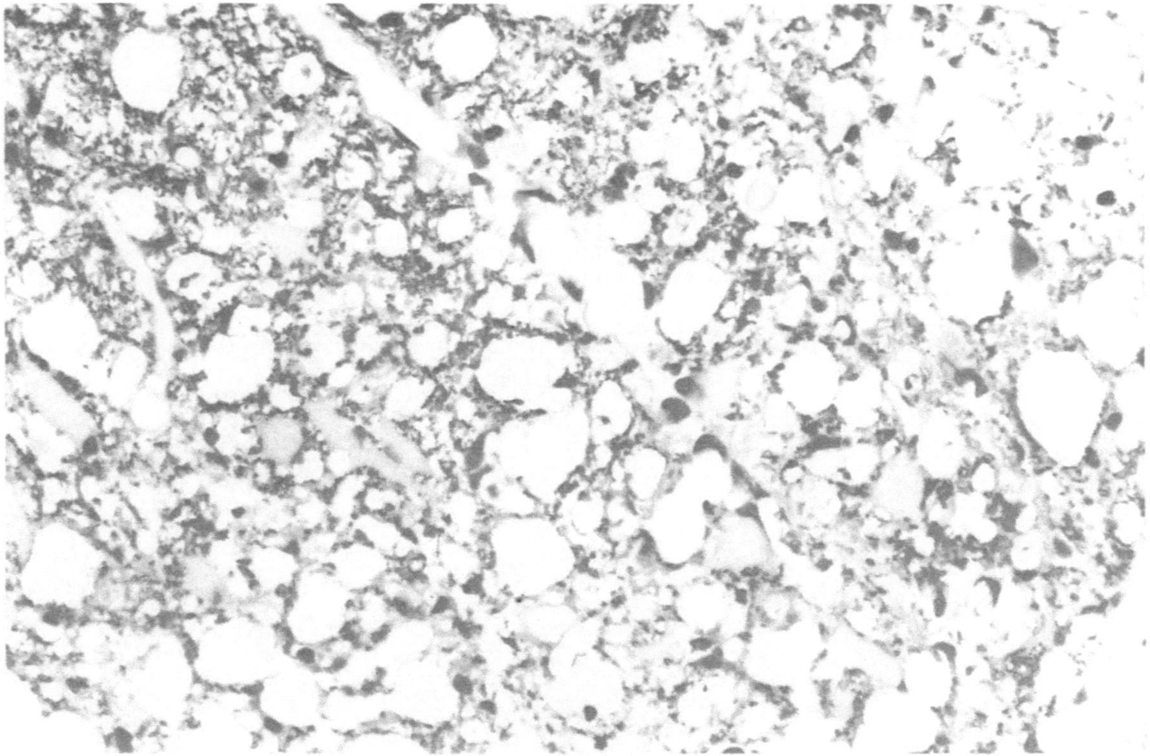

FIGURE 4.133. Severely affected lateral column of the spinal cord with swollen astrocytes, distended axon sheaths, absence of nerve axons, and occasional gitter cells. (PAS - Luxol Fast Blue; magnification × 500.) (From Burek, J. D., Van der Kogel, A. J., and Hollander, C. F., *Vet. Pathol.*, 13, 321, 1976. With permission of S. Karger AG, Basel.)

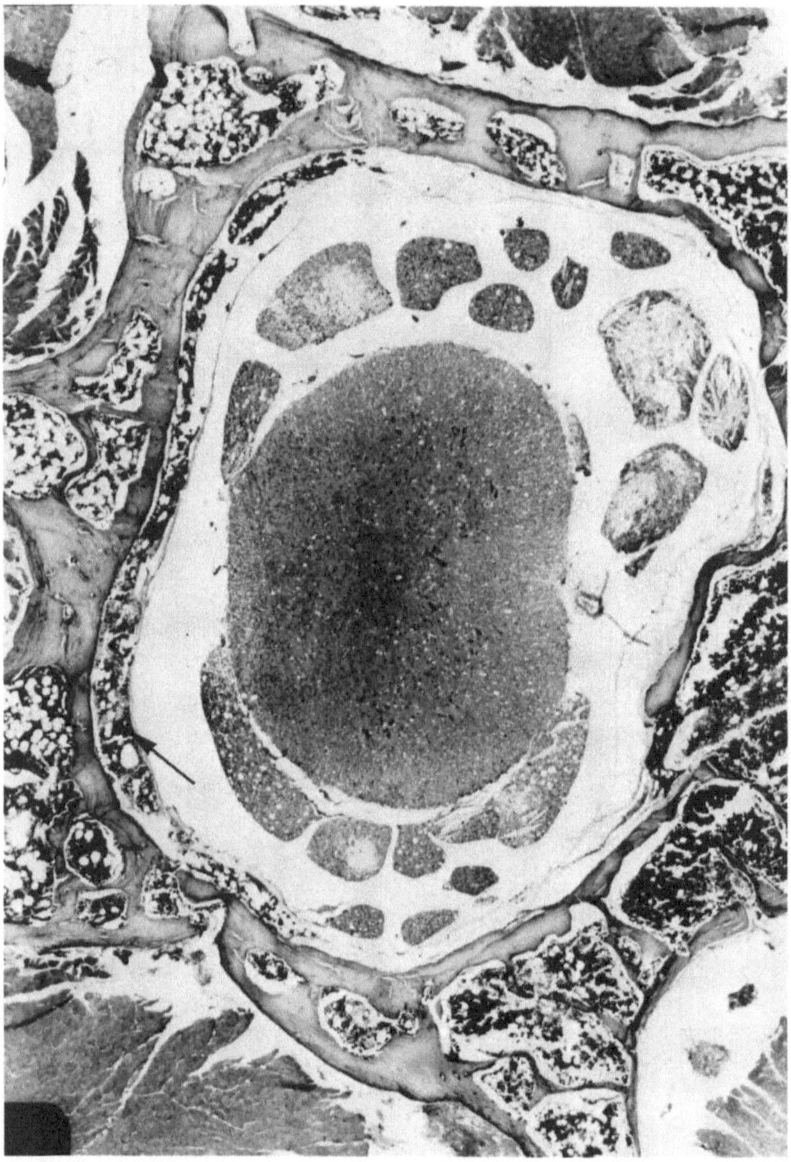

FIGURE 4.134. Cross section through the lumbar spinal cord with severe degeneration of nerve roots, relatively normal spinal cord, and extramedullary hematopoiesis in the dura (↑). (HPS; magnification × 32.)

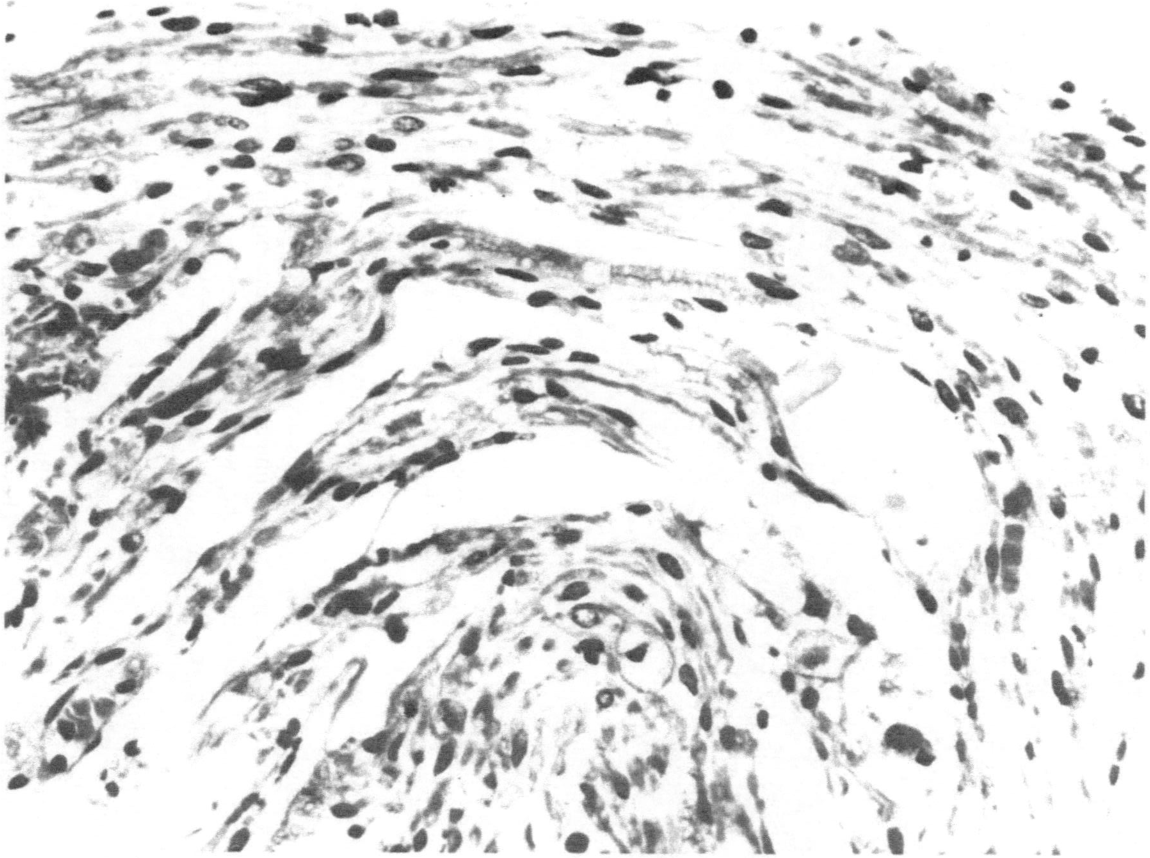

FIGURE 4.135. Wallerian degeneration in ventral nerve root. (PAS - Luxol Fast Blue; magnification × 210.) (From Burek, J. D., Van der Kogel, A. J., and Hollander, C. F., *Vet. Pathol.*, 13, 321, 1976. With permission of S. Karger AG, Basel.)

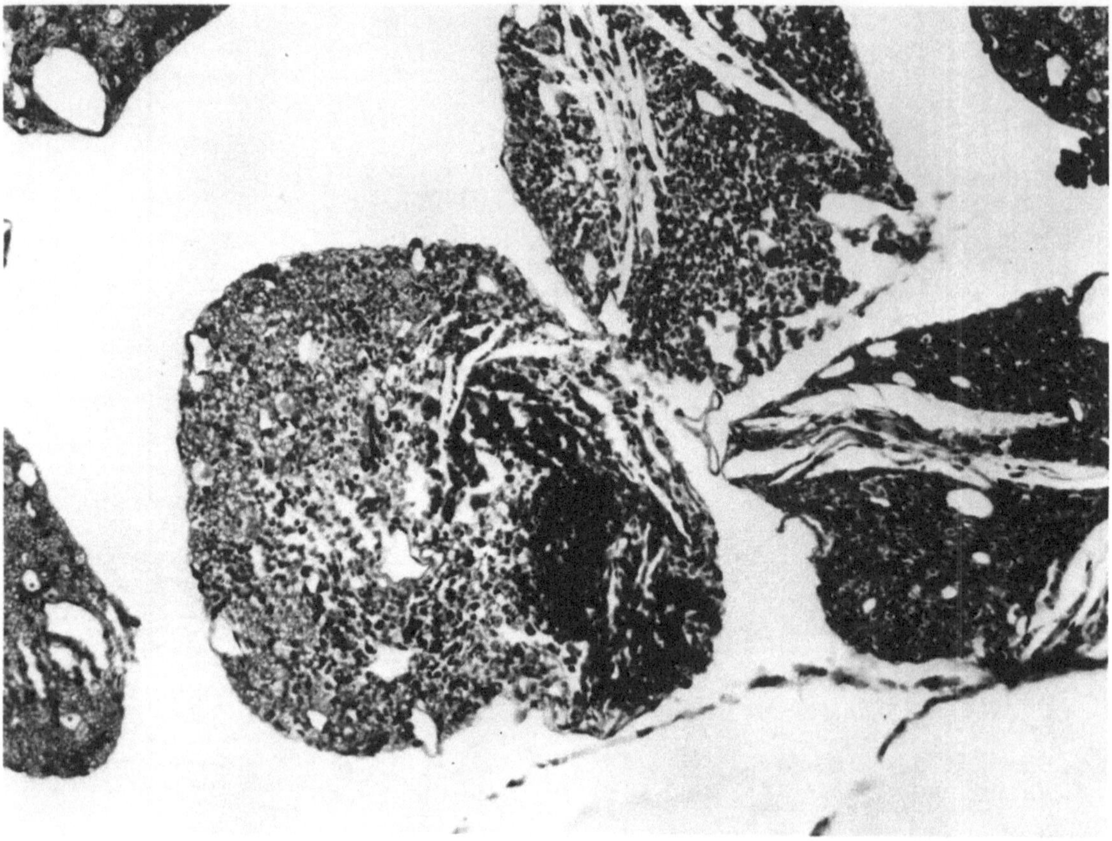

FIGURE 4.136. Rhomboid clefts and recent hemorrhage in nerve roots. (HPS; magnification × 240.) (From Burek, J. D., Van der Kogel, A. J., and Hollander, C. F., *Vet. Pathol.*, 13, 321, 1976. With permission of S. Karger AG, Basel.)

replaced by fibrous tissue. Cartilage, bone, mineral deposits, and even bone marrow elements (Figure 4.134) were occasionally present. Many of the intervertebral foramina were also replaced by or contained increased fibrous tissue, and fibrosis of nerve root sleeves was observed.

When present in the routine sections, intervertebral disks often had degenerative changes. Protrusion or rupture of the disks, resulting in compression of the spinal cord, was not common, but it was occasionally seen.

Peripheral nerves were not included in the routine material for histopathology. Therefore, they could not be evaluated. Additional studies were done, however, and these are reported in Chapter 7.

2. Neoplastic Lesions

Neoplasms originating from the spinal cord or spinal nerve roots were not observed. A few neural tumors arising from peripheral nerves, such as neurofibrosarcomas, were seen and are documented under the organ system in which they were found.

3. Discussion

Burek et al.[27] originally reported the incidence and described the lesions of spontaneously occurring posterior paralysis in aging BN/Bi, WAG/Rij, and (WAG × BN)F_1 rats. They reported a high incidence in (WAG × BN)F_1 males and also reviewed the literature on this disease. Since that study was published, additional studies have further characterized the pathogenesis of the disease and the observed lesions. Therefore, the additional studies, some results from the initial publication,[27] and a more in-depth discussion of this disease are presented in Chapter 7.

Spontaneous tumors of the spinal cord and spinal nerve roots were not observed in the rats of this study. However, van der Kogel[177] has done extensive research on the late effects of irradiation on the spinal cords of BN/Bi, WAG/Rij, and (WAG × BN)F_1 rats. He has observed numerous irradiation-induced tumors as well as degenerative changes and has reviewed the literature concerning these lesions in rats. Others have also reviewed the current literature on neoplastic lesions of the rat spinal cord.[1,163]

Q. Musculoskeletal System

1. Nonneoplastic Lesions

The muscular and skeletal systems were not evaluated systematically. Therefore, age-associated patterns for the observed lesions could not be established. As a result, only a few brief comments about the observed lesions are given in this section.

Generalized atrophy of the skeletal muscles occurred in some rats as a part of generalized cachexia and wasting. Such changes were accompanied by decreased amounts of body fat, roughened hair coat, and an overall unkempt appearance. Rats with this appearance usually had a large tumor or debilitating disease somewhere in the body that may have contributed to the loss of body condition and muscle atrophy.

Some rats had atrophy of the muscles of the lumbar back and hind limbs (Figure 4.137). These changes were invariably associated with severe lesions of the spinal cord and spinal nerve roots as described in Section II. P, with male (WAG × BN)F_1 rats being the most frequently affected.

Bone lesions were sporadic. The most common lesions included focal aseptic necrosis of vertebral bone (Figure 4.138) and focal chondromucoid degeneration of bone (Figure 4.139). Additional bone sections were available from rats with paralysis, and in these sections, lesions were often seen such as degeneration of intervertebral disks, focal prolapse (herniation) of interverterbral disks, and marginal osteophytes in the spinal canal.

2. Neoplastic Lesions

Osteogenic sarcomas occurred in five rats. The rats affected included a 29-month-old (WAG × BN)F_1 male, a 13-month-old BN/Bi male, two BN/Bi females 26 and 30 months of age, and a 38-month-old female WAG/Rij. All five tumors had osteoid formation, marked cellular pleomorphism, mitotic figures, extensive local invasion, and severe destruction of involved tissues. Metastases were only observed in the two BN/Bi female rats.

The only recognized tumor in the muscle was a lipoma in a 33-month-old male BN/Bi rat.

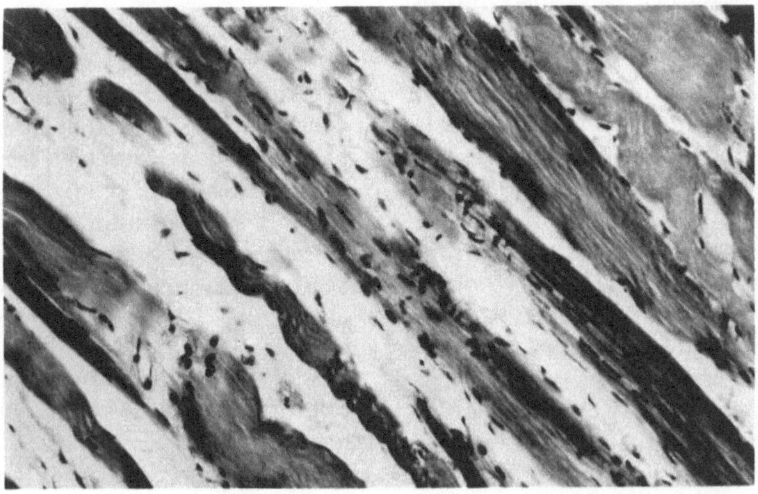

FIGURE 4.137. Atrophy of skeletal muscle in the lumbar back of a paralyzed rat. (HPS; magnification × 210.)

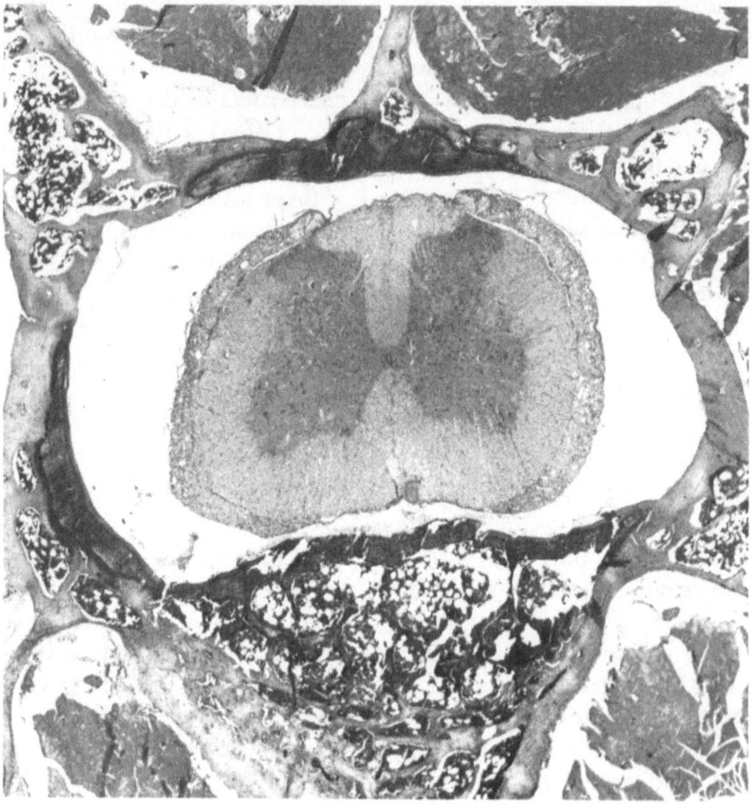

FIGURE 4.138. Aseptic bone necrosis of a vertebral body. (HPS; magnification ×17.)

3. Discussion

Nonneoplastic lesions in bones and joints of mice and rats have been described and the literature reviewed by Sokoloff.[158–160] In addition, degenerative joint disease has also been studied in the Mastomys (*Praomys natalensis*).[160] Lesions documented in these investigations included aseptic necrosis of bone, chondromucoid degeneration of articular cartilage, osteoarthritis, osteophytes at the margins of intervertebral disks, kyphosis, and herniation of intervertebral discs. The most common change in the rat was cystic chondromucoid degeneration of the articular cartilage.

The above-cited references provide a much greater in-depth study of bone lesions of rats and small rodents than could be done in the present study. Despite these excellent studies, there is still a need for more data on the specific age-associated patterns of the various lesions and how they relate to the aging of the rats as a whole.

and how they relate to the aging of the rats as a whole.

Lesions of the skeletal muscles of rats have also been described.[9,26,61,62,158,159,166] The most common lesion was atrophy either secondary to debilitating disease and apparently independent of any obvious cause[9] or secondary to nerve root degeneration.[26,166]

The interrelationship of posterior paralysis, muscle atrophy, and nerve root degeneration is dealt with in greater detail in Chapter 7 where data from additional experiments are presented.

Spontaneous neoplastic lesions of the muscular and skeletal systems are relatively rare in rats,[1,112,163] but tumors have been experimentally induced in both bone[111] and skeletal muscle.[35]

Five osteogenic sarcomas occurred in the 670 rats of this study. One was in a 13-month-old rat, but four were in rats older than 26 months of age.

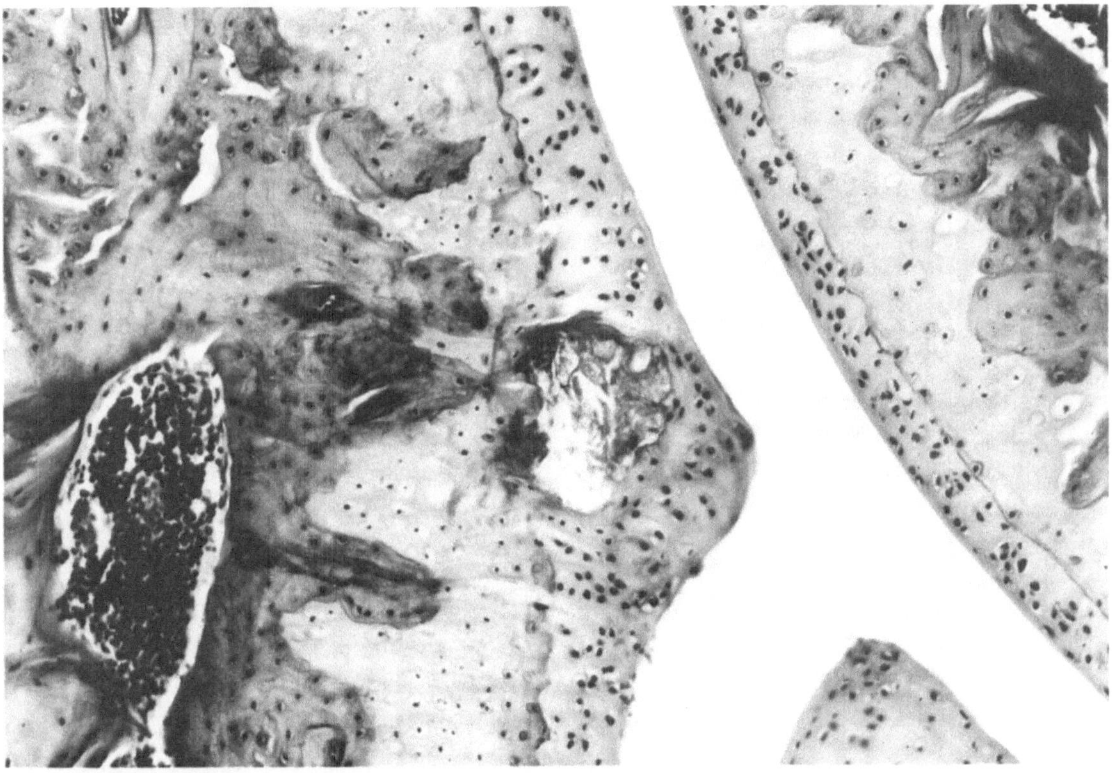

FIGURE 4.139. Cystic chondromucoid degeneration of articular cartilage. (HPS; magnification × 195.)

R. Skin and Subcutaneous Tissues (Excluding Mammary Tissue)

1. Nonneoplastic Lesions

Nonneoplastic lesions were recognized in the skin of some aging rats, but they were usually incidental or nonspecific changes. Such findings included roughened and dirty hair coat, slight alopecia, focal traumatic lesions, and focal mild inflammation.

2. Neoplastic Lesions

Tumors of the skin and subcutaneous tissue were relatively common, as shown in Table 4.33. Epithelial neoplasms consisted of both benign and malignant tumors. The benign lesions included squamous papillomas and inverted papillomas, a basal cell tumor, and epidermal inclusion cysts. Squamous papillomas were typical wart-like growths composed of papillary structures lined by acanthotic and hyperkeratotic squamous epithelium. Inverted papillomas were similar, but they grew down into the dermis as well as some growth out from the surface, thus forming a shallow cavity filled with keratin and debris (Figure 4.140). Epidermal inclusion cysts were found in three female BN/Bi rats. They were located in the subcutaneous tissue along the dorsal midline of the back. They consisted of a cystic structure lined by stratified squamous epithelium and were filled with keratinaceous debris. The last benign tumor, a basal cell tumor composed of cords and ribbons of well-differentiated basal cells, was found in the skin of a 36-month-old (WAG × BN)F_1 rat.

Two malignant epithelial tumors, both squamous cell carcinomas, were also found. Both were small, locally invasive neoplasms. Metastases were not observed with either tumor.

Benign nonepithelial tumors included three fibromas composed of parallel collagen bundles which had a wavy appearance with increased numbers of fibroblasts, two lipomas consisting of nodules composed of normal appearing fat cells, and two hemangioendotheliomas arising in the subcutaneous tissue and composed of numerous endothelial-lined and blood-filled spaces.

Six male rats died with large fibrosarcomas

TABLE 4.33

Occurrence of Tumors in the Skin and Subcutaneous Tissues of Aging BN/Bi, WAG/Rij, and (WAG × BN)F_1 Rats

Type of tumor	Strain	Sex	No. with tumor/ no. examined	%	Mean age (range) in months
Squamous papillomas and inverted papillomas	Bn/Bi	Male	4/74	5	30 (24—37)
	WAG/Rij	Male	4/124	3	25 (21—27)
	F_1	Male	2/67	3	28, 42
	WAG/Rij	Female	2/101	2	33, 36
Basal cell tumor	F_1	Male	1/67	1	36
Epidermal inclusion cyst	BN/Bi	Female	3/236	1	29 (23—34)
Squamous cell carcinoma	BN/Bi	Female	1/236	<1	27
	F_1	Male	1/67	1	35
Fibroma	BN/Bi	Male	1/74	1	26
	WAG/Rij	Male	1/124	1	21
	F_1	Male	1/67	1	36
Fibrosarcoma	WAG/Rij	Male	3/124	2	21 (19—23)
	F_1	Male	3/67	4	34 (32—36)
Lipoma	WAG/Rij	Male	1/124	1	27
	F_1	Female	1/68	1	40
Hemangioendothelioma	BN/Bi	Male	1/74	1	23
	F_1	Male	1/67	1	36
Malignant melanoma	BN/Bi	Female	5/236	2	31 (26—34)
	BN/Bi	Male	3/74	4	31 (29—35)

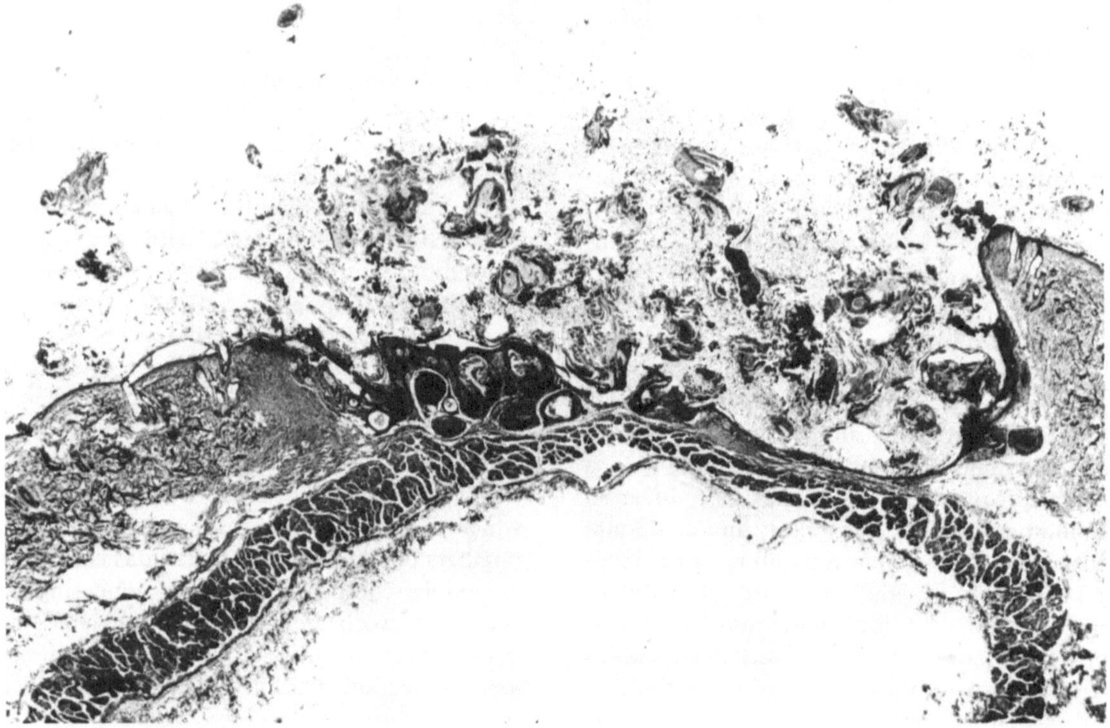

FIGURE 4.140. Inverted papilloma in the skin. (HPS; magnification × 36.)

composed of numerous spindle-shaped cells, mitotic figures, occasional giant cells, and local invasion of surrounding tissues. None had distant metastases. however.

Finally, eight malignant melanomas were present in BN/Bi rats. All were located on the extremities, that is, on the tail, the tip of the ear, the eyelid, or the lip. Grossly, all were relatively small dark to black tumors and were less than 8 mm in diameter. Histologically, the cells were heavily pigmented melanocytes that had extensively invaded the epidermis and surrounding musculature. Three of the eight melanomas had spread to regional lymph nodes. Grossly, the lymph nodes were enlarged and dark in color. Histologically, metastatic tumor was present in these nodes. Five of the eight did not have obviously enlarged regional lymph nodes at the time of gross necropsy. As a result, the regional lymph nodes were not included for routine histopathology, and it is not known whether any of the five rats without gross metastases had microscopic metastases of the malignant melanomas.

3. Discussion

Zackheim,[189] as well as Squire and Goodman[163] and Altman and Goodman,[1] have alluded to the infrequency of spontaneous tumors of the skin and subcutaneous tissues in rats. In the present study, skin tumors were not common, but on the other hand, they were not rare. Of the 670 rats evaluated, 16 had benign epithelial tumors, 2 had malignant epithelial tumors, 13 had mesenchymal tumors in the subcutaneous tissues, and 8 had malignant melanomas. Nearly all of these tumors were in rats older than 24 months of age.

The BN/Bi rat has brown hair. It has heavily pigmented skin and also develops melanomas, as reported previously.[80] Eight (Table 4.33) of these rats developed malignant melanomas of the extremities. Although the number of these tumors is very small, it would appear that they readily metastasize since three of the eight (38%) had spread to regional lymph nodes. The risk of these tumors metastasizing may even be somewhat higher because the regional lymph nodes were not examined histologically from the five rats without metastases. Regardless, melanomas in the BN/Bi rat are malignant tumors and they have a high risk of metastasizing.

S. Mammary Gland

1. Nonneoplastic Lesions

All females had some increased prominence of mammary ducts or lobules suggestive of stimulation or hyperplasia. The amount of mammary tissue hyperplasia was variable and a spectrum of changes was recognized. Most of the changes could be lumped into three major categories. The first was characterized by hyperplasia of the acinar tissue (Figure 4.141). The second was a mild ectasia of ducts and ductules (Figure 4.141). The ducts were dilated and lined by cuboidal epithelium and their lumina usually contained a pink, proteinaceous, acellular material. The third consisted of large cystic ducts that were similar to the ectatic ducts but were greatly dilated cyst-like structures lined by flat epithelium. Some glands had only one of the three changes, while others had various combinations.

2. Neoplastic Lesions

The incidence of mammary gland tumors was relatively high for the rats of the three strains; 42% of female WAG/Rij, 22% of female (WAG × BN)F_1, and 18% of female BN/Bi rats died with mammary tumors. The BN/Bi and (WAG × BN)F_1 rats had only one tumor per rat. Of the 43 WAG/Rij rats, 11 (26%) had two neoplasms while the other 32 had only one mammary tumor.

The age-associated incidence was calculated for all mammary tumors. The calculations were based on the presence of a tumor, irrespective of the histological diagnosis. As is seen in Figure 4.142. the risk of female rats dying with a mammary tumor increased in all age groups, with the peak incidence occurring in rats of 37 months or older. The peak incidence was 40% for (WAG × BN)F_1 females, 38% for WAG/Rij females, and 15% for BN/Bi females. The specific diagnoses of the tumors found are listed in Table 4.34 and their descriptions are summarized in the following paragraphs.

Fibroadenomas were firm, white to tan, nodular masses that ranged in diameter from a few millimeters to several centimeters. Microscopically, they were composed of fibrous connective tissue and epithelial cell structures in varying proportions (Figure 4.143). In some, the fibrous elements predominated, while in others it was the epithelial component. The epithelial cells formed lobules composed of aci-

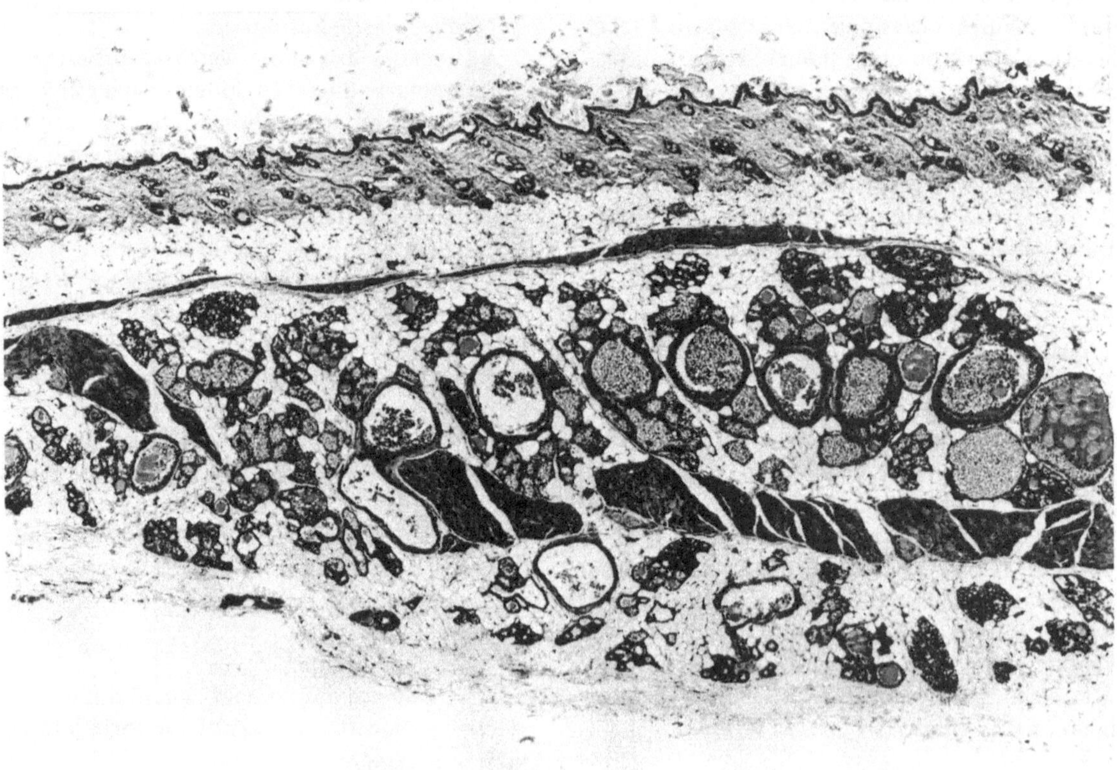

FIGURE 4.141. Slight lobular (acinar) hyperplasia and ductal ectasia of the mammary gland. (HPS; magnification × 32.)

nar-like structures including ducts and ductules. The epithelial structures were surrounded by the fibrous elements that often formed wide bands or trabeculae between the epithelial structures forming a pericanalicular fibroadenoma. Some of the fibroadenomas had focal areas of atypia or "piling-up" of epithelial cells and increased mitoses.

Such tumors were seen in 22% of female WAG/Rij, 16% of female (WAG × BN)F_1, 11% of female BN/Bi, 1% (1 of 74) of male BN/Bi, and 1% (1 of 67) of male (WAG × BN)F_1 rats. The age-associated incidence (Figure 4.144) showed a peak incidence during the period of 31 to 36 months for both the female BN/Bi and WAG/Rij rats. The peak incidence was 20% in WAG/Rij females, but dropped to 11% in the oldest rats (≥37 months). Similarly, the peak incidence in BN/Bi females was 12%, but dropped to 6% in the oldest group (≥37

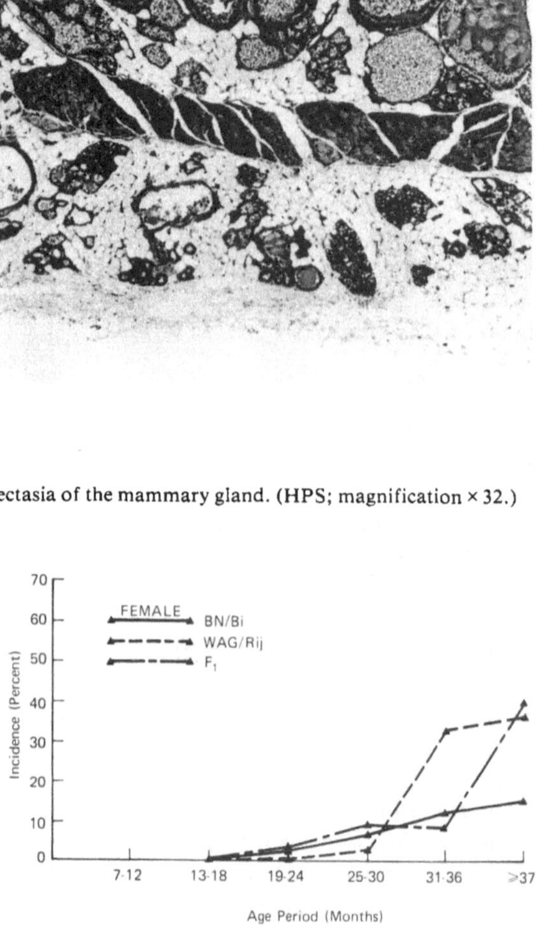

FIGURE 4.142. Percent incidence with age of all tumors arising in the mammary glands of 236 BN/Bi, 101 WAG/Rij, and 68 (WAG × BN)F_1 female rats.

months). The (WAG × BN)F_1 rats, on the other hand, showed a trend for an increased incidence with age so that the peak incidence of 20% occurred in the oldest rats (≥37 months).

TABLE 4.34

Occurrence of Tumors in the Mammary Glands of Aging BN/Bi, WAG/Rij, and (WAG × BN)F_1 Rats

Strain	Sex	No. examined	Type of tumor	No. of each type	%	Mean age (range) in months
BN/Bi	Female	236	Fibroadenoma	26	11	30 (23—40)
			Adenocarcinoma	14	6	30 (19—37)
			Sarcoma	2	<1	36, 37
			Adenoma	1	<1	34
	Male	74	Fibroadenoma	1	1	36
WAG/Rij	Female	101	Fibroadenoma	22	22	34 (24—42)
			Adenocarcinoma	23	23	37 (27—44)
			Fibroma	2	2	31, 31
			Adenoma	2	2	34, 36
			Sarcoma	1	1	34
			Carcinosarcoma	1	1	34
	Male	124	None	—	—	—
F_1	Female	68	Fibroadenoma	11	16	30 (22—38)
			Adenocarcinoma	4	6	35 (26—38)
	Male	67	Adenocarcinoma	1	1	26

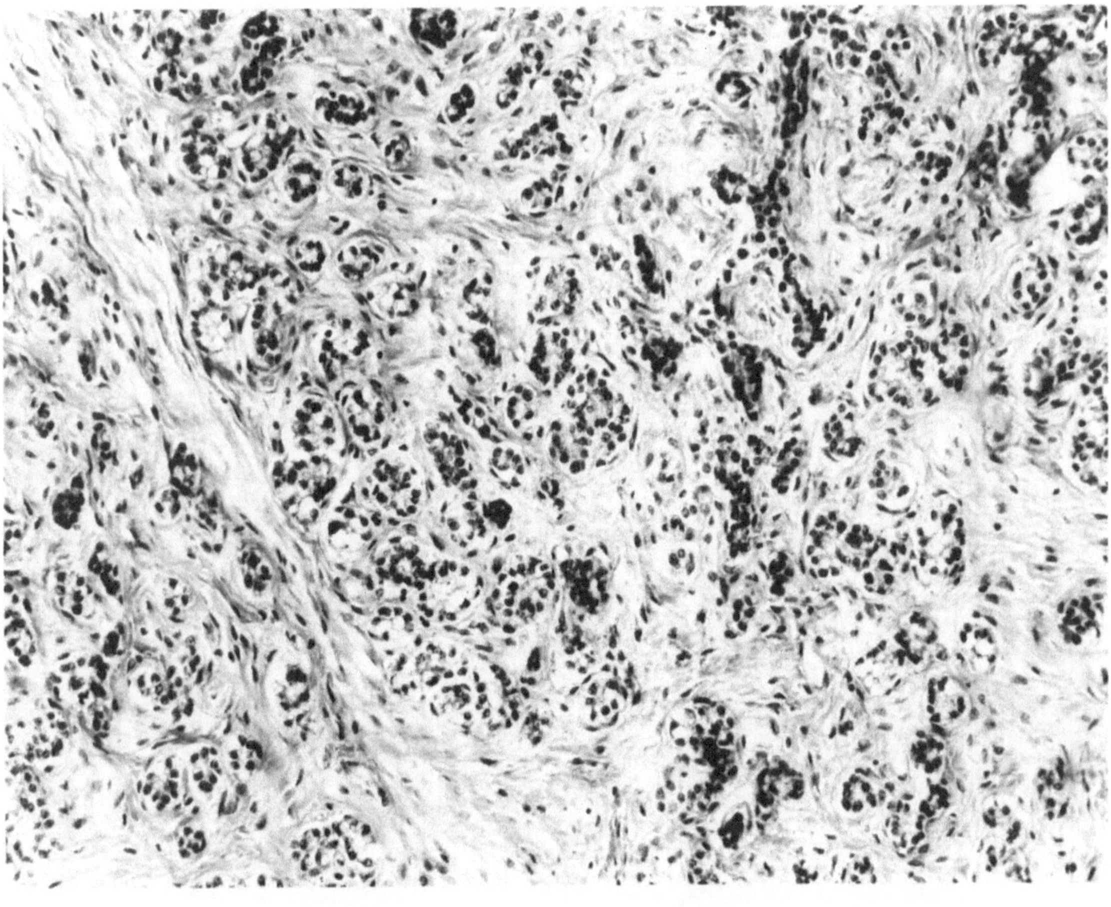

FIGURE 4.143. Fibroadenoma of the mammary gland. (HPS; magnification × 210.)

Fibromas were recognized in two WAG/Rij females, both 31 months of age. They were both large, firm, white tumors that arose in mammary gland tissues. They were composed of well-differentiated fibrocytes and contained abundant collagen. They resembled the fibrous tissue component of the fibroadenomas, but epithelial structures could not be found.

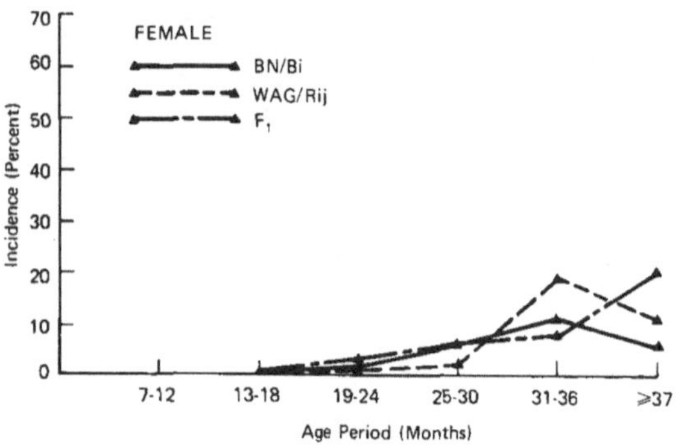

FIGURE 4.144. Percent incidence with age of fibroadenomas arising in the mammary glands of 236 BN/Bi, 101 WAG/Rij, and 68 (WAG × BN)F_1 female rats.

Three adenomas occurred in the 411 female rats of this study. All were well-differentiated epithelial tumors with approximately normal amounts of epithelium, myoepithelium, and stroma, but they were circumscribed masses that were larger than and were clearly demarcated from the normal mammary tissue.

Adenocarcinomas were composed of epithelial cells that were derived from glandular or duct epithelium. The most common form had papillary patterns (Figure 4.145), but follicular patterns were also recognized. The tumors were generally surrounded by a fibrous capsule, but some showed invasion into or beyond the capsule. Most were composed of well-differentiated epithelial cells, often with vacuolated

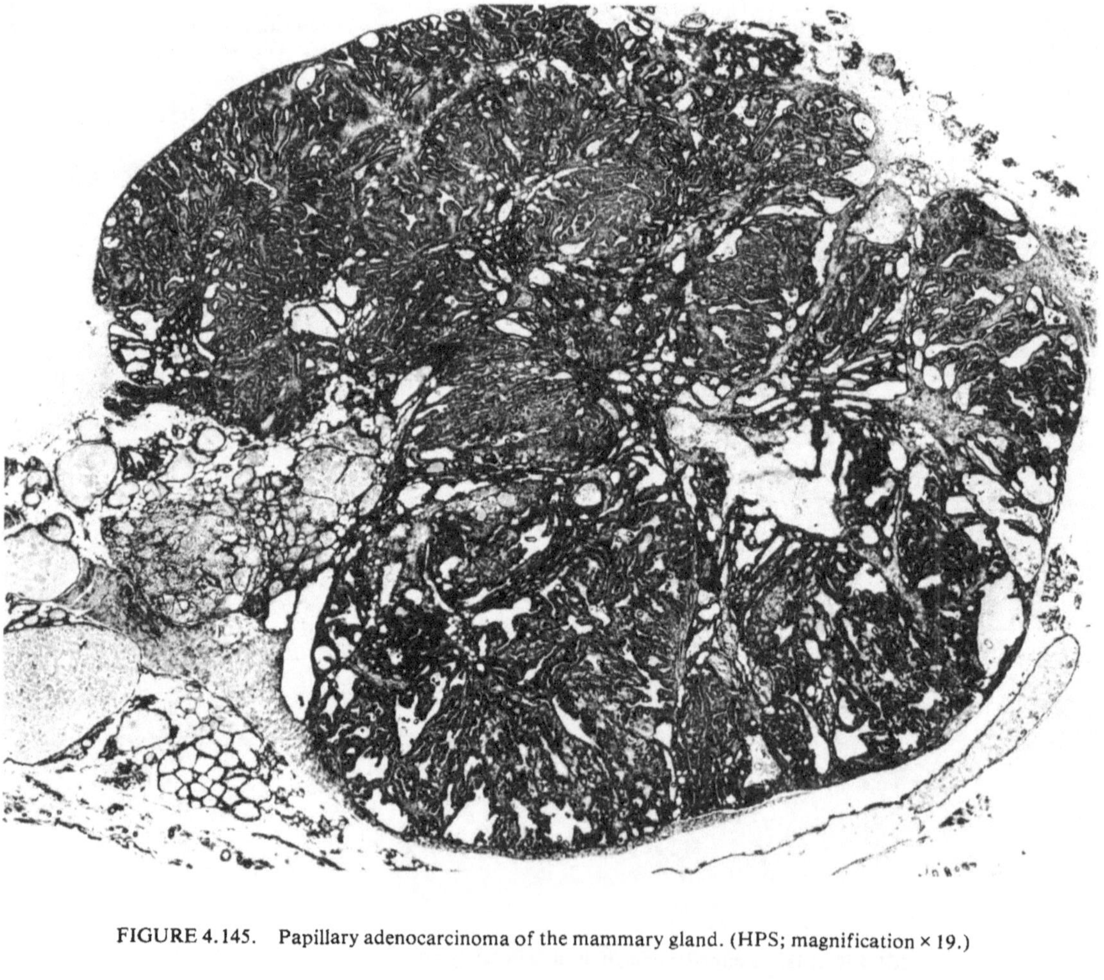

FIGURE 4.145. Papillary adenocarcinoma of the mammary gland. (HPS; magnification × 19.)

pink cytoplasm. Most had areas with secretory material that was pink or basophilic, homogeneous or granular. Mitoses were found in most carcinomas. Such tumors were found in 23% of female WAG/Rij, 6% of female (WAG × BN)F_1 and 6% of female BN/Bi rats. A carcinoma was also seen in a (WAG × BN)F_1 male rat. Only one of the carcinomas metastasized to the lung in a 27-month-old female BN/Bi rat.

Unlike the incidence for fibroadenomas, the age-associated incidence for BN/Bi and WAG/Rij females shows the older the rat, the greater its risk of dying with a carcinoma (Figure 4.146). The peak risk period occurred in the oldest rats (>37 months), reaching 33% for female WAG/Rij and 8% for female BN/Bi rats. Too few carcinomas (only four tumors) were observed in (WAG × BN)F_1 females to calculate the age-associated incidence, but the mean age for these carcinomas was 35 months.

Three sarcomas arose in the region of the mammary gland. They were composed of spindle-shaped cells that had indistinct cell boundaries, with elongated spindle-shaped nuclei and abundant collagen in the surrounding stroma. They were highly invasive and contained numerous mitoses, but distant metastases were not found.

A carcinosarcoma was found in one 34-month-old female WAG/Rij rat. It was composed of intermingled components of epithelial and mesenchymal cells. It was locally invasive, but distant metastases were not found.

3. Discussion

Hyperplasia of the acinar tissue, duct ectasia, or both occurred in nearly all aging female rats of this study. The cause could not be determined, however. It was found that 95% of the WAG/Rij females and 83% of (WAG × BN)F_1 females died with pituitary gland tumors, some of which might have been secreting a hormone such as prolactin. On the other hand, only 26% of BN/Bi females had pituitary gland tumors, but they still exhibited mammary gland stimulation similar to that seen in WAG/Rij or (WAG × BN)F_1 females.

A discussion of the many hormonal factors that can affect rat mammary tissue is not within the scope of this thesis. However, a brief review of normal mammary development, some hormonal influences, as well as some methods of

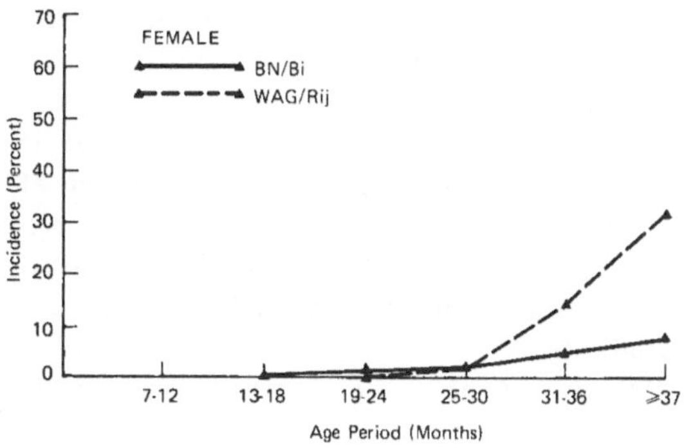

FIGURE 4.146. Percent incidence with age of adenocarcinomas arising in the mammary glands of 236 BN/Bi and 101 WAG/Rij female rats.

experimental tumor induction has been carried out by Young and Hallowes.[188]

Mammary tumors in rats have been reviewed in several recent publications.[1,163,188] In addition, extensive data are being collected on mammary tumors in WAG/Rij, BN/Bi, and Sprague-Dawley rats following various combinations of X-ray irradiation, neutron irradiation, castration, and hormone treatment. The results of the experimentally induced tumors are being compared to those that have spontaneously arisen,[25] and some preliminary findings have been reported.[181]

The risk that a BN/Bi or WAG/Rij rat would die with a mammary tumor seemed to increase with age. However, when fibroadenomas and adenocarcinomas were evaluated separately, a difference was found. Fibroadenomas had a peak risk period between 31 to 36 months old, while those that were 37 months or older had less of a risk. In contrast, the percent incidence of adenocarcinomas continued to increase in each age group. A similar phenomenon was shown by Boorman and Hollander,[17] who evaluated 290 female WAG/Rij rats. Their rats consisted of both virgins and retired breeders, however. They showed a peak risk period for fibroadenomas at 38 months of age and a slightly lower risk at 42 months of age.

III. GENERAL DISCUSSION

Every organ and organ system from rats of this study had age-associated lesions. In some organs, such as the pituitary gland, neoplastic lesions were the most common morphologic alteration. In other organs, like the heart, nonneoplastic lesions predominated. Still other organs had various combinations of both neoplastic and nonneoplastic lesions. The lesions observed, their age of onset, and their effect on life span varied among the different strains and sexes. In addition, individual rats of one sex and strain also showed a considerable variability.

The age of the onset of lesions in the population of rats occurred at about the same age as the bend in their respective survival curves, that is, the time or age that the lesions began to appear in a population was the same time that rats began to die and the percent of surviving animals decreased. Some rats developed lesions earlier in life and died at relatively young ages of 12 to 18 months. Others did not acquire similar lesions until later in life and did not die until much older (greater than 30 months of age). In general, most age-associated neoplastic and nonneoplastic lesions began after the population was 10 to 12 months of age. The number of rats with specific lesions and the severity of these lesions tended to increase as the population aged. The increase was greatest in the population after it reached 24 months of age.

Even though the incidence of most lesions continued to increase as the population aged, there were exceptions, so that several patterns were recognized. Some tumors, for example, peaked at a relatively young age so that older rats were at a much lower risk than their younger cohorts. Other lesions, both neoplastic and nonneoplastic, reached a peak incidence followed by a plateau in the incidence curves. Therefore, many possible combinations were observed in the incidence patterns from these aging rats.

All of the observed lesions will influence the function of organs and cells and thereby can affect experimental results. The impact that such morphological alterations will have on any given study will depend on the strain and sex of rat used and the specific parameters that are to be studied by an individual investigator.

Chapter 5

EVALUATION OF THE AGE-ASSOCIATED RISK FOR MULTIPLE TUMORS AND TUMOR METASTASES IN AGING RATS

I. INTRODUCTION

As the populations of rats aged, it became apparent that the animals died with multiple lesions. This appeared to be true for neoplastic as well as nonneoplastic lesions. In addition, the age-associated risk for cancer to metastasize seemed to be greater in older than younger rats. Therefore, the purpose of this chapter is to evaluate the occurrence of multiple tumors in rats and their occurrence in rats of different ages.

II. AGE-ASSOCIATED RISK FOR MULTIPLE NEOPLASTIC LESIONS

The total tumors were tabulated irrespective of tumor type or their apparent benign or malignant morphology. Also, each separate tumor was counted. For example, a rat with three mammary gland fibroadenomas was tabulated as a rat with three tumors. Similarly, if a pheochromocytoma was found in each adrenal gland, they were counted as two tumors in that rat. As most pathologists know, this type of breakdown can be extremely difficult. However, an intense effort was made to assure uniformity of criteria for tabulating tumor numbers per rat for each of the three rat strains.

The numbers of rats that died with 0, 1, or multiple tumors in each age group are listed in Table 5.1* for BN/Bi females, Table 5.2 for BN/Bi males, Table 5.3 for WAG/Rij females, Table 5.4 for WAG/Rij males, Table 5.5 for (WAG × BN)F_1 females, and Table 5.6 for (WAG × BN)F_1 males.

In general, there was a trend for the average number of tumors per rat to increase in each age group. This was true for the average number of tumors in all dead rats and the average in tumor-bearing rats. The two exceptions to this statement were the BN/Bi and WAG/Rij males which did not show an increase in the average number of tumors in all dead rats. However, the BN/Bi males did show a trend for the average number of tumors to increase in tumor-bearing rats. For the WAG/Rij females, there was nearly a twofold increase in the average number of tumors per rat with increasing age. Other groups such as the BN/Bi females had an increase in average number of tumors, but less than the WAG/Rij females.

The average number of tumors in all dead rats was evaluated by linear regression analysis. This evaluation was conducted to test the statistical significance of the data shown in Tables 5.1 to 5.6. The results are shown in Figure 5.1. There was a clear trend for all groups to develop increased numbers of tumors with age except the male BN/Bi and WAG/Rij rats. Some rats, however, had a much lower risk of developing cancers and died without tumors, even in the oldest age groups. This was especially true for the BN/Bi rats.

III. AGE-ASSOCIATED RISK FOR MALIGNANT NEOPLASMS TO METASTASIZE

Malignant neoplasms were relatively common, as cited in Chapter 4. Some were considered to be malignant as judged by local invasion or direct extension from the primary site to surrounding tissues. Others were classified as malignant based on their cellular morphology. Some primary tumors also had distant metastases.

Tables 5.7 to 5.9 list tumors in male and female BN/Bi, WAG/Rij, and (WAG × BN)F_1 rats, respectively, that had distant metastases. Excluded were all tumors where distant metastases were not found. Also excluded were lymphoreticular tumors. BN/Bi females had the greatest number of metastatic tumors, with 38 in the 236 (16%) rats. The other groups had metastatic cancer in between 3 to 9% of their total populations.

The number of BN/Bi females that had metastatic neoplasms (38) was sufficiently large to permit an evaluation of the age-associated inci-

* All tables and figures appear at the end of this chapter.

dence. The results are shown in Figure 5.2. The older the rat, the greater was its risk of dying with a metastatic cancer.

The other rat groups had fewer tumors with distant metastases, with numbers ranging from only 3 to 7 (3 to 9% of the total rats per group). Therefore, an assessment of the age-associated incidence is more difficult. In these groups, 26 metastatic tumors were found. Only 4 were in rats younger than 24 months, 12 were in rats between 24 to 30 months, 7 were found in rats at 31 to 36 months, and only 3 were in rats 37 months of age or older.

One reason for the large numbers of tumor metastases in the BN/Bi females was the frequency of adrenal cortical carcinomas. Metastases of these tumors accounted for about one third of the total metastatic tumors in that strain. Table 5.10 lists the adrenal cortical carcinomas in these females and the number during the various age ranges. There was a trend for the metastatic adrenal cortical carcinomas to be age related.

IV. DISCUSSION

In general, with increasing age the rats of this study had an increased risk to develop cancer, and those with one were at a greater risk to develop multiple tumors. Evaluation of the risk to develop multiple tumors per rat with increasing age was done using linear regression analysis. The females of all groups and the F_1 males showed a statistically significant increase in the numbers of tumors with age, with the slope of the lines being greatest for the WAG/Rij females. BN/Bi and WAG/Rij males, however, did not show a statistically significant increased number of tumors per rat. This fact suggests that the risk of multiple tumors per rat does not necessarily increase with age, but is relatively constant throughout the life of the population. If this is true, then the risk that rats will develop increased numbers of tumors with age is probably strain and sex dependent.

Some rats are at a low risk for all cancers, irrespective of their age. For example, some male and female BN/Bi rats died without tumors during each age period, even in the oldest rats of 37 months of age or older. Therefore, even in a highly inbred rat strain, individual risks for cancers are variable.

Female BN/Bi rats showed a clear trend for the incidence of metastatic tumor to increase with age. This trend was not as clear for the other groups of rats. In fact, if any trend was seen, it seemed to suggest a lack of any increased risk for the oldest rats (equal to or greater than 37 months) to develop tumor metastases compared to the younger cohorts.

TABLE 5.1

Numbers of Tumors in Female BN/Bi Rats

Age (in months)	Total no. of rats dying during this time	Total no. of dead rats with a tumor	No. of rats with the following no. of tumors									Total no. of tumors	Average no. of tumors per rat	
			0	1	2	3	4	5	6	7	8			
3—12	2	0	2	—	—	—	—	—	—	—	—	0	0 [a]	0 [b]
13—18	9	7	2	7	—	—	—	—	—	—	—	7	0.8	1.0
19—24	26	22	4	14	5	2	1	—	—	—	—	34	1.3	1.5
25—30	73	65	8	32	19	13	—	—	—	1	—	116	1.6	1.8
31—36	98	90	8	33	31	15	5	6	—	—	1	198	2.0	2.2
37—54	28	24	4	7	5	10	2	—	—	—	—	55	2.0	2.3
Total	236	208	28	93	60	40	8	6	0	1	1	410		
% of total		88	12	39	25	17	3	3	—	—	—			

[a] Calculated by dividing the total number of tumors by the total number of rats that died during each time period.

[b] Calculated by dividing the total number of tumors by the number of tumor bearing rats that died during each time period.

TABLE 5.2

Numbers of Tumors in Male BN/Bi Rats

Age (in months)	Total no. of rats dying during this time	Total no. of dead rats with a tumor	No. of rats with the following no. of tumors: 0	1	2	3	4	5	Total no. of tumors	Average no. of tumors per rat	
3—12	3	0	3	—	—	—	—	—	0	0 [a]	0 [b]
13—18	4	4	0	2	2	—	—	—	6	1.5	1.5
19—24	12	11	1	6	2	3	—	—	19	1.6	1.7
25—30	21	18	3	9	6	2	—	1	32	1.5	1.8
31—36	25	18	7	9	3	5	2	—	38	1.5	2.1
37—54	9	6	3	3	1	1	1	—	12	1.3	2.0
Total	74	57	17	29	14	11	3	1	107		
% of total		77	23	40	19	14	4	—			

[a] Calculated by dividing the total number of tumors by the total number of rats that died during each time period.

[b] Calculated by dividing the total number of tumors by the number of tumor bearing rats that died during each time period.

TABLE 5.3

Numbers of Tumors in Female WAG/Rij Rats

Age (in months)	Total no. of rats dying during this time	Total no. of dead rats with a tumor	No. of rats with the following no. of tumors: 0	1	2	3	4	5	6	7	Total no. of tumors	Average no. of tumors per rat	
3—12	0	—	—	—	—	—	—	—	—	—	—	— [a]	— [b]
13—18	0	—	—	—	—	—	—	—	—	—	—	—	—
19—24	3	3	0	1	1	1	—	—	—	—	6	2.0	2.0
25—30	13	12	1	3	5	2	0	1	1	—	30	2.3	2.5
31—36	60	60	0	4	14	20	13	8	1	—	190	3.2	3.2
≥37	25	25	0	0	7	5	6	4	1	2	93	3.7	3.7
Total	101	100	1	8	27	28	19	13	3	2	319		
% of total		99	1	8	27	28	19	13	3	2			

[a] Calculated by dividing the total number of tumors by the total number of rats that died during each time period.

[b] Calculated by dividing the total number of tumors by the number of tumor bearing rats that died during each time period.

TABLE 5.4

Numbers of Tumors in Male WAG/Rij Rats

Age (in months)	Total no. of rats dying during this time	Total no. of dead rats with a tumor	No. of rats with the following no. of tumors: 0	1	2	3	4	5	Total no. of tumors	Average no. of tumors per rat	
3—12	1	1	0	0	0	1	—	—	3	3.0[a]	3.0[b]
13—18	11	10	1	6	3	—	—	1	17	1.5	1.7
19—24	74	74	0	42	22	8	1	1	119	1.6	1.6
25—31	38	38	0	13	16	9	—	—	72	1.9	1.9
Total	124	123	1	61	41	18	1	2	211		
% of total		99	1	49	33	14	1	2			

[a] Calculated by dividing the total number of tumors by the total number of rats that died during each time period.
[b] Calculated by dividing the total number of tumors by the number of tumor bearing rats that died during each time period.

TABLE 5.5

Numbers of Tumors in Female (WAG × BN)F_1 Rats

Age (in months)	Total no. of rats dying during this time	Total no. of dead rats with a tumor	No. of rats with the following no. of tumors: 0	1	2	3	4	5	6	7	8	Total no. of tumors	Average no. of tumors per rat	
3—12	0	0	—	—	—	—	—	—	—	—	—	—	—[a]	—[b]
13—18	3	3	0	2	1	—	—	—	—	—	—	4	1.3	1.3
19—24	13	13	0	6	4	3	—	—	—	—	—	23	1.8	1.8
25—30	29	29	0	16	8	3	2	—	—	—	—	49	1.7	1.7
31—36	13	13	0	5	5	2	0	1	—	—	—	26	2.0	2.0
≥37	10	8	2	0	2	4	1	1	—	—	—	25	2.5	3.1
Total	68	66	2	29	20	12	3	2	—	—	—	127		
% of total		97	3	43	29	18	4	3	—	—	—			

[a] Calculated by dividing the total number of tumors by the total number of rats that died during each time period.
[b] Calculated by dividing the total number of tumors by the number of tumor bearing rats that died during each time period.

TABLE 5.6

Numbers of Tumors in Male (WAG × BN)F_1 Rats

Age (in months)	Total no. of rats dying during this time	Total no. of dead rats with a tumor	No. of rats with the following no. of tumors					Total no. of tumors	Average no. of tumors per rat	
			0	1	2	3	4			
3—12	0	0	—	—	—	—	—	—	—[a]	—[b]
13—18	0	0	—	—	—	—	—	—	—	—
19—24	6	6	0	4	1	1	—	9	1.5	1.5
25—30	18	16	2	6	6	3	1	31	1.9	1.7
31—36	23	21	2	4	8	6	3	50	2.4	2.2
≥37	20	19	1	5	5	4	5	47	2.5	2.4
Total	67	62	5	19	20	14	9	137		
% of total		92	8	28	30	21	13			

[a] Calculated by dividing the total number of tumors by the total number of rats that died during each time period.

[b] Calculated by dividing the total number of tumors by the number of tumor bearing rats that died during each time period.

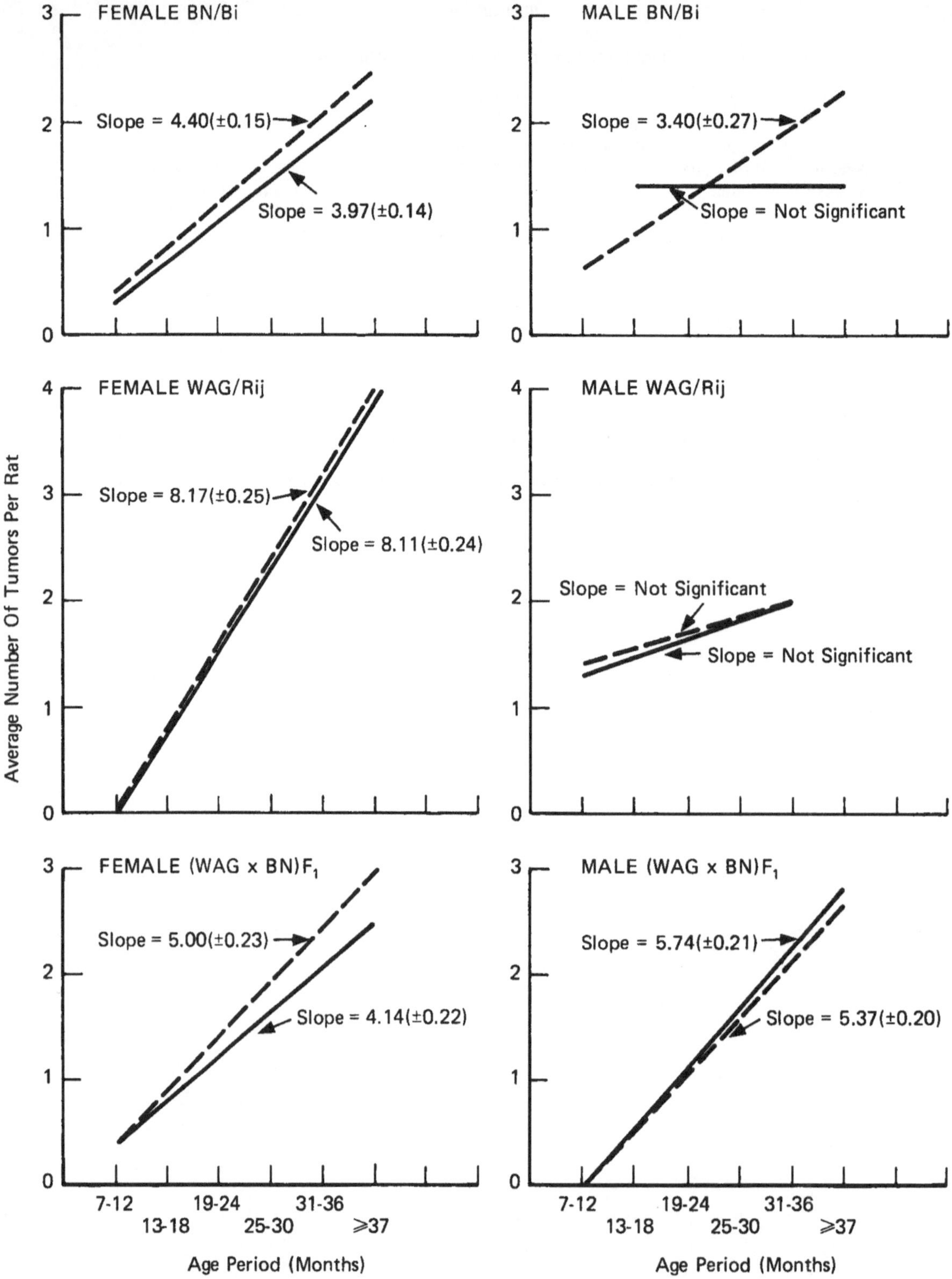

FIGURE 5.1. Summary of the linear regression analysis to determine the line of best fit for the average number of tumors per rat (—) and the average number of tumors in tumor bearing rats (---) for female and male BN/Bi, WAG/Rij, and (WAG × BN)F_1 rats. The term "Not Significant" for BN/Bi and WAG/Rij male rats means that a valid line could not be drawn through the data points and that the apparent increase or decrease in the slope of these lines was not significant.

TABLE 5.7

Total Number of Tumors in 236 Female and 74 Male BN/Bi Rats where Distant Metastases could be Found

Location and type of tumors	No. of males	Ages (months)	No. of females	Ages (months)
Adrenal cortical carcinoma	0		13	19, 25, 27, 31, 31, 32, 32, 34, 35, 35, 37, 38, 40
Carcinoma of ureter	2	25, 26	4	26, 31, 31, 33
Carcinoma of bladder	1	27	1	27
Sarcoma of cervix/vagina	—		5	26, 26, 31, 34, 36
Melanoma of skin	1	29	2	32, 34
Squamous cell carcinoma of mouth	0		2	31, 38
Medullary thyroid carcinoma	0		2	35, 38
Pheochromo-cytoma	1	20	1	39
Mammary carcinoma	0		1	27
Intestinal sarcoma	1	33	0	
Osteogenic sarcoma of vertebrae	0		2	26, 30
Sarcoma of neck	0		1	13
Carcinoma of lung	0		1	29
Sarcoma of ovary (?)	0		1	39
Lymph nodes with squamous cell carcinoma, origin *undetermined*	0		2	18, 36
Total (%)	6 (8%)		38 (16%)	

TABLE 5.8

Total Number of Tumors in 101 Female and 124 Male WAG/Rij Rats with Distant Metastases

Location and type of tumors	No. of males	Ages (months)	No. of females	Ages (months)
Medullary thyroid carcinoma	1	29	5	32, 33, 35, 35, 39
Adrenal cortical carcinoma	2	24, 25	0	—
Adenocarcinoma of salivary gland	1	22	0	—
Inguinal lymph node and lung with metastatic adenocarcinoma origin undetermined	—	—	1	34
Leiomyosarcoma of uterus	—	—	1	19
Total (%)	4 (3%)		7 (7%)	

TABLE 5.9

Total Number of Tumors in 68 Female and 67 Male (WAG × BN)F_1 Rats with Distant Metastases

Location and type of tumors	No. of males	Ages (months)	No. of females	Ages (months)
Medullary thyroid carcinoma	3	28, 30, 38	3	25, 28, 37
Adrenal cortical carcinoma	0	—	2	26, 34
Sarcoma in subcutis of skin	0	—	1	22
Total (%)	3 (4%)		6 (9%)	

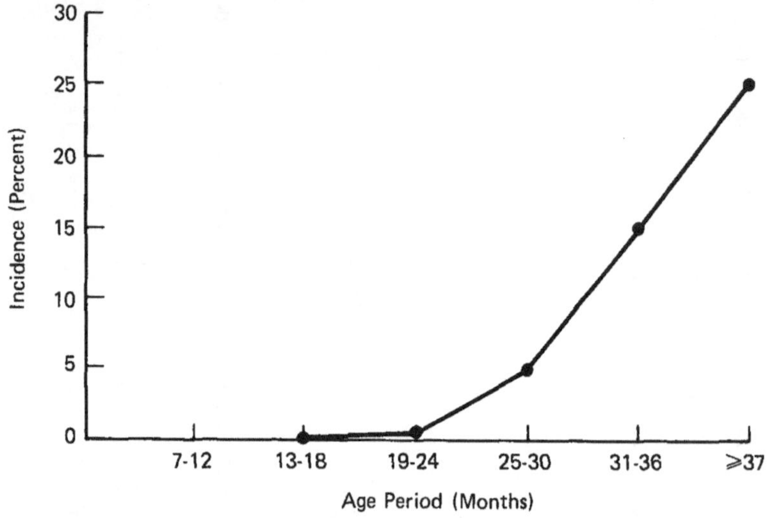

FIGURE 5.2. Percent incidence with age of tumors with distant metastases in 236 female BN/Bi rats.

TABLE 5.10

Adrenal Cortical Carcinomas in Female BN/Bi Rats: Comparison of Number of Tumors and Number of Metastases with Age

Age (months)	Total no. rats alive at beginning of period	Total no. rats dying during period	Total no. of dead rats with adrenal carcinoma	Total no. of rats with metastatic adrenal carcinoma	Total metastatic adrenal carcinomas × 100 / Total adrenal carcinomas
3—12	236	2	0	0	—
13—18	234	9	0	0	—
19—24	225	26	3	1	33%
25—30	199	73	4	2	50%
31—36	126	98	11	7	64%
37—54	28	28	4	3	75%
Total	0	236	22	13	59%

Chapter 6

COMPARISONS OF THE INCIDENCE OF LESIONS IN RATS THAT DIED SPONTANEOUSLY TO THOSE KILLED AT DIFFERENT AGES

I. INTRODUCTION

Age-associated incidence patterns of lesions in Chapter 4 were calculated by using life table techniques as described by Sachs.[148] The calculations were based on the number of rats that were alive at the beginning of a period, divided into the number of rats with a specific lesion that died during that period. The result was multiplied by 100, yielding a percentage, and it was this percentage that was plotted to show the incidence of different lesions with age. All lesions were based on their presence at the time of death and in no way implied that they were the cause of death; rather, they reflected the risk that a rat would die with each lesion.

Such methods of plotting data are useful for evaluating trends or patterns of lesions with age in the population, but caution must be exercised in the interpretation of the percentages because they are based on lesions found in dead rats. The real incidence of a lesion in the living population can be determined only by killing animals at specific ages.

There were differences observed in the percent of dead rats with a lesion as compared to the percent calculated by the life table methods. This chapter compares the percentages of selected lesions in rats that died as part of the aging study with the percentage of lesions in rats killed at two ages, namely, 14 and 21 months. Comparisons will be based on three types of findings, namely, the percent of lesions in dead rats during each age period, the age-associated percent based on the population at risk during each age period as calculated by life table techniques, and the percent in rats killed at specific ages.

This chapter will try to answer the following questions:

1. Will the percent incidence of a lesion, as calculated by life table methods, give a true indication of the incidence of that lesion in the living population?
2. Will the percent of a lesion in spontaneously dead rats indicate the percentage of that lesion in the living population?
3. Is the percent of a lesion in rats that died spontaneously similar to the percent in the population at risk as calculated by life table methods?

II. COMPARISONS OF THE PERCENTAGE OF SELECTED LESIONS IN DEAD RATS VS. PERCENTAGE IN THE POPULATION AT RISK AS DETERMINED BY LIFE TABLE CALCULATIONS

Before specific examples are presented, a brief description of the methods used to determine the data is needed. First, the ages of rats were divided into age periods. The number that were alive at the beginning of each period, the number that died during the period, and the number of dead rats that had lesions were listed for each period. The findings were then used to calculate the percent of dead rats and the percent of the population at risk that had the lesion. As an example, Table 6.1 shows the data tabulated for biliary cysts in aging BN/Bi females with the percentage of dead rats and the percentage of the population at risk in the two right-hand columns. Once tabulated, the data were then graphed as shown in Figure 6.4 for biliary cysts and Figures 6.1 to 6.5 were graphed in the same manner.

A. Medullary Thyroid Carcinomas

These tumors were typical of most neoplastic lesions and are illustrated in Figure 6.1 for male and female WAG/Rij and female BN/Bi rats. The curves for the percent in dead rats and the percent in the population at risk both increased with age. The BN/Bi females represented a low incidence strain while the WAG/Rij rats were a relatively high incidence strain. The peak incidence was reached in the oldest rats. Therefore, as the population aged, the risk of a rat dying with a medullary thyroid tumor increased

TABLE 6.1

Biliary Cysts in the Liver of Aging Female BN/Bi Rats

Age period (months)	No. alive at beginning of period	No. dead	No. with cysts	% of dead	% of population at risk
3—6	236	0	0	0	0
7—12	236	2	0	0	0
13—18	234	9	2	22	1
19—24	225	26	9	35	4
25—30	199	73	33	45	17
31—36	126	98	64	65	51
⩾37	28	28	21	75	75

and the percentage of rats dying with these lesions also increased.

B. Pituitary Tumors

Pituitary tumors gave different shaped curves than the medullary thyroid carcinomas, as illustrated in Figure 6.2 for male and female WAG/Rij and female BN/Bi rats. The percent of pituitary tumors in the population at risk increased with age. However, the percent in dead rats was relatively constant throughout the life of the population. Between 90 to 100% of the female and 80 to 100% of the male WAG/Rij rats that died died with a pituitary tumor regardless of their age. The BN/Bi females represented a low incidence strain for this lesion and again similar curves were seen, with 20 to 30% of the dead rats having pituitary tumors. The percent in the population at risk shows that a much lower percentage should be found in the population at risk. For example, for the age period of 19 to 24 months, approximately 3% of WAG/Rij females should die with the lesion, but 100% of the dead rats had tumors. Therefore, which of the two curves provides the best means of predicting the real incidences in the living population? There is no way of knowing without killing rats at selected ages.

C. Urothelial Tumors

Some tumors had a peak incidence in younger aged rats and a decreased incidence in the oldest. As examples, ureter and bladder tumors in male and ureter tumors in female BN/Bi rats are given in Figure 6.3. All three showed peaks in both curves (i.e., the percent of dead and percent of population at risk). Rats that survived the peak risk periods for these tumors apparently stood a much lower risk of developing the lesion in later life. Therefore, it would appear that most susceptible rats developed these tumors before 30 months of age.

D. Nonneoplastic Lesions

Nonneoplastic lesions were all similar; that is, the percent of dead rats with a lesion and the percent of the population at risk both increased with each age period. This is illustrated by two examples, namely, biliary cysts (Figure 6.4) and pancreatic atrophy (Figure 6.5) in BN/Bi females. Here, as was true with the medullary thyroid carcinomas, the percent in dead rats and the percent in the population at risk increased with age and the peak incidence occurred in the oldest age groups.

III. COMPARISONS OF THE PERCENTAGE OF SELECTED LESIONS IN RATS KILLED VS. THOSE THAT DIED SPONTANEOUSLY

The incidence of lesions in rats that lived out their life spans was presented in Chapter 4. The data were summarized as (1) the percent of the total population and (2) the age-associated incidence in the population at risk during different age periods as determined using life table techniques.

In Section II of this chapter, selected lesions were studied by comparing (1) the percent in rats that died spontaneously during different age periods and (2) the age-associated incidence in the population at risk as determined by life table calculations. Several different patterns of curves were seen in Figures 6.1 to 6.5 when the

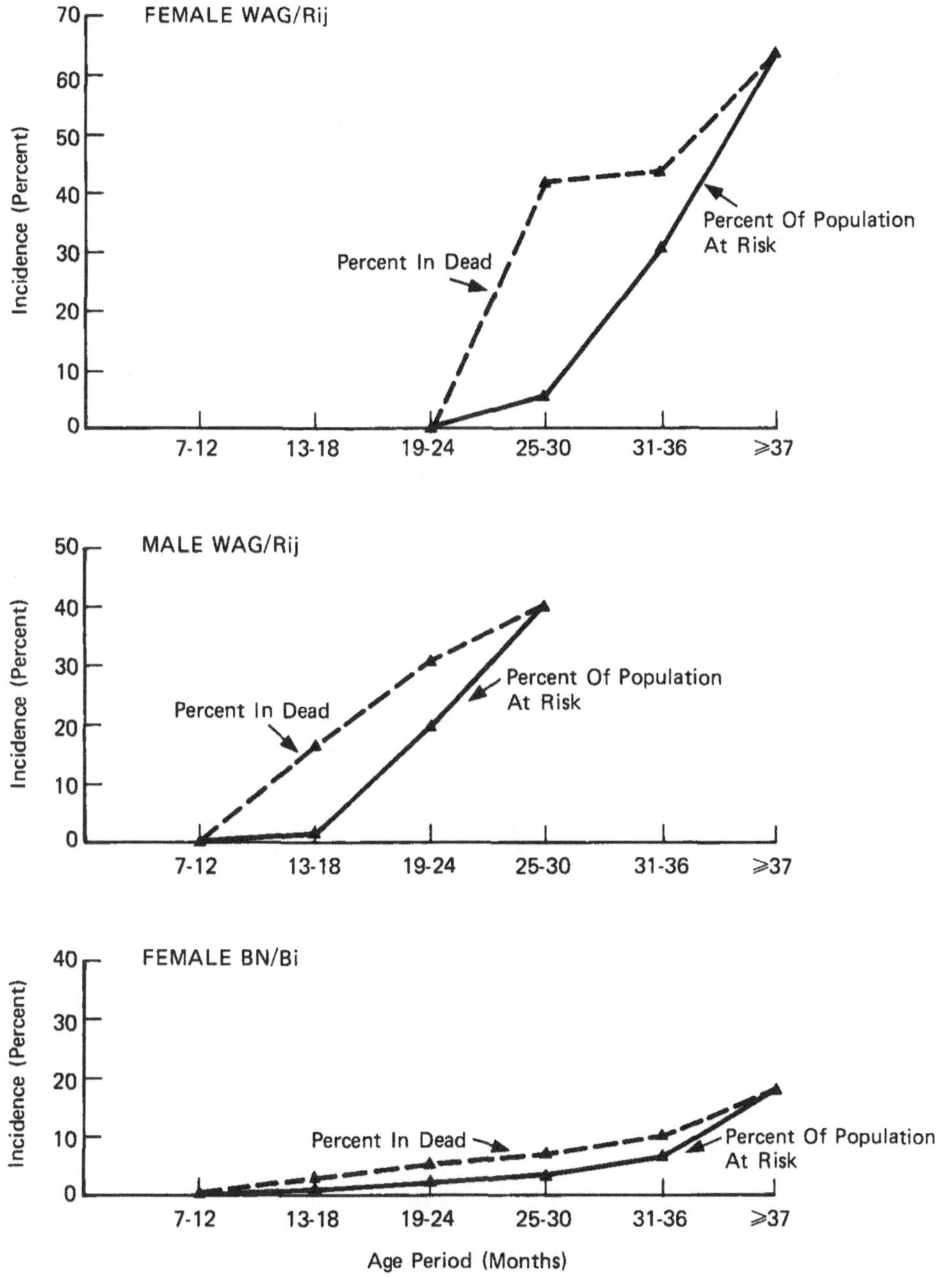

FIGURE 6.1. The percent with age of medullary thyroid tumors in dead rats and the percent of the population at risk as determined by life table calculations for female and male WAG/Rij and female BN/Bi rats.

two types of data were plotted. In order to determine which of the two curves comes closest to predicting the incidence in the living population, rats must be killed during the study. The incidence (percent) in the killed rats can then be compared to the percent in spontaneously dead and to the percent of the population at risk. The purpose of this section is to compare the previous findings with the incidence (percent) actually found in rats killed at selected ages.

Thirty female BN/Bi and thirty female WAG/Rij rats of 14 and 21 months of age were randomly selected from existing cohorts in the aging colony. These animals were killed with

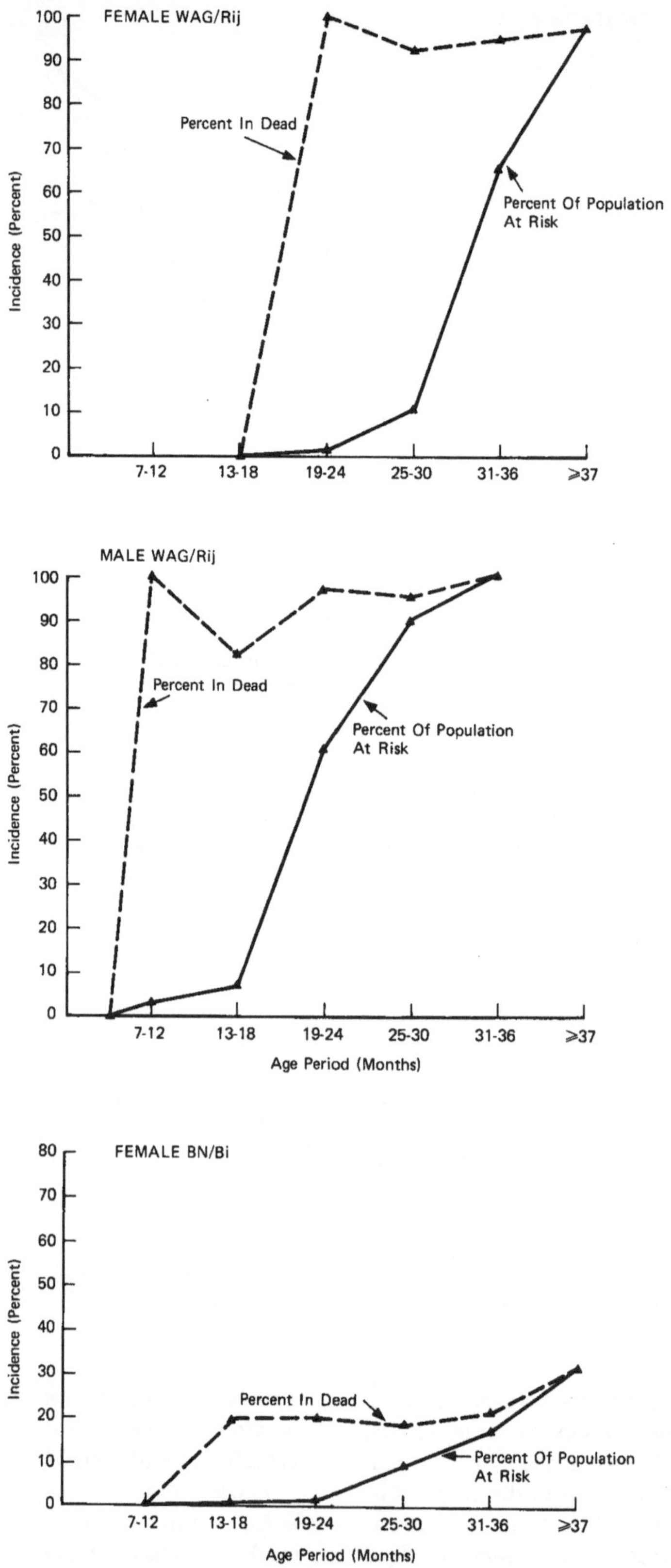

FIGURE 6.2. The percent with age of pituitary gland tumors in dead rats and the percent of the population at risk as determined by life table calculations for female and male WAG/Rij and female BN/Bi rats.

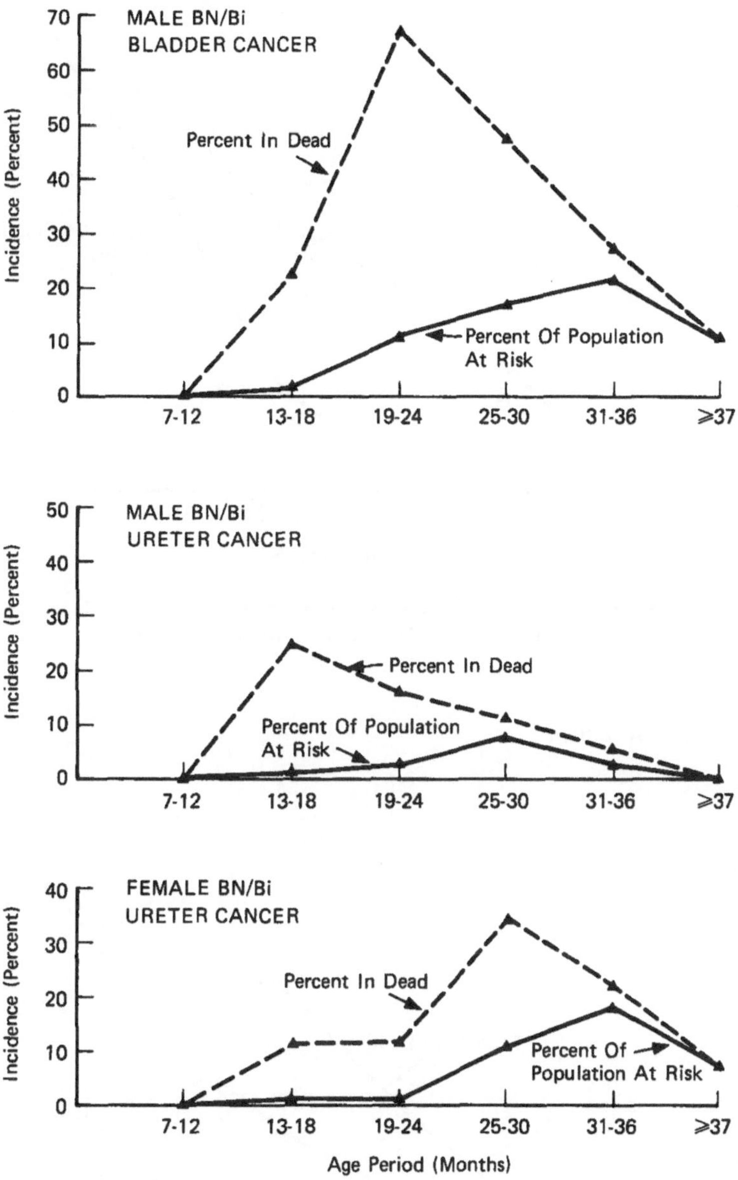

FIGURE 6.3. The percent with age of ureter and bladder cancers in dead rats and the percent of the population at risk as determined by life table calculations for male and female BN/Bi rats.

ether and a partial necropsy was performed. The organs examined from the WAG/Rij females included pituitary gland, thyroid glands, adrenal glands, mammary glands, and liver. The organs examined from the BN/Bi rats included pituitary gland, thyroid, adrenal glands, mammary glands, liver, left and right ureters, and urinary bladder. All tissues were grossly examined and processed using the same protocol for the tissues of the aging rats as described in Chapter 2.

Table 6.2 shows the results for selected lesions obtained from the BN/Bi females and Table 6.3 for WAG/Rij females. The percentages shown in the tables were used to plot the percentage of lesions in killed rats in Figures 6.6 to 6.10. The percent of each lesion in the spontaneously dead rats and percent of the population at risk were plotted as previously described in Section II.

There are only two data points for killed rats, and they were plotted in Figures 6.6 to 6.10.

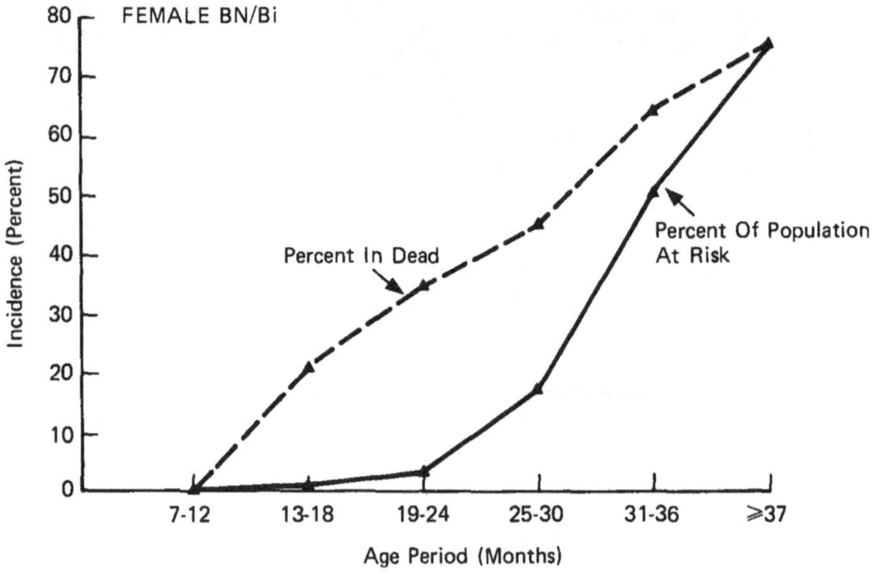

FIGURE 6.4. The percent with age of biliary cysts in the livers of dead rats and the percent of the population at risk as determined by life table calculations for female BN/Bi rats.

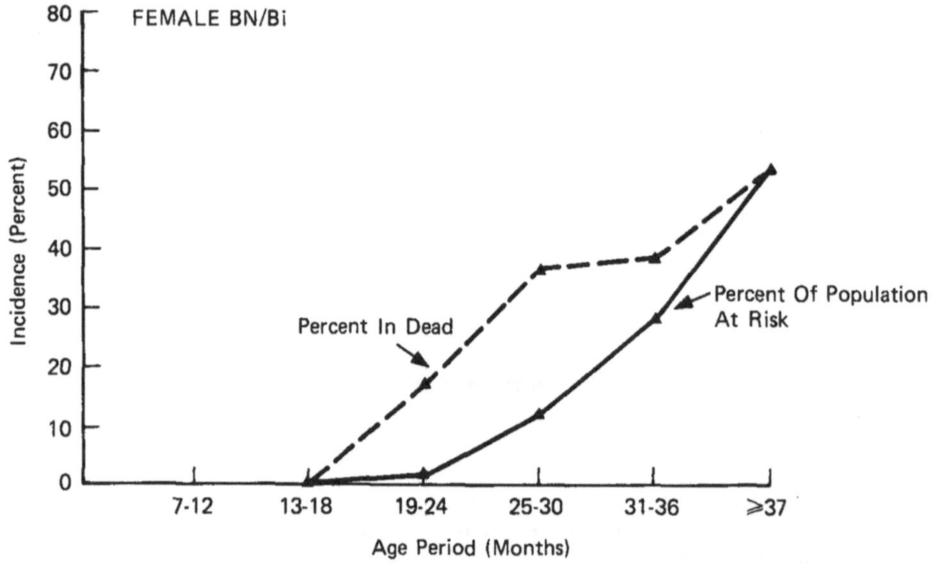

FIGURE 6.5. The percent with age of pancreatic atrophy in dead rats and the percent of the population at risk as determined by life table calculations for female BN/Bi rats.

Curves were drawn through these points, however, in order to connect them with the y-axis and the percentage in the oldest rats. It is possible that such curves may not reflect the percentage of these lesions at all age groups, but it does permit a better visualization of the data.

A. Pituitary Tumors

In Figure 6.6, the percent of pituitary tumors in the killed rats most closely corresponded to the percent found in the population at risk. The killed WAG/Rij females had more tumors than predicted by the percent in the population at risk, but clearly not 100% as seen in the spontaneously dead. The percentage of tumors in the killed BN/Bi females was virtually identical to that predicted by life table calculations at both the 14- and 21-month periods.

TABLE 6.2

Occurrence of Selected Lesions in 30 Female BN/Bi Rats Killed at 14 and 21 Months of Age

Lesion	Age (months)	No. with lesion/ no. killed	%
Pituitary tumor	14	0/30	0
	21	1/30	3
Medullary thyroid carcinoma	14	0/30	0
	21	0/30	0
Ureter carcinoma	14	2/30	7
	21	9/30	30
Urinary bladder carcinoma	14	0/30	0
	21	0/30	0
Liver-biliary cysts	14	1/30	3
	21	8/30	27
Liver foci and areas	14	0/30	0
	21	0/30	0
Liver nodules	14	0/30	0
	21	0/30	0
Mammary gland tumor	14	0/30	0
	21	0/30	0

TABLE 6.3

Occurrence of Selected Lesions in 30 Female WAG/Rij Rats Killed at 14 Months Old and 30 Killed at 21 Months

Lesion	Age (months)	No. with lesion/ no. killed	%
Pituitary tumor	14	2/30	7
	21	8/30	27
Medullary thyroid carcinoma	14	3/30	10
	21	8/30	27
Liver-biliary cysts	14	0/30	0
	21	0/30	0
Liver foci and areas	14	3/30	10
	21	27/30	90
Liver nodules	14	0/30	0
	21	1/30	3
Mammary gland tumor	14	0/30	0
	21	1/30	3

The pituitary lesions in the killed rats differed in size from those seen in the rats that died spontaneously. To demonstrate this difference, the lesions in pituitary glands were graded as follows:

1. Foci of cells with clearly different morphology than the normal pituitary, but without compression
2. Microtumors which resembled foci but having zones of compressed normal pituitary cells along the margins
3. Pituitary tumors that were seen grossly as enlarged pituitary glands and were confirmed histologically as tumors.

The results in the killed rats are given in Table 6.4. Eight 21-month-old WAG/Rij females had tumors, but only three of the eight (38%) were recognized grossly. In contrast, the tumors from the 101 aging WAG/Rij females that died spontaneously were graded using the same criteria and 91 of 96 (95%) tumors were seen during the gross necropsy. Microfoci were recog-

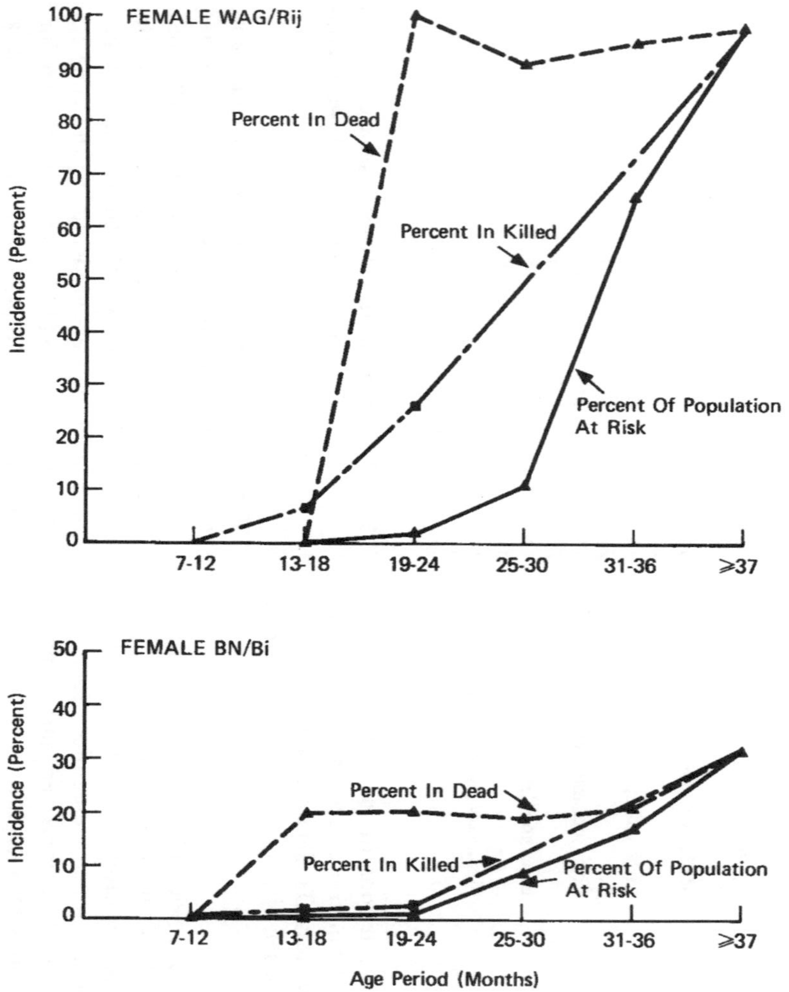

FIGURE 6.6. The percent with age of pituitary gland tumors in dead rats, the percent of the population at risk as determined by life table calculations, and the percent in 30 rats killed at 14 and 21 months of age for female WAG/Rij and BN/Bi rats.

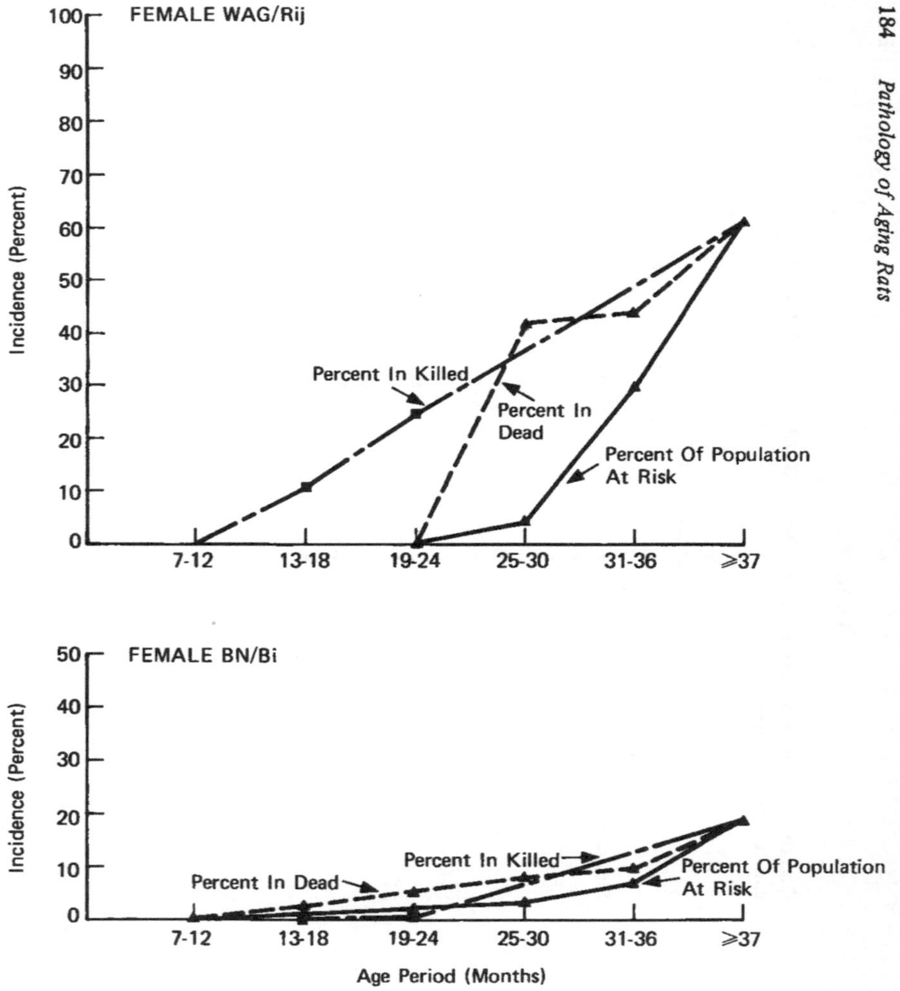

FIGURE 6.7. The percent with age of medullary thyroid carcinomas in dead rats, the percent of the population at risk as determined by life table calculations, and the percent in 30 rats killed at 14 and 21 months of age for female WAG/Rij and BN/Bi rats.

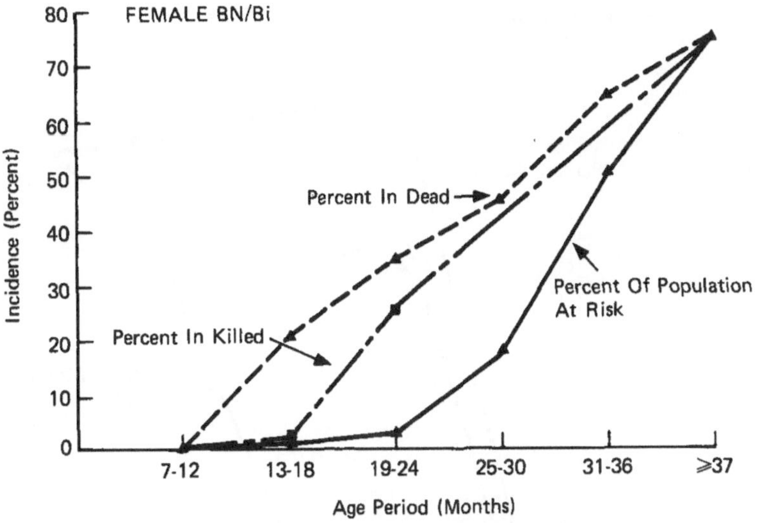

FIGURE 6.8. The percent with age of ureter carcinomas in dead rats, the percent of the population at risk as determined by life table calculations, and the percent in 30 rats killed at 14 and 21 months of age for female BN/Bi rats.

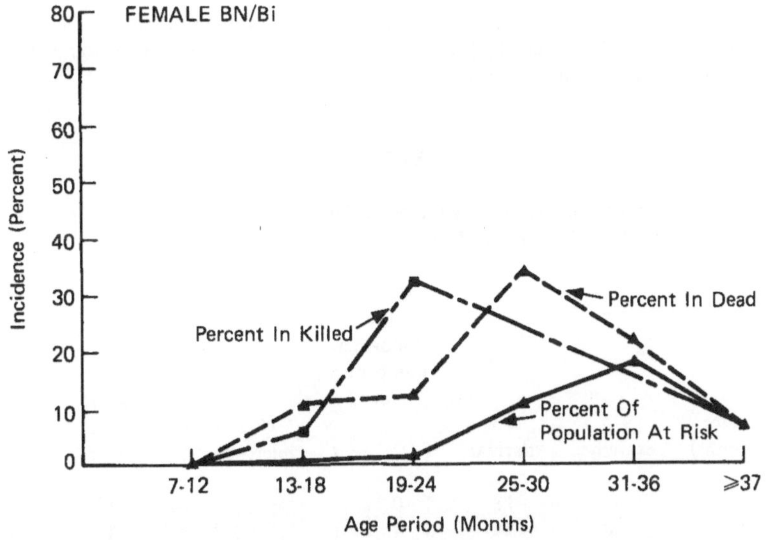

FIGURE 6.9. The percent with age of biliary cysts in the livers of dead rats, the percent of the population at risk as determined by life table calculations, and the percent in 30 rats killed at 14 and 21 months of age for female BN/Bi rats.

nized in 8 of the 30 killed rats, but in only 1 of the 101 spontaneously dead rats. In view of the 95% incidence of pituitary tumors in this strain, it seems likely that the observed foci represent early pituitary tumors or, at the very least, "precancerous" lesions of which some progress into pituitary neoplasms.

B. Medullary Thyroid Carcinomas

Medullary thyroid tumors in killed WAG/Rij females appeared to fall on, or closest to, the curve for the percent in dead rats (Figure 6.7). The BN/Bi females killed at 14 and 21 months, however, yielded results that were identical to the life table curves.

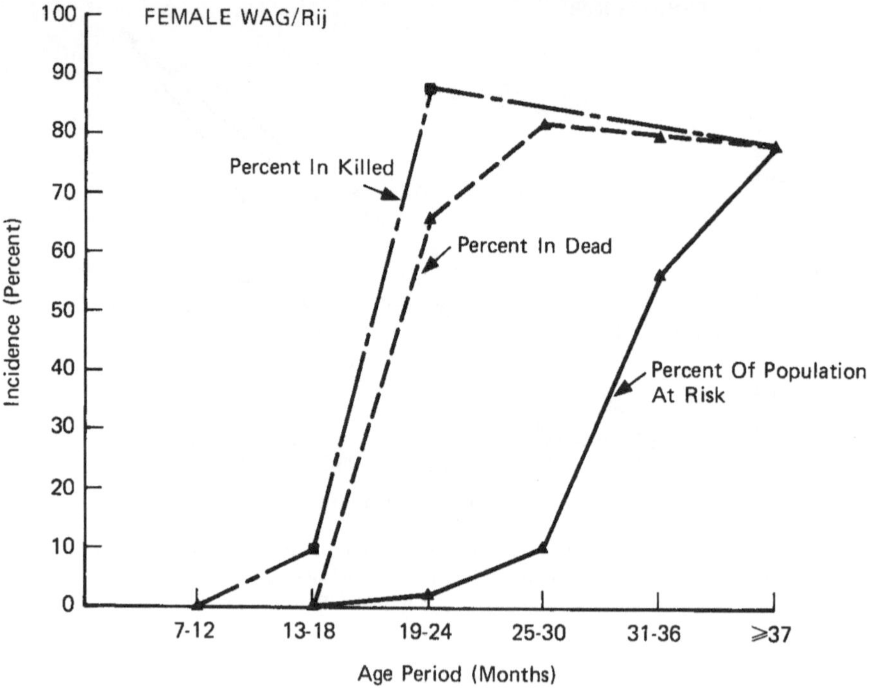

FIGURE 6.10. The percent with age of foci and areas in the livers of dead rats, the percent of the population at risk as determined by life table calculations, and the percent in 30 rats killed at 14 and 21 months of age for female WAG/Rij rats.

TABLE 6.4

Foci of Altered Cells, Microtumors, and Grossly Observed Tumors in the Pituitary Glands from BN/Bi and WAG/Rij Females Killed at 14 and 21 Months of Age

Strain	Age (months)	No. examined	No. normal	No. with the following number of foci				No. with microtumors	No. with gross tumors	Total tumors	%
				1	2	3	4				
WAG/Rij	14	30	28	1	0	0	0	0	1	1	7
	21	30	14	3	4	0	1	5	3	8	27
BN/Bi	14	30	30	0	0	0	0	0	0	0	0
	21	30	28	1	0	0	0	0	1	1	3

Similar to the pituitary lesions, the medullary thyroid tumors were graded as (1) microtumors which were found only histologically and (2) those tumors that were recognized grossly as enlarged thyroid glands and confirmed histologically. The results of the killed rats are given in Table 6.5. All three tumors in 14-month-old WAG/Rij females and five of eight (62%) in the 21-month-old group were microtumors. The lesions in the 101 aging female WAG/Rij that died spontaneously were similarly graded and 24 of 47 (51%) were microtumors. Therefore, the sizes of the medullary thyroid tumors in killed rats were relatively similar to those in rats that died spontaneously. This is in contrast to the pituitary tumors, which were larger in dead rats than in those killed.

C. Ureter Carcinoma

Ureter carcinomas in BN/Bi females were

TABLE 6.5

Microtumors and Grossly Recognized Medullary Thyroid Carcinomas in BN/Bi and WAG/Rij Females Killed at 14 and 21 Months of Age

Strain	Age (months)	No. examined	No. with microtumors	No. With gross tumors	Total tumors	%
WAG/Rij	14	30	3	0	3	10
	21	30	5	3	8	27
BN/Bi	14	30	0	0	0	0
	21	30	1	0	1	3

evaluated because they represented an example of a lesion that had a higher incidence in younger rats. In Figure 6.8 the percent of tumors in the ureters of 14- and 21-month-old killed females is shown along with the percent in dead and the percent in the population at risk as determined from spontaneous deaths. It would appear that the percent in the dead came closest to the incidence in the living population. Based on the data from killed rats, the actual age of the peak incidence is earlier than that indicated by either of the other two curves.

The lesions were graded and it was seen that the tumors in killed rats were smaller or less invasive than the cancers found in the spontaneously dead. For example, over one half of the tumors found in the spontaneously dead females were graded as P2 to P4 cancers. In the 21-month-old killed females, on the other hand, eight of the nine tumors were P1S or P1 and only one was graded as P3. This supports the finding that most of the rats at risk for such tumors develop them before 2 years of age. The ureter cancers in the older rats (greater than 2 years) probably represent expansion or continued growth of tumors that began earlier in life. Rats that survived until about 2 years of age without developing one of these lesions were apparently at a very low risk to develop one later in life.

Stones were found in the ureter or bladder of approximately 60% of the rats that died spontaneously (Chapter 4, Section II.K). Most were in the ureter, but some were found in the bladder. Stones were also present in BN/Bi females that were killed. Of the 30 14-month-old rats, 6 (20%) had stones in the ureter (2 in left ureter, 2 in right ureter, and 2 bilateral). At 21 months, 12 of the 30 (40%) had stones (5 in left ureter, 4 in right ureter, 2 bilateral, and 1 in the urinary bladder). It seemed, therefore, that a positive correlation existed between the presence of stones and the development of tumors. To test this hypothesis, the data for killed rats with ureter tumors only, stones only, and tumors with stones are summarized in Table 6.6. The findings suggest the lack of any direct correlation between stones and the development of cancer in these rats.

Urinary bladders were also examined from these rats, but no tumors were recognized in the 30 BN/Bi females killed at 14 or 21 months of age.

D. Biliary Cysts

The percent of biliary cysts in the livers of female BN/Bi rats killed at 14 months of age was identical to the percent of the population at risk between the ages of 13 and 18 months (Figure 6.9). The percent in those killed at 21 months was closest to the percent in the spontaneously dead rats. Therefore, the age of onset of this lesion was between 12 to 18 months in the population and rapidly increased after 18 months of age.

Like the tumors discussed earlier, the biliary cysts in killed rats were not identical in size to the cysts in the rats that died spontaneously. The cysts in rats that died spontaneously were greater than 5 mm, and most were larger than 1 cm in diameter. In addition, they were often multiple. In killed rats of 14 and 21 months of age, all were solitary and less than 5 mm in size. This confirms that the biliary cysts are proliferative lesions that increased in number and size with age. Therefore, the older the rat, the greater was the risk of dying with one or more

TABLE 6.6

The Occurrence of Tumors, Stones, and Tumors with Stones in the Ureters of 30 Female BN/Bi Rats Killed at 14 and 21 Months of Age

Age (months)	No. with tumor only	No. with stones only	No. with tumors and stones
14	2	6	0
21	4	7	5

cysts, and the older the rat, the larger the lesion.

E. Liver Foci, Areas, and Nodules

BN/Bi females had a very low background incidence (<5%) of liver foci and areas (Chapter 4, Section II.F). This was confirmed in the killed BN/Bi females where none had lesions at either 14 or 21 months old (Table 6.2).

WAG/Rij females, on the other hand, had a high incidence of liver foci and areas (Chapter 4, Section II.F). In the females killed at 14 and 21 months, 10 and 90%, respectively, had these cellular changes (Table 6.3). This is shown in Figure 6.10. The percent in killed rats indicated a rapid increase in the incidence between 14 and 21 months of age. This suggests that the percent in the dead rats was a much better prediction of the incidence in the living population than the percent of the population at risk as determined by life table calculations.

Ninety percent of the 21-month-old killed rats had these changes, and the peak incidence in the dead rats was 80%. This suggests that the incidence in the population older than 21 months must remain relatively constant (80 to 90%) through the remaining life of the population.

Although 90% of the 21-month-old WAG/Rij females had liver foci and areas, the severity was clearly less than that observed in the older rats that died. These changes were focal and relatively mild in the killed females. Those that died usually had many more such lesions that involved a much greater area of the liver sections. Therefore, 80 to 90% of these rats acquire focal hepatocellular alterations by 21 months, and this incidence remains constant for the rest of the population life. The severity of these lesions continues to increase with age, however, resulting in more extensive lesions with advancing age.

Of 30 21-month-old WAG/Rij females, 1(3%) had a small hepatocellular neoplastic nodule (Table 6.3).

F. Mammary Tumors

None of the killed BN/Bi and only one of the 21-month-old killed WAG/Rij females had a mammary tumor. The incidence in these killed rats was 0 to 3%. This was similar to the data present in Figure 4.142 which showed that the percent of mammary tumors in these two strains was between 1 to 4% in the age period of 19 to 24 months as determined by life table calculations.

IV. DISCUSSION

Neither the percent of a lesion in spontaneously dead rats nor the percent of the population at risk predicted the incidence of all lesions in the living population at a given age. However, based on the results of the selected examples presented in this chapter and the data obtained from the calculations used to plot the life table data in Chapter 4, several conclusions can be made.

First, the incidence patterns of most lesions show that the percent of the population at risk and the percent of the spontaneously dead rats increased with age; that is, the older the rat, the greater was its risk to die with a specific lesion. In general, the percent of lesions in killed rats was somewhere in between these two values.

Second, exceptional lesions such as pituitary gland tumors showed a different pattern. The percent of pituitary tumors in dead rats was constant throughout the life of the population, while, in contrast, the percent in the population at risk increased with each age group. In this example, the percent in killed rats was closest to the value as determined from the population at risk, especially in younger age groups. A pattern such as this may indicate that these tumors are rapidly growing. Once they arise, they progress rapidly and contribute to the death of the rat. It should be emphasized that only pituitary gland tumors showed such curves. However, pituitary tumors in all groups had similar shaped curves, suggesting a tumor-specific pattern. The same pattern was seen in the WAG/Rij strain with a 95% incidence and in the BN/Bi strain which had less than a 25% incidence. The curves for liver foci and areas were similar but not identical to those for pituitary gland tumors. The percent of these changes increased rapidly and seemed to reach a peak by 24 months. From 24 months on, the percent incidence remained relatively constant, with 80 to 90% in the living population.

Third, a few lesions had a peak incidence in rats during one age period, but a lower incidence in the older age animals. The actual onset of such lesions is probably earlier than indicated by the percent of the population at risk. Curves with such shapes show that the percent of tumors in dead rats is a better indicator of the incidence in the living population than the percent based on the life table calculations. This was certainly true for the ureter carcinomas as illustrated in this chapter, but additional studies are needed to determine if this is true for all tumors that show peak incidences in younger aged rats. Such curves may also indicate that some of the lesions observed in the 21-month-old rats may be reversible and as such they disappear rather than progress.

Finally, patterns such as seen for medullary thyroid carcinomas in the female WAG/Rij rats (Figure 6.6) may mean that such lesions are slowly growing. Such lesions are present in the population long before the first deaths occur. They grow slowly and contribute to the death of the animal only in later life.

In general, the percent of the population at risk was the best predictor of the percent in the living population in rats less than 18 months old. The values for the percent of the population at risk and the percent of the dead were nearly identical in the oldest groups of rats (greater than 30 months old). The greatest variability was seen in rats between 18 and 30 months of age. The percent of some lesions in killed rats at 21 months was closest to the percent of the population at risk, others in-between, and still others closest to the percent in the dead rats. Therefore, interim kills at selected time intervals during a long-term or life span rat study are essential in order to monitor the change, over a period of time, of the incidence of a lesion in the living population. Such interim kills are important for any life span study, but they are especially needed in aging, chronic toxicity, and carcinogenesis studies.

Chapter 7

SPONTANEOUSLY OCCURRING POSTERIOR PARALYSIS: FURTHER STUDIES TO DETERMINE THE AGE OF ONSET AND PROGRESSION OF THE SYNDROME

I. INTRODUCTION

By studying the lesions in rats that had severe paresis or paralysis at the time of death (Chapter 4, Section II.P), it was possible to determine several features of this syndrome. First, the different incidences in the various strains and sexes was determined (Table 4.32). Second, the age-associated incidence in male F_1 rats was established (Figure 4.131). Finally, the lesions that were present in the spinal cords, spinal nerve roots, vertebrae, intervertebral disks, and skeletal muscle were described.

On the other hand, several features of this syndrome could not be determined without additional studies. As examples peripheral nerves were not available so it could not be determined if they had lesions; as complete semiserial sections through the entire length of the spinal cord were not available, specific lesions could not be localized; as spinal cords were not available from the clinically normal rats, comparisons of paralyzed and nonparalyzed rats could not be done; finally, it was impossible to determine the age of onset of each lesion and the interrelationships of the various lesions because serial killings of younger rats would be needed. To answer some of these questions, additional studies were performed. Some of the findings were previously published,[26] but are reported here again for completeness along with some new findings.

II. SEMISERIAL SECTIONS OF SPINAL CORDS

Twelve male F_1 rats with clinical signs of severe posterior paresis or paralysis were killed to study the entire length of the spinal cord, as previously reported.[26] Six rats were first fixed by whole body perfusion with formalin, as previously described,[26] and six were formalin fixed without perfusion. The entire vertebral column, including the spinal cord and nerve roots, was studied by taking multiple, semiserial sections (approximately 25 to 30 blocks per rat).

All 12 male F_1 rats had similar lesions and a similar distribution as shown in Table 7.1. Lesions in the region of cervical (C) 1 to 3 were minimal or moderate. Severe lesions were present in the lateral and ventral columns in the region of C4-7 and were most severe between thoracic (T) 1 and 4. Caudal from T4, the lesions were less severe, and by the lumbar segments they were minimal or absent. Nerve root lesions were observed caudal to T3, with the ventral roots in this region affected and the dorsal roots either normal or minimally involved. In the lumbosacral region, both the dorsal and ventral roots were affected, but the severity varied among the individual roots.

III. MINERAL DEPOSITS IN THE VERTEBRAL CANAL OF MALE (WAG × BN)F_1 RATS

Nodular deposits were observed in the vertebral canal of a few male F_1 rats that died with paralysis. In the decalcified sections, they appeared as barely visible, soft white to tan nodules on the dura. In most rats, they were incidental findings occurring at one or two sites in the available material. Rarely, such deposits were extensive and were seen in all levels of the spinal canal (Figure 7.1). Histologically, they were amorphous or crystalline, stained positive with Alizerin red, and occasionally seemed to compress the spinal cord or nerve roots.

Scanning electron microscopy was done on decalcified cross sections and longitudinal sections of spinal cord within the vertebral column. X-ray diffraction analysis was done on undecalcified deposits from three different rats to determine their composition and general structure. All studies were done by and with the complete cooperation of the Central Laboratory TNO in Delft, The Netherlands. The scanning electron microscopy on decalcified sections showed that they were often localized on the dura where they appeared as irregular, convoluted masses (Figure 7.2). Individual undecalcified deposits that were first removed from

TABLE 7.1

Distribution and Severity[a] of Lesions in the Spinal Cord and Nerve Roots of (WAG × BN)F_1 Male Rats

Rat number	Cervical 1 to 3			Cervical 4 to 7			Thoracic 1 to 4			Thoracic 5 to lumbosacral			Spinal nerve roots in lumbo sacral region
	D[b]	L	V	D	L	V	D	L	V	D	L	V	
1	−	−	+ +	±	+	+ + +	±	+ +	+ + +	−	−	±	+ + +
2	−	−	+	+ +	+ + +	+ +	−	±	+	−	−	−	+ + +
3	+	+	+ +	−	+ +	+ + +	−	+ + +	+ + +	−	−	−	+ + +
4	−	±	+	−	+ +	+ +	−	−	+	−	−	−	+ +
5	−	−	+	−	+	+ +	−	+ +	+ +	−	−	−	+ + +
6	−	−	+	−	+	+ +	−	+ + +	+	−	−	±	+ + +
7	−	−	+	−	+ +	+	±	+ +	+ +	−	−	−	+ + +
8	−	−	−	−	+	+	−	+ +	+ +	−	−	+	+ + +
9	−	−	+	−	+ +	+ +	−	+ + +	+	−	+	+	+ + +
10	−	−	±	−	+	+	−	+ + +	+	−	−	±	+ + +
11	−	±	+	−	+	+ +	+	+ + +	+ +	−	−	−	+ + +
12	−	±	+	−	+	+ +	−	+ + +	+ +	−	−	−	+ + +

[a] Range of severity: (−) normal, (±), questionable, (+) mild, (+ +) moderate, (+ + +) severe.
[b] D = dorsal funiculi; L = lateral funiculi; V = ventral funiculi.

From Burek, J. D., Van der Kogel, A. J., and Hollander, C. F., *Vet. Pathol.*, 13,321, 1976. With permision of S. Karger AG, Basel.

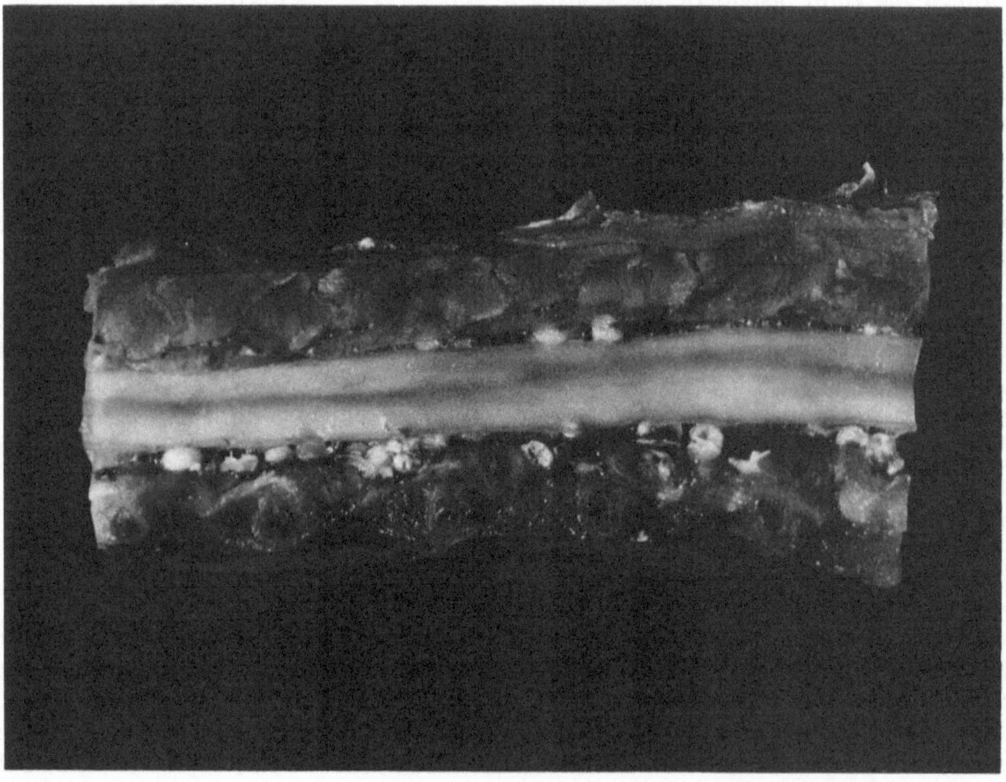

FIGURE 7.1. Multifocal mineral deposits in the spinal canal of a male (WAG × BN)F_1 rat.

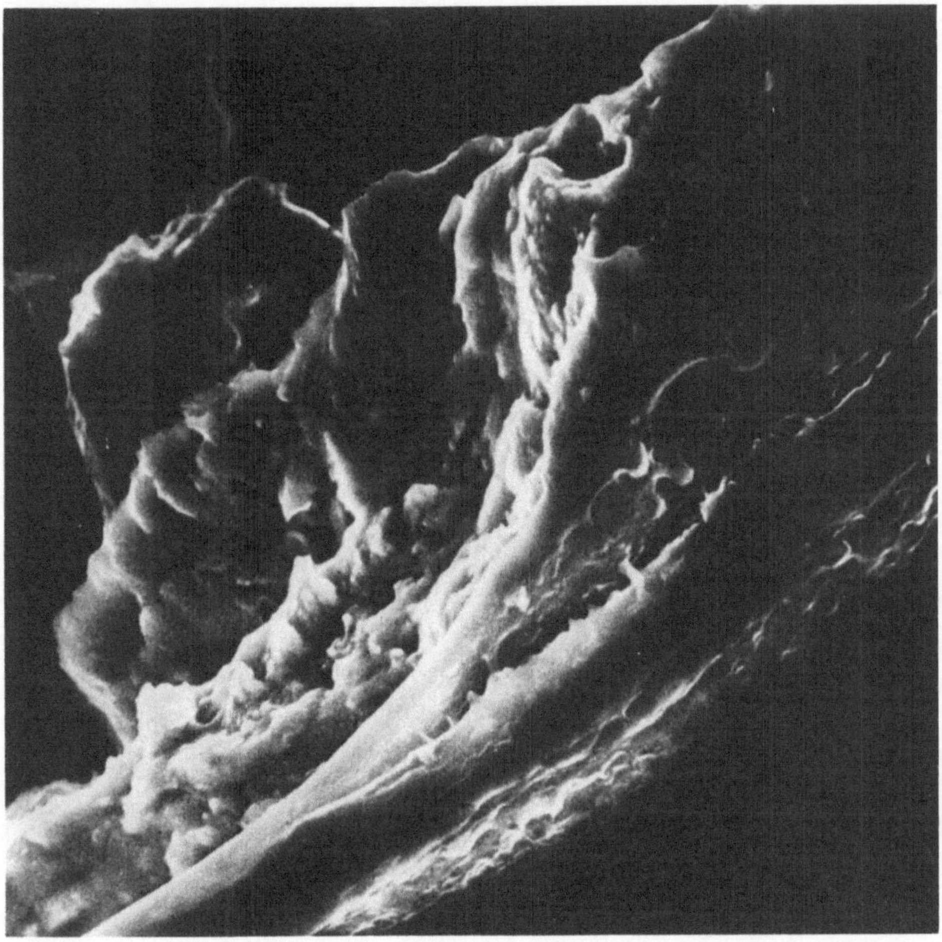

FIGURE 7.2. Scanning electron photomicrograph of a decalcified irregularly convoluted mineral deposit on the dura of the spinal cord from a male (WAG × BN)F_1 rat. (Magnification × 440.)

the spinal canal and then studied had a similar appearance (Figure 7.3). X-ray diffraction analysis of these individual, undecalcified masses showed that most were composed of calcium and phosphate. However, a few deposits appeared to be composed predominantly of organic material with little or no calcium and phosphate found.

IV. SERIAL KILLING OF CLINICALLY NORMAL MALE AND FEMALE (WAG × BN)F_1 RATS

Five male and five female (WAG × BN)F_1 rats of 6, 12, 18, and 22 months were randomly selected from existing cohorts in the aging colony. They were killed in order to study the age of onset of the earliest spinal cord and nerve root lesions, to determine if peripheral nerves were also affected, to determine if males and females both had similar lesions, and to determine if the muscle atrophy observed in paralyzed rats preceded or followed the cord and nerve root changes. Therefore, in addition to the routine necropsy material, additional tissue was obtained from all 40 rats in this study. A new protocol was established so that tissues from each rat were all handled in the same way and their sections were similar.

The additional material obtained from these rats included: (1) the entire length of the spinal column with the spinal cord, (2) the right and left brachial plexus and medial nerves from the front legs, (3) the right and left sciatic nerves from the hind legs, and (4) extensor and flexor muscles from the right and left front and right

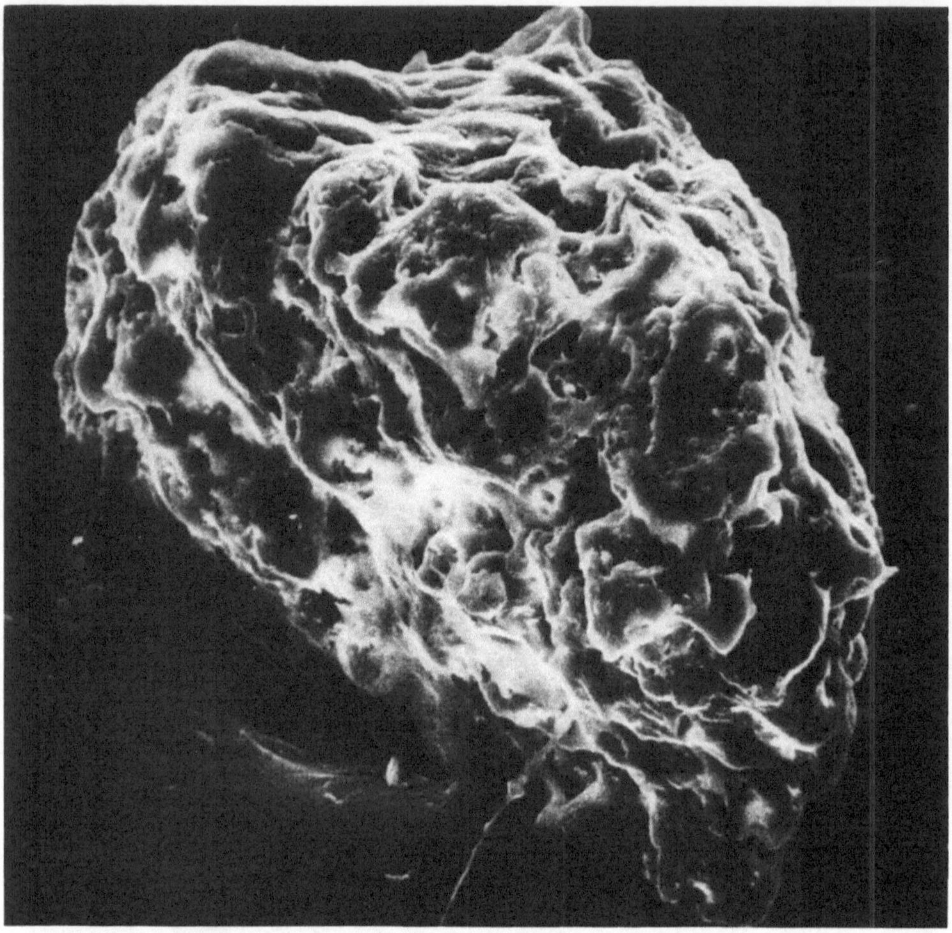

FIGURE 7.3. Scanning electron photomicrograph of an undecalcified mineral deposit similar to the one shown in Figure 7.2. This one was removed from the spinal canal for examination. (Magnification × 330.)

and left hind legs. The spinal column was first fixed in formalin and decalcified. It was trimmed, so that a cross section of spinal cord was obtained at about the level of C2, T1, T7, and L5. The remaining portions of spinal cord were longitudinally sectioned. All muscles were trimmed in both cross and longitudinal sections.

In contrast to the finding in the paralyzed rats, lesions, when present, were mild, and several findings in the aging animals were not seen in those killed at younger ages. For example, intervertebral disk herniations, aseptic necrosis of vertebral bone, dural fibrosis, fibrosis of nerve root sleeves, and skeletal muscle lesions were not observed in any of these younger rats. Nerve root lesions were recognized in the 22-month-old males and females. The lesions were mild and consisted of dilated nerve sheaths with absent or swollen axons and a few lipid-filled macrophages. In contrast to the finding in the aging rats, areas of recent hemorrhage, hemosiderin pigment, and cholesterol clefts were not seen in these younger animals.

In addition to the mild nerve root lesions, peripheral nerve degeneration was also seen. When present, the changes were similar to those seen in the roots and consisted of dilated nerve sheaths, focal demyelinization, swollen or absent axons, and one or a few lipid-filled macrophages.

As shown in table 7.2, lesions in the spinal cords, spinal nerve roots and peripheral nerves were found only in the 22 month old male and female rats. In these, five of five females had nerve root changes in the cervical, thoracic and lumbar regions. Three of the 5 had lesions in the roots in the sacral region which consisted of

TABLE 7.2

Spinal Cord, Spinal Nerve Root, and Peripheral Nerve Lesions in Male and Female (WAG × BN)F_1 Rats that were Killed at Various Ages

			Spinal cords				Nerve roots				Peripheral nerves	
Age (months)	Sex	No. examined	C[a]	T	TL	LS	C	T	L	S	Fore limb	Hind limb
6	Male	5	0	0	0	0	0	0	0	0	0	0
	Female	5	0	0	0	0	0	0	0	0	0	0
12	Male	5	0	0	0	0	0	0	0	0	0	0
	Female	5	0	0	0	0	0	0	0	0	0	0
18	Male	5	0	0	0	0	0	0	0	0	0	0
	Female	5	0	0	0	0	0	0	0	0	0	0
22	Male	5	0	1	0	0	0	2	5	2	0	1
	Female	5	0	0	0	0	5	5	5	3	5	3

[a] C = cervical; T = thoracic; TL = thoracic-lumbar; L = lumbar; LS = lumbosacral; S = sacral.

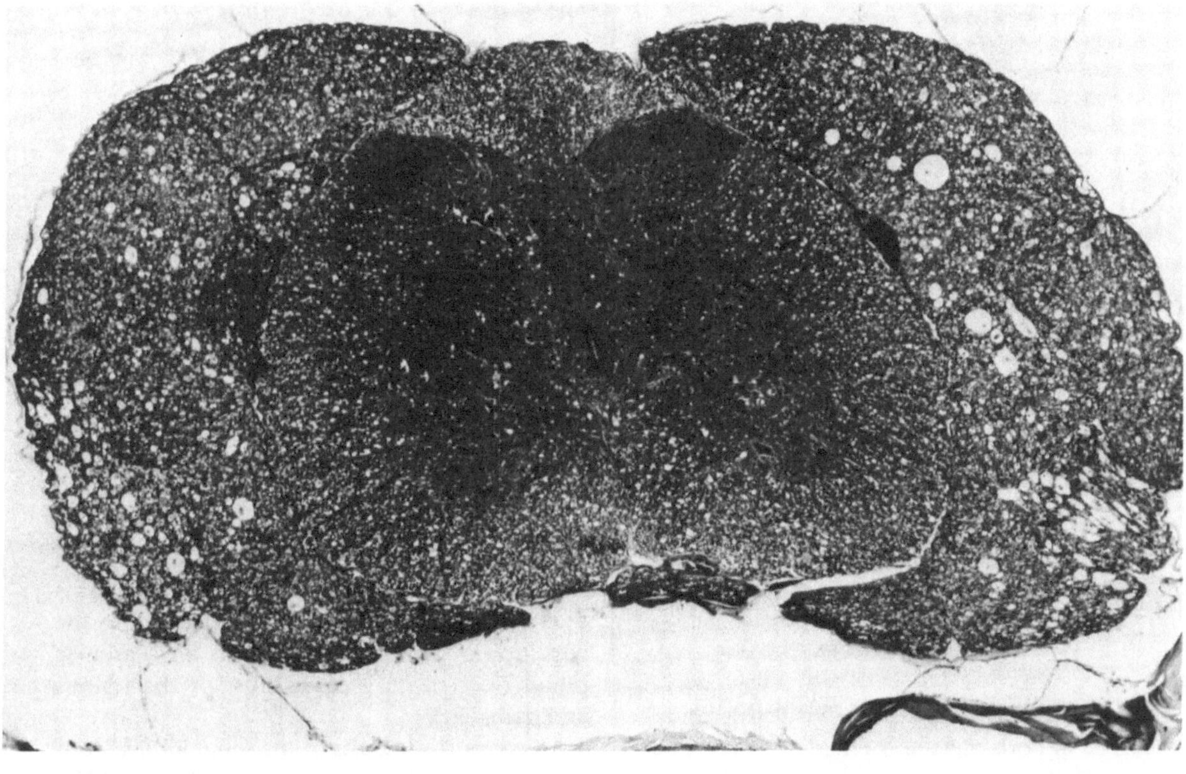

FIGURE 7.4. Lumbar nerve roots from a male (WAG × BN)F_1 rat showing degenerative changes. (HPS; magnification × 40.)

very mild focal lesions that were less severe than those in the lumbar roots of the same rat. All five had distinct degenerative changes in the nerves of the brachial plexus. Three also had lesions in the peripheral nerves of the hind legs, but these were focal and very mild.

The 22-month-old males had fewer lesions than the females. None had cervical root changes and only two had thoracic root changes. All five showed root lesions in the lumbar region. Sacral root changes were present in only two of the five. In the two rats with sacral nerve root degeneration, the lesions were less extensive than those in the lumbar roots of the same rat (Figures 7.4 and 7.5). Not only were the root lesions less extensive than in the

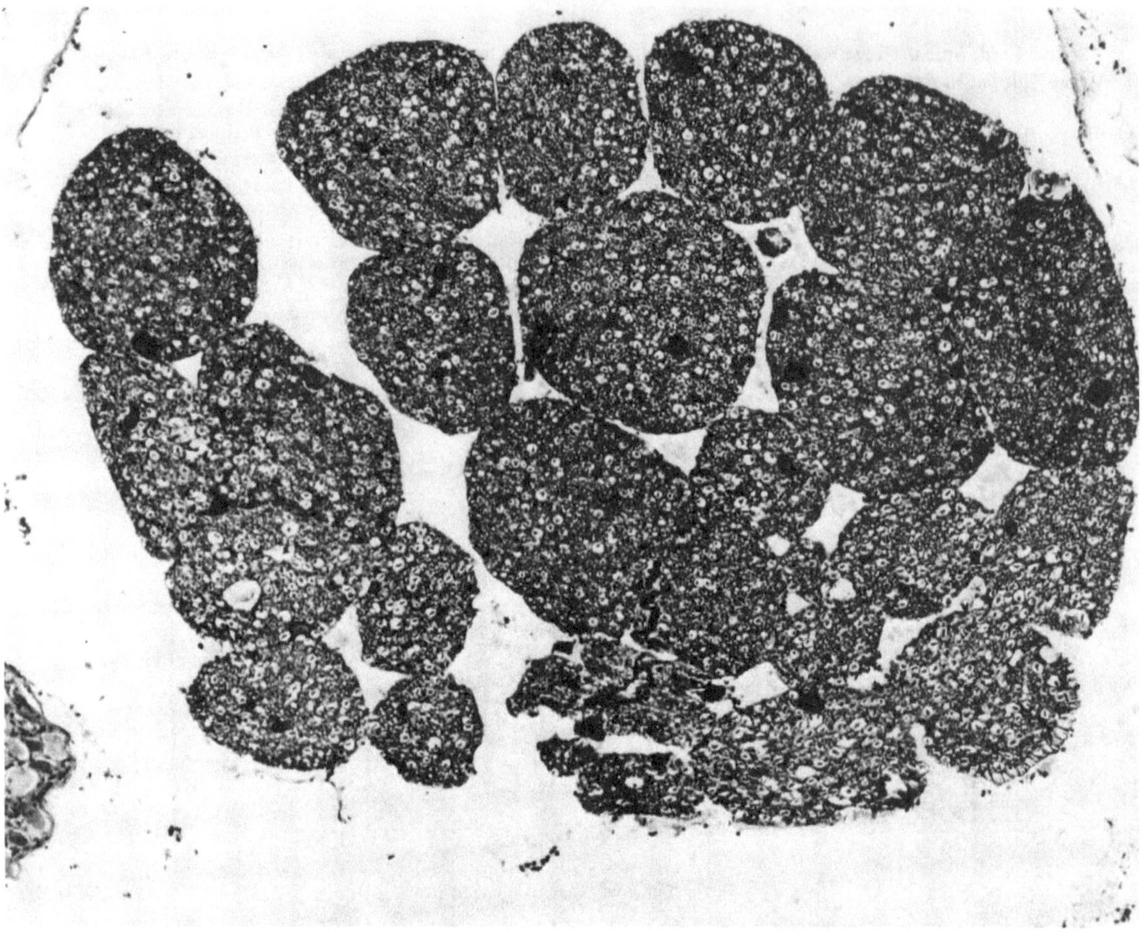

FIGURE 7.5. Nerve roots in the sacral region from the same rat illustrated in Figure 7.4. Degenerative changes are less extensive than those observed in the lumbar nerve roots. (HPS; magnification × 110.)

females, but peripheral nerve changes were seen in only one of the five males.

V. DISCUSSION

Nerve root and peripheral nerve degeneration occurred in both male and female F_1 rats. The earliest lesions began in rats between 18 to 22 months old and, surprisingly, these seemed to be more frequent in females than in the males (Table 7.2). Only in the males did the syndrome progress and paralysis result (Chapter 4, Section II.P).

It is interesting that, in contrast to the rats that died spontaneously, none of the rats killed at 6, 12, 18, or 22 months had any of the following lesions: intervertebral disk protrusions, dural fibrosis, fibrosis of intervertebral foramina, aseptic necrosis of bone, osteophytes, or exostoses. This is significant, because such changes were frequently identified in those rats dying with paralysis. These findings led to the conclusion that obstruction of the vascular blood flow could explain most of the lesions and paralysis.[26]

However, the results of the serial killing studies show that this is not completely true. The earliest lesions, namely the degeneration of spinal nerve roots and peripheral nerves, preceded any evidence of vascular obstruction. Therefore, obstruction of the blood supply may lead to the advanced lesions and subsequent paralysis, but it cannot account for the initial lesions.

All of the rats that died spontaneously with paralysis had atrophied skeletal muscles. None of the rats killed at 6, 12, 18, or 22 months old had muscle atrophy, even though all 22-month-old males and females had nerve root lesions and some had peripheral nerve lesions. This seems to confirm that the muscle lesions are secondary to the nerve lesions. These findings are similar to those of van Steenis and Kroes,[166] who considered skeletal muscle atrophy to be neurogenic and secondary to lesions in the peripheral nerves in paralyzed Wistar rats. In contrast, Berg et al.[12] thought that skeletal muscle atrophy occurred independent of spinal nerve root and peripheral nerve degeneration in Sprague-Dawley rats.

One of five 22-month-old males had degeneration of the thoracic region of the spinal cord (Table 7.2). In contrast, all 12 rats showing evidence of severe paresis had spinal cord lesions (Table 7.1). Therefore, it would appear that spinal cord degeneration develops after the onset of the nerve root lesions.

It is difficult to explain why paralysis is common in male F_1 rats and not females even though both have spinal nerve root and peripheral nerve degeneration. It may be that the severity of the degeneration stabilizes in females, but progresses in males. On the other hand, aging males may be more likely to develop more intervertebral disk protrusions, dural fibrosis, and mineral deposits in the spinal canal. Such lesions in combination could result in vascular obstruction which could cause the degenerative lesions that are already present to progress, thus resulting in paralysis.

Part IV
General Discussion

Chapter 8

THE F_1 HYBRID: COMPARISON WITH THE PARENT STRAINS

No published studies were found that compared F_1 hybrid rats with their parent strains with regard to longevity, body weights, neoplastic lesions, and nonneoplastic lesions. Therefore, a brief discussion is given in this chapter to emphasize the differences observed between the F_1 hybrid and the parent strains. These differences will be shown by presenting a few selected examples from Chapters 2, 3, and 4.

The cross between WAG/Rij females with white hair and BN/Bi males with brown hair yielded offspring that had brown and white hair. The body weights of female F_1 rats were similar to WAG/Rij and BN/Bi females, regardless of age (Figure 2.2). Similarly, the body weights of F_1 males were also similar to the WAG/Rij and BN/Bi males.

Despite the fact that the F_1 rats were hybrids, there was no evidence of so-called "hybrid vigor". As summarized in Table 3.1, the longevity of F_1 males and females was nearly identical to that of BN/Bi males and females and WAG/Rij females.

Several variations in the incidence patterns of neoplastic and nonneoplastic lesions were recognized in F_1 rats compared to the parent strains. The incidence of some lesions in F_1 rats was intermediate with respect to the incidence of the same lesions in the two parent strains; that is, one parent strain had a high incidence of a certain lesion, the second parent a relatively low incidence, and the F_1 had an incidence somewhere in-between. This is illustrated by the occurrence of foci and areas of hepatocellular alteration. This cellular change occurred in the liver of 83% of the female WAG/Rij rats and in less than 5% of the male BN/Bi rats. Both the male and female F_1 rats had incidences of these lesions of 37 and 47%, respectively. Similarly, pituitary tumors were present in about 95% of WAG/Rij females. Of the male BN/Bi rats, 14% had pituitary tumors. The F_1 rats were intermediate, with 83% in females and 64% in males.

Another variation was the higher incidence of certain other lesions in the F_1 rats as compared to either parent strain. This variation is illustrated by the finding of severe paresis and paralysis, which occurred in less than 2% of the BN/Bi or WAG/Rij rats. Similarly, it was uncommon in F_1 females. The F_1 males, on the other hand, had a high incidence of this syndrome. Over one half of those dying had this syndrome and over 80% of those older than 30 months developed it.

Some lesions were common in one or both of the parent strains, but were uncommon in the F_1 animals. For example, 20 to 35% of BN/Bi males and females had urothelial cancer. In contrast, about 1% of the F_1 rats had similar lesions.

From these few examples, it is clear that one cannot predict what changes will occur in F_1 rats compared to their parent strains. The incidence of certain lesions can be higher, lower, intermediate, or the same as that in the parent strains.

Chapter 9

SUMMARY OF POTENTIAL MODELS FOR AGING AND CARCINOGENESIS RESEARCH USING BN/Bi, WAG/Rij, AND (WAG × BN)F_1 RATS

Several pathological processes were documented in Chapters 4 to 6 that either warrant further studies or represent diseases that can be developed as animal models. Obviously, nearly every observed pathological change could be considered as a model; however, a few were unique enough to justify special interest.

Among the many cancers that were found, the following should be investigated further: medullary thyroid carcinoma in all groups, but especially in aging WAG/Rij females; carcinomas of the ureter and bladder in aging female and male BN/Bi rats; multiple endocrine tumors in all groups of aging rats; aortic body lesions and tumors in aging WAG/Rij females; granular cell tumors in the brain of aging BN/Bi, WAG/Rij and (WAG × BN)F_1 rats; sarcomas and to a lesser extent carcinomas of the cervix and vagina in aging BN/Bi females; "myelomonocytic" leukemia in BN/Bi males; and the relatively high incidence of age-associated metastatic cancers in BN/Bi females compared to the other groups.

Nonneoplastic findings also warrant further investigations to determine their possible significance with aging. Among these are the earlier mortality of WAG/Rij males and the possibility that it is the result of "premature" aging; the earlier onset of thymic atrophy in WAG/Rij males and its possible link to the earlier onset of age-associated lesions and earlier mortality; the epithelial cell proliferation in the thymus of aging BN/Bi females and the possibility that these epithelial cells are producing a thymic hormone or hormones; the possibility that degenerative myelopathy and nerve root degeneration may be a "lesion of aging" in the nervous system; and the multiplicity and increased variability of age-associated lesions as markers for a pathological aging syndrome.

These are but a few possible models that need additional studies. Now that the background data are available on these rat strains, they may be ideally suited to use for other studies, especially in the field of aging research. It would now be possible to try to modify the background incidence of the various lesions and thereby try to alter the longevity of these strains. By so doing, it may be possible to alter the aging process in these animals.

Chapter 10

IMPORTANCE OF LIFE SPAN LONGEVITY AND PATHOLOGY DATA FROM RATS

The life span and pathologic lesions of an individual rat or population of rats are determined by their genetic background and external environmental factors. Two obvious genetic variables are the strain or stock of rats and their sex. Imposed upon the genetic background are many external factors such as the husbandry conditions, diet, infectious diseases, and breeding conditions. The many combinations of interacting factors determine the longevity and pathologic findings in rats maintained under specific conditions. If one or more factors is changed, the longevity and pathologic lesions may also be changed. Such varying combinations of factors easily explain why rats derived from common stock can be so different with respect to the lesions of aging and to longevity.

Baseline longevity data are important for several reasons. Complete survival data make it possible for an investigator to select a long-lived or short-lived strain of rats. With survival curves, it is possible to determine what ages constitute the adult and aged portions of a population. This is a critical point for gerontological studies. The literature contains many descriptions of physiological, behavioral, and biochemical differences between young and "old" rats which are, in fact, differences between young and adult rats. It is impossible, therefore, to conduct gerontologic studies without knowing the life span of the experimental animal.

Longevity data are needed to compare the life spans of different rat strains and sexes. Most published data indicate that the mean survival for most rat strains is about 28 to 30 months. The males of some, but not all, strains have significantly shorter life spans than females. Since the mean or 50% survival age is about 28 to 30 months for most rat strains, it would be inadvisable to consider 18- to 24-month-old rats of these strains to be "old." At best, they could only be considered "middle-aged" and could still expect to live another 6 to 24 months.

In addition to longevity data, the baseline neoplastic and nonneoplastic lesions must be determined for each strain of rats. There are many reasons why this information is needed, some of which have been discussed by others.[29,41,79,167] In a colony free of life-shortening infectious diseases, the age-associated lesions are the most common factors associated with death. Their onset and progression directly affect mortality and thereby influence the shape of the survival curves.

Lesions, both neoplastic and nonneoplastic, are pathologic processes that alter the morphology and physiology of each organ. Not only is one specific organ affected, but a lesion in one part of the body can cause changes in other distant organs. A rat that dies from kidney failure, for example, has severe lesions in the kidneys. Because the kidneys fail to function, the rat becomes uremic and develops many secondary lesions. The rat loses weight; parathyroid glands become hyperplastic; the stomach and cecum become edematous and inflamed; mineral deposits appear in the stomach wall, cecum, arteries, and lungs; the lungs are often edematous and hydrothorax occurs; and bones become soft. The rat eventually dies. Similarly, a large mammary tumor may be localized in the subcutaneous tissue under the skin. The center of the tumor may become necrotic and hemorrhagic, so that the rat becomes toxic, anemic, and loses weight. Death may result.

Kidney disease and mammary tumors are only two examples of how age-associated lesions in one part of the body can influence distant organs. The reason for emphasizing these examples should be clear. Rats are living organisms similar to man; they develop many age-associated lesions. These age-associated lesions can influence the physiology, biochemistry, and morphology of aging rats.

Baseline pathology data are useful in selecting rats for future studies. A knowledge of the major age-associated lesions helps to select those rats best suited for certain studies. For example, the WAG/Rij rat livers had many areas of altered hepatocytes. The BN/Bi livers had apparently normal hepatocytes. It would seem, therefore, that the BN/Bi would be a better strain to use to study changes in hepatocytes with age.

Equally important, baseline pathology data can be used to eliminate some rats because they are unsuited for certain studies. The BN/Bi rat would be a poor choice for aging studies on kidneys for two reasons. First, this strain has a high incidence of congenital hydronephrosis. Second, it also has a high incidence of urothelial cancers that result in severe hydronephrosis secondary to the obstruction of urine flow. Therefore, if one had to choose between the BN/Bi and WAG/Rij rats for aging kidney research, the WAG/Rij would probably be the best choice.

Another result of determining baseline pathologic lesions is the selection of animal models for future research. Such findings are "spin offs", but can be very useful for biomedical research. Examples of such "spin offs" were discussed in detail in Chapter 9. Tumor models included such findings as medullary thyroid carcinomas, urothelial cancers, cervix and vaginal cancers, and granular cell (myoblastomas) tumors. Nonneoplastic models included age-associated spinal cord and nerve root degeneration and age-associated changes in the thymuses of the different strains and sexes of rats. Another observation was the possibility of developing the male WAG/Rij rat as an example of "premature" aging.

Chapter 11

THE ONSET OF AGE-ASSOCIATED LESIONS AND AGING IN RATS

Researchers in gerontology have been devoting increased efforts into studying functional and subcellular alterations with age in the hope of finding the cause or causes of aging. However, one problem is that rats, as well as man, usually die with multiple pathological processes. Therefore, gerontologists have been trying to differentiate what constitutes a "true" aging process and what is a disease that incidentally happens to be age associated.

In rats that die spontaneously, age-associated lesions are multiple and are both neoplastic and nonneoplastic. The combined result of all changes is death. Some rats die at a relatively young chronological age (about 12 months), while others die relatively late (older than 48 months). This variability in the age of death is associated with the onset of lesions. This finding was observed in the so-called inbred BN/Bi and WAG/Rij rat strains and in their F_1 hybrid. This variability in the age of death has also been shown in outbred rats as well.

In rats and man, the incidence of cancer increases significantly with age. However, most investigators in aging research try to separate cancer from the aging process. Cancers can kill rats and are extremely important in shaping the survival curves for a rat population. Since this is true, it may be that cancer, rather than being separate from aging, is actually one manifestation of the aging process. The onset and progression can be modified by many external factors and these may even cause some cancers. However, the genetic background of the individual animal sets the stage for the age of onset and for the rate at which many cancers, and probably most nonneoplastic lesions, will occur. It is these multiple lesions that finally kill the individual rats. Therefore, if one can understand what initiates cancer, some insight may be gained into the causes of cell and organ aging. Similarly, elucidation of the aging process within a cell may also help in understanding the processes which lead to cancer. This concept is not new, and Pitot[125] has recently extensively reviewed this topic.

Many studies have tried to modify the aging process. However, dietary manipulation is the only method that has significantly altered longevity of various rat strains, as illustrated in the studies by Ross et al.,[140,141,144,145] Nolen,[123] and Berg and Simms.[10,11,13] The altered longevity seems to be caused by an alteration in the course or the onset of pathological lesions. For example, changing the protein or carbohydrate in the diet of rats appeared to decrease the age of onset of tumors and thereby increased the longevity. Similarly, decreased protein in the diet resulted in increased longevity that appeared to result from rats showing a slowing of the onset of chronic kidney disease.

Therefore, pathologic lesions, either neoplastic or nonneoplastic, cannot easily be separated from the aging process. It is extremely important to keep these age-associated lesions in proper perspective when doing aging research with rats. Every effort should be made to correlate morphological changes with functional changes in order to better understand the biology of the aging process.

Summary

SUMMARY

There is a large gap in our understanding of the normal biological variability that occurs in experimental animals used in biomedical research. One of the reasons for this gap is our lack of knowledge about the normal background patterns of age-associated lesions and how these lesions influence the shape of survival curves over the life span of a population of animals. The purpose of this study was to help narrow this gap by evaluating two strains of aging rats and their F_1 hybrid and to compare the age-associated pathological findings in these rats that had been maintained under similar and well-controlled laboratory conditions with the longevity data obtained from the same rats. In addition, groups of rats were killed at selected ages in order to compare the lesions in the killed rats vs. those from aging cohorts that had died spontaneously.

Spontaneous age-associated patterns of neoplastic and nonneoplastic lesions were studied in 670 rats, and survival data were compiled on 791 rats that completed their normal life spans. Rats were of the BN/Bi (Brown Norway) and WAG/Rij (Wistar Strains) and their (WAG × BN)F_1 hybrid and included both males and females. They were maintained under similar conditions, were free of mycoplasmal pneumonia, and one half were older than 30 months at the time of their death.

The life span study showed that male and female BN/Bi, male and female (WAG × BN)F_1, and female WAG/Rij rats had 50% survival ages of about 30 months. This was similar to most other rat strains where a 50% survival was reached at 28 to 30 months. In a few strains, the male does not live as long as the female, and this was also true for the male WAG/Rij rats which had a 50% survival of about 24 months. In addition, retired breeder females do not necessarily have shorter life spans than virgins. In fact, the BN/Bi retired breeders had slightly better survival than virgin BN/Bi rats. Also, there was no evidence for F_1 hybrid vigor if longevity is used as a criterion. In this study, the F_1 males and females both had 50% survival ages that were nearly identical to their parents.

The studies to determine the age-associated pathological lesions resulted in several findings which correlated with the longevity data. The results of the pathologic evaluation of these animals can be summarized as follows:

1. Rats died with or from multiple age-associated lesions, not necessarily from one specific lesion.
2. In some organs or tissues, neoplastic lesions predominated, in others nonneoplastic lesions predominated, and in still others various combinations of both neoplastic and nonneoplastic lesions were present.
3. The multiple age-associated lesions usually resulted in mortality and thereby determined the shape of the survival curves.
4. The age of onset of most lesions was between 12 to 18 months in the aging population. However, the age of onset in an individual rat varied from earlier than 18 months in some rats to later than 30 months in other rats.
5. The incidence of some lesions increased with time, others reached a peak and plateaued, and still others reached a peak followed by a decrease. To determine such variable patterns in the incidence of lesions, life span studies are clearly needed.
6. The types and patterns of age-associated lesions were different for each strain of rat. Despite this, the survival curves were similar.
7. The older the population, the greater the variability of lesions within the population and within an individual rat.

Major differences were found when the percentage or incidence of lesions in rats that died spontaneously was compared to rats that were killed at selected ages. The percentage of some lesions in killed rats was closest to that found in the population at risk as calculated by life table techniques, whereas the incidence of other lesions in the killed rats was closest to that found in animals that had died spontaneously. Finally, the percentage of still other lesions in the killed rats was found to be somewhere in-between the percentage determined by life table calculations and the percentage in the dead rats at different time periods. Results such as these clearly indicate the need for serial or multiple sacrifices for aging studies. Similarly, serial or multiple sacrifices would be equally important for chronic toxicity and carcinogenesis studies.

In conclusion, to adequately evaluate aging rats in long-term research, whether the studies are on the population, on the whole animal, on an organ, on an individual cell, or even at the subcellular level, one needs rats that have survived until at least the 50% survival age for the particular strain. It is also critically important to know the natural incidence of age-associated lesions and the patterns of these lesions with age in the animal as a whole and in the particular organ to be studied. This information is important because these lesions reflect pathological processes that directly affect the experimental results obtained from aging animals.

References

REFERENCES

1. **Altman, N. H. and Goodman, D.**, Neoplastic diseases of the laboratory rat, in *The Biology of the Laboratory Rat*, Baker, H. J., Lindsey, J. R., and Weisbroth, S., Eds., Academic Press, New York, in press.
2. **Andrew, W. and Pruett, D.**, Senile changes in the kidneys of Wistar Institute rats, *Am. J. Anat.*, 100, 51, 1957.
3. **Antoon, J. W. and Gregg, R. V.**, The influence of body temperature on the population of ulcers of restraint in the rat, *Gastroenterology*, 70, 747, 1976.
4. **Anver, M. and Cohen, B. J.**, Lesions associated with aging, in *The Biology of the Laboratory Rat*, Baker, H. J., Lindsey, J. R., and Weisbroth, S., Eds., Academic Press, New York, in press.
5. **Anver, M. and Cohen, B. J.**, Neoplastic and nonneoplastic lesions in Sprague-Dawley rats, in preparation.
6. **Arakaw, M.**, A scanning electron microscopy of the glomerulus of normal and nephrotic rats, *Lab. Invest.*, 23, 489, 1970
7. **Arakaw, M. and Tokunaga, J.**, A scanning electron microscope study of the glomerulus. Further consideration of the mechanism of the fusion of podocyte terminal processes in nephrotic rats, *Lab. Invest.*, 27, 366, 1972.
8. **Baba, N. and Von Haam, E.**, Tumours of the uterus, in *Pathology of Tumours in Laboratory Animals*, Vol. 1, Part 2, Turusov, V. S., Ed., IARC Scientific Publ. No. 6, World Health Organization, Geneva, 1976, 161.
9. **Berg, B. N.**, Muscular dystrophy in aging rats, *J. Gerontol.*, 11, 134, 1956.
10. **Berg, B. N. and Simms, H. S.**, Nutrition and longevity in the rat. II. Longevity and the onset of disease with different levels of food intake, *J. Nutr.*, 71, 255, 1960.
11. **Berg, B. N. and Simms, H. S.**, Nutrition and longevity in the rat. III. Food restriction beyond 800 days, *J. Nutr.*, 74, 23, 1961.
12. **Berg, B. N., Wolf, A., and Simms, H. S.**, Degenerative lesions of spinal roots and peripheral nerves in aging rats, *Gerontologia*, 6, 72, 1962.
13. **Berg, B. N. and Simms, H. S.**, Nutrition, onset of disease and longevity in the rat, *Can. Med. Assoc. J.*, 93, 911, 1965.
14. **Berg, B. N.**, Longevity studies in rats. II. Pathology of aging rats, in *Pathology of Laboratory Rats and Mice*, Cotchin, E. and Roe, F. J. C., Eds., Blackwell Scientific, Oxford, 1967, 749.
15. **Berlin, M. and Wallace, R. B.**, Aging and the central nervous system, *Exp. Aging Res.*, 2, 125, 1976.
16. **Bolton, W. K., Benton, F. R., Maclay, J. G., and Sturgill, B. C.**, Spontaneous glomerular sclerosis in aging Sprague-Dawley rats, *Am. J. Pathol.*, 85, 277, 1976.

16a. **Boorman, G. A.**, unpublished observations.

17. **Boorman, G. A. and Hollander, C. F.**, Occurrence of spontaneous cancer with aging in an inbred strain of rats, *TNO Nieuws*, 27, 692, 1972.
18. **Boorman, G. A., van Noord, M. J., and Hollander, C. F.**, Naturally occurring medullary thyroid carcinoma in the rat, *Arch. Pathol.*, 94, 35, 1972.
19. **Boorman, G. A. and Hollander, C. F.**, Spontaneous lesions in the female WAG/Rij (Wistar) rat, *J. Gerontol.*, 28, 152, 1973.
20. **Boorman, G. A., Hollander, C. F., and Feron, V. J.**, Naturally occurring endocardial disease in the rat, *Arch. Pathol.*, 96, 39, 1973.
21. **Boorman, G. A., Heersche, J. N. M., and Hollander, C. F.**, Transplantable calcitonin-secreting medullary carcinomas of the thyroid in the WAG/Rij rat, *J. Natl. Cancer Inst.*, 53, 1011, 1974.
22. **Boorman, G. A. and Hollander, C. F.**, Medullary carcinoma of the thyroid in the rat. Animal model of human disease, *Am. J. Pathol.*, 83, 237, 1976.
23. **Boorman, G. A. and Hollander, C. F.**, High incidence of spontaneous urinary bladder and ureter tumors in the Brown Norway rat, *J. Natl. Cancer Inst.*, 52, 1005, 1974.
24. **Boorman, G. A., Burek, J. D., and Hollander, C. F.**, Carcinoma of the ureter and urinary bladder. Animal model of human disease, *Am. J. Pathol.*, 88, 251, 1977.
25. **Broerse, J. J., Hollander, C. F., and van Zwieten, M. J.**, personal communication.
26. **Burek, J. D., van der Kogel, A. J., and Hollander, C. F.**, Degenerative myelopathy in three strains of aging rats, *Vet. Pathol.*, 13, 321, 1976.
27. **Burek, J. D., Zurcher, C., and Hollander, C. F.**, High incidence of spontaneous cervical and vaginal tumors in an inbred strain of Brown Norway rats (BN/Bi), *J. Natl. Cancer Inst.*, 57, 549, 1976.
28. **Burek, J. D. and Hollander, C. F.**, Incidence patterns of spontaneous tumors in BN/Bi rats, *J. Natl. Cancer Inst.*, 58, 99, 1977.
29. **Burek, J. D. and Hollander, C. F.**, Uses in aging research, in *The Biology of the Laboratory Rat*, Baker, H. J., Lindsey, J. R., and Weisbroth, S., Eds., Academic Press, New York, in press.
30. **Burek, J. D. and Meihuizen, S. P.**, Age-related characteristics of thymuses from BN/Bi rats, in *Proc. 5th Eur. Symp. Basic Research in Gerontology*, Schmidt, U. J., Brüschke, G., Lange, E., Viidik, A., Platt, D., Frolkis, V. V., and Schulz, F. H., Eds., Verlag Dr. med. D. Straube, Erlangen, 1977, 167.
31. **Burek, J. D., Zurcher, C., Van Nunen, M. C. J., and Hollander, C. F.**, A naturally occurring epizootic caused by Sendai virus in breeding and aging rodent colonies. II. Sendai virus infection in rats, *Lab. Anim. Sci.*, 27, 963, 1977.

31a. **Burek, J. D.**, unpublished observations.

32. **Bullock, B. C., Banks, K. L., and Manning, P. J.,** Common lesions in the aged rat, in *The Laboratory Animal in Gerontological Research,* Publ. No. 1591, National Academy of Sciences, Washington, D.C., 1968, 62.
33. **Byfield, P. G. H., Matthews, E. W., Heersche, J. N. M., Boorman, G. A., Girgis, S. I., and MacIntyre, I.,** Isolation of calcitonin from rat thyroid medullary carcinoma, *FEBS Lett.,* 65, 238, 1976.
34. **Byfield, P. G. H., McLaughlin, L., Matthews, E. W., and MacIntyre, I.,** A proposed structure for rat calcitonin, *FEBS Lett.,* 65, 242, 1976.
35. **Carter, R. L.,** Tumours of soft tissues, in *Pathology of Tumours in Laboratory Animals,* Vol. 1, Part 1, Turusov, V. S., Ed., IARC Scientific Publ. No. 5, World Health Organization, Geneva, 1973, 151.
36. **Carter, R. L. and Ird, E. A.,** Tumours of the ovary, in *Pathology of Tumours of Laboratory Animals,* Vol. 1, Part 2, Turusov, V. S., Ed., IARC Scientific Publ. No. 6, World Health Organization, Geneva, 1976, 189.
37. **Cherry, C. P., Eisenstein, R., and Glücksmann, A.,** Epithelial cords and tubules of the rat thymus: effects of age, sex, castration, of sex, thyroid and other hormones on their incidence and secretory activity, *Br. J. Exp. Pathol.,* 48, 90, 1967.
38. **Chesky, J. A. and Rockstein, M.,** Life span characteristics in the male Fischer rat, *Exp. Aging Res.,* 2, 399, 1976.
39. **Cohen, B. J.,** Effects of environment on longevity in rats and mice, in *The Laboratory Animal in Gerontological Research,* Publication No. 1591, National Academy of Sciences, Washington, D.C., 1968, 21.
40. **Cohen, B. J., de Bruin, R. W., and Kort, W. J.,** Heritable hydronephrosis in a mutant strain of Brown Norway rats, *Lab. Anim. Care,* 20, 489, 1970.
41. **Cohen, B. J. and Anver, M. R.,** Pathological changes during aging in the rat, in *Special Review of Experimental Aging Research. Progress in Biology,* Eleftheriou, M. F. and Elias, P. K., Eds., Experimental Aging Research, Bar Harbor, Maine, 1977, 379.
42. **Coleman, G. L., Jonas, A. M., Hoffman, H., Barthold, S., Osbaldiston, G., and Foster, S.,** Pathological changes during aging in barrier reared Fischer 344 male rats, *J. Gerontol.,* 32, 258, 1977.
43. **Comfort, A.,** *Ageing, The Biology of Senescence,* Routledge and Kegan Paul Ltd., London, 1964.
44. **Cotchin, E. and Roe, F. J. C.,** *Pathology of Laboratory Rats and Mice,* Blackwell Scientific, Oxford, 1967.
45. **Couser, W. G. and Stilmant, M. M.,** Mesangial lesions and focal glomerular sclerosis in the aging rat, *Lab. Invest.,* 33, 491, 1975.
46. **Couser, W. G. and Stilmant, M. M.,** The immunopathology of the aging rat kidney, *J. Gerontol.,* 31, 13, 1976.
47. **Crain, R. C.,** Spontaneous tumors in the Rochester strain of Wistar rat, *Am. J. Pathol.,* 34, 311, 1958.
48. **Crumeyrolle-Arias, M., Scheib, D., and Aschheim, P.,** Light and electron microscopy of the ovarian interstitial tissue in the senile rat: normal aspect and response to HCG of "deficiency cells" and "epithelial cords", *Gerontology,* 22, 185, 1976.
49. **de Estable-Puig, R. F. and de Estable-Puig, J. F.,** Vacuolar degeneration in neurons of aging rats, *Virchows Arch. B,* 17, 337, 1975.
50. **de Leeuw-Israël, F. R.,** Aging Changes in the Rat Liver. An Experimental Study of Hepatocellular Function and Morphology, Ph.D. thesis, University of Leiden, The Netherlands, 1971.
51. **DeLellis, R. A., Nunnemacher, G., and Wolfe, H. J.,** C-cell hyperplasia. An ultrastructural analysis, *Lab. Invest.,* 36, 237, 1977.
52. **Elema, J. D. and Arends, A.,** Focal and segmental glomerular hyalinosis and sclerosis in the rat, *Lab. Invest.,* 33, 554, 1975.
53. **Fairweather, F. A.,** Cardiovascular disease in rats, in *Pathology of Laboratory Rats and Mice,* Cotchin, E. and Roe, J. C., Eds., Blackwell Scientific, Oxford, 1967, 213.
54. **Farber, E.,** Hyperplastic liver nodules, in *Methods in Cancer Research,* Vol. 7, Busch, H., Ed., Academic Press, New York, 1973, 345.
55. **Festings, M. and Staats, J. N.,** Standardized nomenclature for inbred strains of rats. Fourth listing, *Transplantation,* 16, 221, 1973.
56. **Finch, C. E.,** Comparative biology of senescence-evolutionary and developmental considerations, in *Animal Models for Biomedical Research,* National Academy of Sciences, Washington, D.C., 1971, 47.
57. **Finch, C. E.,** Enzyme activities, gene function and ageing in mammals (review), *Exp. Gerontol.,* 7, 53, 1972.
58. **Flodh, H.,** Pulmonary foam cells in rats of different ages, *Z. Viersuchstierk,* 16, 299, 1974.
59. **Franks, L. M. and Maldague, P.,** Tumours of the accessory male sex gland, in *Pathology of Tumours in Laboratory Animals,* Vol. 1, Part 2, Turusov, V. S., Ed., IARC Scientific Publ. No. 6, World Health Organization, Geneva, 1976, 151.
60. **Frith, C. H., Farris, H. E., and Highman, B.,** Endocardial fibromatous proliferation in a rat, *Lab. Anim. Sci.,* 27, 114, 1977.
61. **Fujisawa, K.,** Some observations on the skeletal musculature of aged rats. I. Histological aspects, *J. Neurol. Sci.,* 22, 353, 1974.
62. **Fujisawa, K.,** Some observations on the skeletal musculature of aged rats. II. Fine morphology of diseased muscle fiber, *J. Neurol. Sci.,* 24, 447, 1975.
63. **Furth, J. and Clifton, K. H.,** Experimental pituitary tumours, in *The Pituitary Gland,* Vol. 2, Harris, G. W. and Donovan, B. T., Eds., Butterworths, London, 1966, 460.
64. **Furth, J., Nakane, P. K., and Pasteels, J. L.,** Tumours of the pituitary gland, in *Pathology of Tumours in Laboratory Animals,* Vol. 1, Part 2, Turusov, V. S., Ed., IARC Scientific Publ. No. 6, World Health Organization, Geneva, 1976, 201.

65. **Gaitz, C. M.,** *Aging and the Brain,* Plenum Press, New York, 1972.
66. **Garner, F. M., Innes, J. R. M., and Nelson, D. H.,** Murine neuropathology, in *Pathology of Laboratory Rats and Mice,* Cotchin, E. and Roe, F. J. C., Eds., Blackwell Scientific, Oxford, 1967, 295.
67. **Giddens, W. E. and Whitehair, C. K.,** The peribronchial lymphocytic tissue in germfree, defined-flora, conventional and chronic murine pneumonia affected rats, in *Germfree Biology,* Plenum Press, New York, 1969, 75.
68. **Gilbert, C. and Gillman, J.,** Spontaneous neoplasms in the albino rat, *S. Afr. J. Med. Sci.,* 23, 257, 1958.
69. **Gillman, J., Gilbert, C., and Spense, I.,** Phaeochromocytoma in the rat. Pathogenesis and collateral reactions and its relation to comparable tumours in man, *Cancer,* 6, 494, 1953.
70. **Glucksmann, A. and Cherry, C. P.,** Tumours of the salivary glands, in *Pathology of Tumours in Laboratory Animals,* Vol. 1, Part 1, Turusov, V. S., Ed., IARC Scientific Publ. No. 5, World Health Organization, Geneva, 1973, 75.
71. **Gore, I. Y.,** Methodology in gerontological research, *Geront. Clin.,* 15, 133, 1973.
72. **Gössner, W., Hollander, C. F., Maisin, J. R., Nilsson, A., and Lutz, A.,** EULEP Pathology Atlas, Report of Workshops, Committee on Pathology Standardization, European Late Effects Project Group, in preparation.
73. **Gould, D. H.,** Mesotheliomas of the tunica vaginalis propria and peritoneum in Fischer rats, *Vet. Pathol.,* 14, 372, 1977.
74. **Gray, J. E., Weaver, R. N., and Purmalis, A.,** Ultrastructural observations of chronic progressive nephrosis in the Sprague Dawley Rat, *Vet. Pathol.,* 11, 153, 1974.
75. **Gsell, D.,** Absterbekurven und Wachstumscharakteristika einer "Alterszucht" von Wistar-Ratten, in *Die Umwelt der Versuchstiere und ihre Standardisierung in biologischen Test,*Welke, W., Ed., *Int. Z. Vitaminforsch.* (Suppl.), 9, 114, 1964.
76. **Hebel, R. and Stromberg, M. W.,** *Anatomy of the Laboratory Rat,* Williams and Wilkins, Baltimore, 1976, 96.
77. **Hicks, R. M., St. Wakefield, J. St. J., Vlasov, N. N., and Pliss, G. B.,** Tumours of the bladder, in *Pathology of Tumours in Laboratory Animals,* Vol. 1, Part 2, Turusov, V. S., Ed., IARC Scientific Publ. No. 6, World Health Organization, Geneva, 1976, 103.
77a. **Hirokawa, K.,** Characteristics of age-associated kidney disease in Wistar rats, *Mech. Age. Dev.,* 4, 301, 1975.
78. **Hollander, C. F.,** Cartilagenous focus at the base of the non-coronary semilunar valve of the aorta in rats of different ages, *Exp. Gerontol.,* 3, 303, 1968.
79. **Hollander, C. F.,** Guest Editorial: Animal models for aging and cancer research, *J. Natl. Cancer Inst.,* 51, 3, 1973.
80. **Hollander, C. F., Boorman, G. A., and Zurcher, C.,** Classification of malignant tumours in rodents, in *Multiple Primary Malignant Tumors,* Proc. 6th Perugia Quadrenmial Int. Conf. Cancer, Severi, L., Ed., Perugia Division of Cancer Research, Perugia, Italy, 1974, 279.
81. **Hollander, C. F.,** Current experience using the laboratory rat in aging studies, *Lab. Anim. Sci.,* 26, 320, 1976.
82. **Hollander, C. F., Burek, J. D., Boorman, G. A., Snell, K. C., and Laqueur, G. L.,** Granular cell tumors of the central nervous system of rats, *Arch. Pathol. Lab. Med.,* 100, 445, 1976.
83. **Hollander, C. F., Zurcher, C., Nooteboom, A., van Zwieten, M. J., and Burek, J. D.,** unpublished observations, 1976.
84. **Hollander, C. F. and Snell, K. C.,** Tumours of the adrenal gland, in *Pathology of Tumours in Laboratory Animals,* Vol. 1, Part 2, Turusov, V. S., Ed., IARC Scientific Publ. No. 6, World Health Organization, Geneva, 1976, 273.
85. **Hruza, Z. and Zbuzkova, V.,** Cholesterol turnover in plasma, aorta, muscles, and erythrocytes in young and old rats, *Mech. Aging Dev.,* 4, 169, 1975.
86. **Huang, H. H. and Meites, J.,** Reproductive capacity of aging female rats, *Neuroendocrinology,* 17, 289, 1975.
87. **Hueper, W. C.,** Cartilaginous foci in hearts of white rats and mice, *Arch. Pathol.,* 27, 446, 1939.
88. **Innes, J. R. M. and Saunders, L. Z.,** *Comparative Neuropathology,* Academic Press, New York, 1962.
89. **Innes, J. R. M.,** Lesions of the respiratory tract of small laboratory animals, in *The pathology of Laboratory Animals,* Ribelin, W. E. and McCoy, J. R., Eds., Charles C Thomas, Springfield, Ill., 1965, 49.
90. **Innes, J. R. M.,** Respiratory diseases in rats, in *Pathology of Laboratory Rats and Mice,* Cotchin, E. and Roe, F. J. C., Eds., Blackwell Scientific, Oxford, 1967, 229.
91. **Inukai, T.,** On the loss of purkinje cells with advanced age from the cerebellar cortex of the albino rat, *J. Comp. Neurol.,* 45, 1, 1928.
92. **Ito, A.,** Pituitary tumors in rats. Animal model of human disease, *Am. J. Pathol.,* 83, 423, 1976.
93. **Ivankovic, S.,** Tumours of the heart, in *Pathology of Tumours in Laboratory Animals,* Vol. 1, Part 2, Turusov, V. S., Ed., IARC Scientific Publ. No. 6, World Health Organization, Geneva, 1976, 313.
94. **Jacobs, B. B. and Huseby, R. A.,** Neoplasms occurring in aged Fischer rats with special reference to testicular, uterine and thyroid tumors, *J. Natl. Cancer Inst.,* 39, 303, 1967.
95. **Jänisch, W. and Schreiber, D.,** *Experimental Tumors of the Central Nervous System,* 1st English ed., Bigner, D. D. and Swenberg, J. A., Eds., The Upjohn Company, Kalamazoo, Mich., 1977.
96. **Jay, G. E., Jr.,** Genetic strains and stocks, in *Methodology in Mammalian Genetics,* Burdette, W. J., Ed., Holden-Day, San Francisco, 1963, 107.
97. **Jonas, A. M.,** Long-term Holding of Laboratory Rodents, *ILAR News,* 19, L1, 1976.
98. **Knook, D. L. and Hollander, C. F.,** Embryology and aging of the rat liver, in *Rat Hepatic Neoplasia,* Proc. Rat Liver Workshop, Newberne, P. M. and Butler, W. A., Eds., MIT Press, Cambridge, Mass., in press.
99. **Kociba, R. J., Keyes, D. G., Jersey, G. C., Ballard, J. J., Dittenber, D. A., Quast, J. F., Wade, C. E., Humiston, C. G., and Schwetz, B. A.,** Results of a Two-Year Chronic Toxicity Study with Hexachlorobutadiene (HCBD) in Rats, Internal Report, Toxicology Research Laboratory, The Dow Chemical Company, Midland, Mich., 1976.

100. **Kohn, R. R.,** *Principles of Mammalian Aging,* Foundations of Developmental Biology Series, Prentice-Hall, Englewood Cliffs, N.J., 1971.
101. **Kraus, B. and Cain, H.,** Über eine Spontane Nephropathie bei Wistar-Ratten: Die Licht und Elektronenmikroskopischen Glomerulum Veränderungen, *Virchow Arch. Pathol. Anat. Histol.,* 363, 343, 1974.
102. **Kroes, R., Berkvens, J. M., de Vries, T., and van Nesselrooy, J. H. J.,** Tumorincidenties en incidenties van andere pathologische veranderingen in Wistar-ratten welke door het Laboratorium Voor Toxicologie van het Rijksinstituut voor de Volksgezondheid worden Gebruikt. Deel I and II, Internal Rep. No. 127/76 Path/Tox, Rijksinstituut voor de Volksgezondheid, Bilthoven, The Netherlands, 1976.
103. **Kruisbeek, A. M., Zijlstra, S. P., Meihuizen, S. P., and Burek, J. D.,** Comparison Between Age-Related Changes in Thymus Morphology and Thymus-Dependent Immune Functions in WAG/Rij Rats, Annual Report of the REP Institutes of the Organization for Health Research TNO, Rijswijk (ZH), The Netherlands, 1976, 293.
104. **Kunstyr, I. and Leuenberger, H. W.,** Gerontological data of C57BL/6J mice. I Sex differences in survival curves, *J. Gerontol.,* 30, 157, 1975.
105. **Lalich, J. J., Faith, G. C., and Harding, G. E.,** Protein overload nephropathy, *Arch. Pathol.,* 89, 548, 1970.
106. **Lamb, D.,** Rat lung pathology and quality of laboratory animals: the user's view, *Lab. Anim.,* 9, 1, 1975.
107. **Lehr, D.,** Lesions of the cardiovascular system, in *The Pathology of Laboratory Animals,* Ribelin, W. E. and McCoy, J. R., Eds., Charles C Thomas, Springfield, Ill., 1965, 124.
108. **Lindsay, S., Nichols, C. W., Jr., and Chaikoff, I. L.,** Naturally occurring thyroid carcinoma in the rat: similarities to human medullary thyroid carcinoma, *Arch. Pathol.,* 86, 353, 1968.
109. **Lindsay, S. and Nichols, C. W.,** Medullary thyroid carcinoma and parathyroid hyperplasia in rats, *Arch. Pathol.,* 88, 402, 1969.
110. **Lindsey, J. R., Baker, H. J., Overcash, R. G., Cassell, G. H., and Hunt, C. E.,** Murine chronic respiratory disease. Significance as a research complication and experimental production with mycoplasma pulmonis, *Am. J. Pathol.,* 64, 675, 1971.
111. **Litvinov, N. N. and Soloviev, J. N.,** Tumours of the bone, in *Pathology of Tumours in Laboratory Animals,* Vol. 1, Part 1, Turusov, V. S., Ed., IARC Scientific Publ. No. 5, World Health Organization, Geneva, 1973, 169.
112. **Machado, E. A. and Beauchene, R. E.,** Spontaneous osteogenic sarcoma in the WI/Ten rat: a case report, *Lab. Anim. Sci.,* 26, 98, 1976.
113. **Mawdesley-Thomas, L. E. and Hague, P. H.,** Mesothelioma of the tunica vaginalis in a rat, *Lab. Anim.,* 4, 29, 1970.
114. **Meier-Ruge, W.,** Experimental pathology and pharmacology in brain research and aging, *Life Sci.,* 17, 1627, 1975.
115. **Meihuizen, S. P. and Burek, J. D.,** The epithelial cell component of thymic tissue in aging female BN/Bi rats, in *Electron Microscopy,* Vol. 2, Ben-Shaul, Y., Ed., Tal International, Israel, 1976, 569.
116. **Meites, J. and Huang, H. H.,** Relation of the neuroendocrine system to loss of reproductive functions in aging rats, in *Neuroendocrine Regulation of Fertility,* Karger, S., Basel, 1976, 246.
117. **Mennel, H. D. and Zülch, E.,** Tumours of the brain, in *Pathology of Tumours in Laboratory Animals,* Vol. 1, Part 2, Turusov, V. S., Ed., IARC Scientific Publ. No. 6, World Health Organization, Geneva, 1976, 295.
117a. **Moloney, W. C., Boschetti, A. E., and King, V. P.,** Spontaneous leukemia in Fischer rats, *Cancer Res.,* 30, 41, 1970.
118. **Mostofi, F. K. and Bresler, V. M.,** Tumours of the testis, in *Pathology of Tumours in Laboratory Animals,* Vol. 1, Part 2, Turusov, V. S., Ed., IARC Scientific Publ. No. 6, World Health Organization, Geneva, 1976, 135.
119. **Nagayo, T.,** Tumours of the stomach, in *Pathology of Tumours in Laboratory Animals,* Vol 1, Part 1, Turusov, V. S., Ed., IARC Scientific Publ. No. 5, World Health Organization, Geneva, 1973, 101.
120. **Napalkov, N. P.,** Tumours of the thyroid, in *Pathology of Tumours in Laboratory Animals,* Vol. 1, Part 2, Turusov, V. S., Ed., IARC Scientific Publ. No. 6, World Health Organization, Geneva, 1976, 239.
121. **Nelson, A. A.,** The recording and reporting of pathology data, in *The Pathology of Laboratory Animals,* Ribelin, W. E. and McCoy, J. R., Eds., Charles C Thomas, Springfield, Ill., 1965, 406.
122. **Newburgh, L. H. and Curtis, A. C.,** Production of renal injury in the white rat by the protein of the diet: dependence of the injury on the duration of feeding and on the amount and kind of protein, *Arch. Intern. Med.,* 42, 801, 1928.
123. **Nolen, G. A.,** Effect of various restricted dietary regimes on the growth, health, and longevity of albino rats, *J. Nutr.,* 102, 1477, 1972.
124. **Paget, G. E. and Lemon, P. G.,** The interpretation of pathology data, in *Pathology of Laboratory Animals,* Ribelin, W. E. and McCoy, J. R., Eds., Charles C Thomas, Springfield, Ill., 1965, 382.
125. **Pitot, H. C.,** Carcinogenesis and aging — two related phenomena? A review, *Am. J. Pathol.,* 87, 444, 1977.
126. **Plagge, J. C.,** Some effects of prolonged massive estrogen treatment on rat, with special reference to thymus, *AMA Arch. Pathol.,* 42, 598, 1946.
127. **Porta, E. A. and Hartroft, W. S.,** Lipid pigments in relation to aging and dietary factors (lipofuscins), in *Pigments in Pathology,* Wolman, M., Ed., Academic Press, New York, 1969, 191.
128. **Potter, V. R.,** Transplantable animal cancer, the primary standard, *Cancer Res.,* 21, 1331, 1961.
129. **Pour, P., Stanton, M. F., Kuschner, M., Laskin, S., and Shabad, L. M.,** Tumours of the respiratory tract, in *Pathology of Tumours in Laboratory Animals,* Vol. 1, Part 2, Turusov, V. S., Ed., IARC Scientific Publ. No. 6, World Health Organization, Geneva, 1976, 1.
130. **Pozharisski, K. M.,** Tumours of the esophagus, in *Pathology of Tumours in Laboratory Animals,* Vol. 1, Part 1, Turusov, V. S., Ed., IARC Scientific Publ. No. 5, World Health Organization, Geneva, 1973, 87.

131. **Pozharisski, K. M.,** Tumours of the intestines, in *Pathology of Tumours in Laboratory Animals,* Vol. 1, Part 1, Turusov, V. S., Ed., IARC Scientific Publ. No. 5, World Health Organization, Geneva, 1973, 119.
131a. **Quast, J. F. and Jersey, G. C.,** unpublished observations.
132. **Ratcliffe, H. C.,** Spontaneous tumors in two colonies of rats of the Wistar Institute of Anatomy and Biology, *Am. J. Pathol.,* 16, 237, 1940.
133. **Reuber, M. D.,** Development of preneoplastic and neoplastic lesions of the liver in male rats given 0.025 percent N-2-fluroenyldiacetamide, *J. Natl. Cancer Inst.,* 34, 697, 1965.
134. **Ribelin, W. E. and McCoy, J. R.,** *Pathology of Laboratory Animals,* Charles C Thomas, Springfield, Ill., 1965.
135. **Robinson, M. R. G., Rigby, C. C., Pugh, R. C. B., Vaughan, L. C., and Dumonde, D. C.,** Experimental cellular allergic reactions in normal canine and malignant human prostate, *Clin. Exp. Immunol.,* 26, 137, 1976.
136. **Rockstein, M. and Sussman, M. L.,** *Development and Aging in the Nervous System,* Academic Press, New York, 1973.
137. **Roe, F. J. C.,** Spontaneous tumors in rats and mice, *Food Cosmet. Toxicol.,* 3, 707, 1965.
138. **Roe, F. J. C. and Roberts, J. D. B.,** Tumours of the pancreas, in *Pathology of Tumours in Laboratory Animals,* Vol. 1, Part 1, Turusov, V. S., Ed., IARC Scientific Publ. No. 5, World Health Organization, Geneva, 1973, 141.
139. **Ross, M. A. and Korenchevsky, V.,** Thymus of rat and sex hormones, *J. Pathol. Bacteriol.,* 52, 349, 1941.
140. **Ross, M. H.,** Length of life and nutrition in the rat, *J. Nutr.,* 75, 197, 1961.
141. **Ross, M. H. and Bras, G.,** Tumor incidence patterns and nutrition in the rat, *J. Nutr.,* 87, 245, 1965.
142. **Ross, M. H., Bras, G., and Ragbeer, M. S.,** The influences of protein and caloric intake upon spontaneous tumor incidence of the anterior pituitary gland of the rat, *J. Nutr.,* 100, 177, 1970.
143. **Ross, M. H.,** Length of life and caloric intake, *Am. J. Clin. Nutr.,* 25, 834, 1972.
144. **Ross, M. H. and Bras, G.,** Influence of protein under and over nutrition on spontaneous tumor prevalence in the rat, *J. Nutr.,* 103, 944, 1973.
145. **Ross, M. H., Lustbader, E., and Bras, G.,** Dietary practices and growth responses as predictors of longevity, *Nature (London),* 262, 548, 1976.
146. **Rowlatt, U. F.,** Neoplasms of the alimentary canal of rats and mice, in *Pathology of Laboratory Rats and Mice,* Cotchin, E. and Roe, F. J. C., Eds., Blackwell Scientific, Oxford, 1967, 57.
147. **Russfield, A. B.,** Pathology of the endocrine glands, ovary and testis of rats and mice, in *Pathology of Laboratory Rats and Mice,* Cotchin, E. and Roe, F. J. C., Eds., Blackwell Scientific, Oxford, 1967, 391.
148. **Sachs, R.,** Life table technique in the analysis of response-time data from laboratory experiments in animals, *Toxicol. Appl. Pharmacol.,* 1, 203, 1959.
149. **Sass, B., Rabstein, L. S., Madison, R., Nims, R. M., Peters, R. L., and Kelloff, G. J.,** Incidence of spontaneous neoplasms in F344 rats throughout the natural life-span, *J. Natl. Cancer Inst.,* 54, 1449,1975.
150. **Saxton, J. A., Jr. and Kimball, G. C.,** Relation of nephrosis and other diseases of albino rats to age and to modifications of diet, *Arch. Pathol.,* 32, 951, 1941.
151. **Schärer, K.,** The effect of chronic underfeeding on organ weights of rats. How to interpret organ weight changes in cases of marked growth retardation in toxicity tests?, *Toxicology,* 7, 45, 1977.
152. **Schlettwein-Gsell, D.,** Survival curves of an old age rat colony, *Gerontologia,* 16, 111, 1970.
153. **Simms, H. S. and Berg, B. N.,** Longevity and the onset of lesions in male rats, *J. Geront.,* 12, 244, 1957.
154. **Simms, H. S.,** Longevity studies in rats. I. Relation between life span and age of onset of specific lesions, in *Pathology of Laboratory Rats and Mice,* Cotchin, E. and Roe, F. J. C., Eds., Blackwell Scientific, Oxford, 1967, 733.
155. **Snell, K. C. and Stewart, H. L.,** Variations in histologic pattern and functional effects of a transplantable adrenal cortical carcinoma in intact, hypophysectomized, and new-born rats, *J. Natl. Cancer Inst.,* 22, 1119, 1959.
156. **Snell, K. C.,** Renal disease of the rat, in *Pathology of Laboratory Rats and Mice,* Cotchin, E. and Roe, F. J. C., Eds., Blackwell Scientific, Oxford,1967, 105.
157. **Sobel, J. H. and Marquet, E.,** Granular cells and granular cell lesions, *Pathol. Annul.,* 9, 43, 1974.
158. **Sokoloff, L.,** Musculoskeletal lesions in experimental animals, in *The Pathology of Laboratory Animals,* Ribelin, W. E. and McCoy, J. R., Eds., Charles C Thomas, Springfield, Ill., 1965, 3.
159. **Sokoloff, L.,** Articular and musculoskeletal lesions of rats and mice, in *Pathology of Laboratory Rats and Mice,* Cotchin, E. and Roe, F. J. C., Eds., Blackwell Scientific, Oxford, 1967, 373.
160. **Sokoloff, L., Snell, K. C., and Stewart, H. L.,** Degenerative joint disease in *Praomys* (Mastomys) *natalensis, Ann. Rheum. Dis.,* 26, 146, 1967.
161. **Squire, R. A.,** A Guide to Infectious Diseases of Mice and Rats, Committee on Laboratory Animal Diseases, Institute of Laboratory Animal Resources, National Academy of Sciences, Washington, D.C., 1971.
162. **Squire, R. A. and Levitt, M. H.,** Classification of specific hepatocellular lesions in rats: Report of a workshop, *Cancer Res.,* 35, 3214, 1975.
163. **Squire, R. A. and Goodman, D. G.,** Tumors of laboratory animals, in *Pathology of Laboratory Animals,* Vol. 2, Jones, T. C., Garner, F. M., and Benirschke, K., Eds., Springer-Verlag, New York, in press.
164. **Stark, O. and Kren, V.,** Five congenic resistant lines of rats differing at the R+ H-1 locus, *Transplantation,* 8, 200, 1969.
165. **Stark, O., Kren, V., Krenova, D., and Frenzl, B.,** Serologically detected antigens in two sublines of the WAG rats, *Folia Biol.,* 15, 259, 1969.

166. **van Steenis, G. and Kroes, R.**, Changes in the nervous system and musculature of old rats, *Vet. Pathol.*, 8, 320, 1971.
167. **Stewart, H. L.**, Comparative aspects of certain cancers, in *Cancer*, Vol. 4, Becker, F. F., Ed., Plenum Press, New York, 1975, 303.
168. **Storer, J. B.**, Longevity and gross pathology at death in 22 inbred mouse strains, *J. Gerontol.*, 21, 404, 1966.
169. **Swaen, G. J. V. and van Heerde, P.**, Tumours of the haematopoietic system, in *Pathology of Tumors in Laboratory Animals*, Vol. 1, Part 1, Turusov, V. S., Ed., IARC Scientific Publ. No. 5, World Health Organization, Geneva, 1973, 185.
170. **Thompson, S. W., Husely, R. A., Fox, M. A., Davis, C. L., and Hunt, R. D.**, Spontaneous tumors in the Sprague-Dawley rat, *J. Natl. Cancer Inst.*, 27, 1037, 1961.
171. **Thung, P. J.**, Ovaria van Oude Muizen, Ph.D. thesis, University of Leiden, The Netherlands, 1958.
172. **Trevino, G. S. and Nessmith, W. B.**, Aortic body tumor in a white rat, *Vet. Pathol.*, 9, 243, 1972.
173. **Tschalargane, C., Goerttler, K., Ule, G., and Volk, B.**, Granularzell tumor des ZNS bei der Ratte nach Intraperitonealer Applikation von Aflatoxin B1 Während der Tragezeit, *Acta Neuropathol.*, 32, 281, 1975.
174. **Turusov, V. S., Ed.**, *Pathology of Tumours in Laboratory Animals*, Vol. 1, Part 1, IARC Scientific Publ. No. 5, World Health Organization, Geneva, 1973.
175. **Turusov, V. S., Ed.**, *Pathology of Tumours in Laboratory Animals*, Vol. 1, Part 2, IARC Scientific Publ. No. 6, World Health Organization, Geneva, 1976.
176. **van Bezooijen, C. F. A., von Noord, M. J., and Knook, D. L.**, The viability of parenchymal liver cells isolated from young and old rats, *Mech. Aging Dev.*, 3, 107, 1974.
177. **van der Kogel, A. J.**, Ph.D. thesis, University of Amsterdam, The Netherlands, in preparation.
178. **van der Waaij, R. and Speltie, T. M.**, Biotyping of Enterobacteriaceae as a test for the evaluation of isolation systems, *J. Hyg.*, 70, 639, 1972.
179. **van Haelst, U. J. G. M.**, De normale Thymus en zijn veranderingen Tijdens de Experimentale, snell Involutie en Regeneratie, Een licht-en electronmicrorcopisch Onderzoek bij de Rat, Ph.D. thesis, University of Nijmegen, The Netherlands, 1967.
180. **van Zwieten, M. J., Burek, J. D., Zurcher, C., and Hollander, C. F.**, Aortic body lesions in the hearts of aging rats, in preparation.
181. **van Zwieten, M. J., Burek, J. D., Nooteboom, A. L., Hollander, C. F., and Broerse, J. J.**, Morphological aspects of mammary tumours in irradiated and non-irradiated rats of different strains, *Int. J. Radiat. Biol.*, 31, 379, 1977.
182. **Walford, R. L.**, When is a mouse "old"? Letters to the Editor, *J. Immunol.*, 117, 352, 1976.
183. **Weisbroth, S. H.**, Pathogen-free substrates for gerontological research — review, sources, and comparison of barrier-sustained vs. conventional rats, *Exp. Gerontol.*, 7, 417, 1972.
184. **Wexler, E. J. and Greenberg, L. J.**, Co-existent arteriosclerosis, PAN and premature aging, *J. Gerontol.*, 25, 373, 1970.
185. **Wilens, S. C. and Sproul, E. E.**, Spontaneous cardiovascular disease in the rat. I. Lesions of the heart, *Am. J. Pathol.*, 14, 177, 1938.
186. **Wilens, S. L. and Sproul, E. E.**, Spontaneous cardiovascular disease in the rat. II. Lesions of the vascular system, *Am. J. Pathol.*, 14, 201, 1938.
187. **Yang, Y. H., Yang, C. Y., and Grice, H. C.**, Multifocal histiocytosis in the lungs of rats, *J. Pathol. Bacteriol.*, 92, 559, 1966.
188. **Young, S. and Hallowes, R. C.**, Tumours of the mammary gland, in *Pathology of Tumours in Laboratory Animals*, Vol. 1, Part 1, Turusov, V. S., Ed., IARC Scientific Publ. No. 5, World Health Organization, Geneva, 1973, 31.
189. **Zackheim, H. S.**, Tumours of the skin, in *Pathology of Tumours in Laboratory Animals*, Vol. 1, Part 1, Turusov, V. S., Ed., IARC Scientific Publ. No. 5, World Health Organization, Geneva, 1973, 1.
190. **Zurcher, C., Burek, J. D., Van Nunen, M. C. J., and Meihuizen, S. P.**, A naturally occurring epizootic caused by Sendai virus in breeding and aging rodent colonies. I. Infection in aging mice, *Lab. Anim. Sci.*, 27, 955, 1977.

Index

INDEX

A

B

D

E

F

G

H

M

N

O

P

R

S

T

U

V

W

X